ETHNOBOTANY
OF THE HERITAGE REGION OF
SHIWALIK HIMALAYA

ETHNOBOTANY
OF THE HERITAGE REGION OF
SHIWALIK HIMALAYA

S.K. Sood • Suresh Kumar
Department of Biosciences
Himachal Pradesh University, Shimla (H.P., India)

Susheel Kumar Bassi • J.C. Rana
National Bureau of Plant Genetic Resources
Phagli, Shimla (H.P., India)

ANAMIKA PUBLISHERS & DISTRIBUTORS (P) LTD.
4697/3, 21A, Ansari Road, Daryaganj, New Delhi 110002
Phones: 011-2328 1655, 011-2327 0239
Fax: 011-2786 8035
e-mail: anamikapublishers@yahoo.co.in

First Published 2012

ISBN 978-81-7975-323-1

PRINTED IN INDIA

Published by Anamika Publishers & Distributors (P) Ltd., 4697/3, 21-A, Ansari Road, Daryaganj, New Delhi 110 002.
Typeset and Printed by Aadil Printographics, Delhi 110 092

CONTENTS

LIST OF FIGURES AND PHOTOGRAPHS

Figures

Photographs

Pl. 7	Ornaments adorned by womenfolk of village Pragpur. (A) Silver bracelet; (B) A pair of large-sized silver ear rings; (C) Necklace of gold; (D) 'Kantha'—A necklace of silver; (E) 'Pajeb'—An ornament for feet; (F) 'Chak'—An adornment for head of married woman; (G) Various types of gold ornaments.
Pl. 8	(A) Radha-Krishna Temple; (B) Close view of monkey god—'Hanuman' resting under the canopy of a tree in Hanuman temple of village Pragpur.
Pl. 9	(A-D) 'Naun'—Water storage reservoirs for drinking and irrigation in Pragpur.
Pl. 10	(A) Sickle; (B) Earthen Pitchers; (C) 'Chakla' and 'Belan'—A device for rolling of 'chappatis' and 'Madani' (wooden churner—a device for churning milk and curd; (D) Bamboo baskets of different sizes for drying and serving purposes.
Pls. 11-61	Ethnobotanically important plants.

PREFACE

Despite the progress of ethnobotany in India since the time of Janaki Ammal (1956), about 70% of Indian plant species are yet to be explored ethnobotanically and ethnopharmacologically. As a matter of fact, bioscientists consider ethnobotanical knowledge system as a first effective means for identifying as well as locating alternative food source, leads for drugs and pharmaceuticals, natural dyes, colours, gums, resins, etc. This is a well-known fact that traditional knowledge survives usually among the indigenous and local community as they maintain a balanced ecological rhythm in their surroundings and needs to be properly documented before it is lost forever due to the assault of modern civilization into tribal areas. This book has, therefore, been necessitated, and is based on intensive ethnobotanical surveys undertaken by the authors for a period spread over more than five years (2001-2006) in the first ever declared heritage region of Shiwalik Himalaya having rich diversity of floristic and economic plant resources. A comprehensive account of 204 plant spp. incorporates information on nomenclature updated after Bennet (1987); Vernacular, English, Sanskrit and Regional Names; distribution; description; flowering and fruiting; habitat ecology; part/s used; folk uses; active constituents; biological activity; uses in literature; major Ayurvedic formulations; adulterant or substitute; and trade. Contents of the book include information on Ethnobotany—A Multidisciplinary Science, Indigenous Societies, Traditional Knowledge and Tradomedicalism, Ethnic Communities of Himachal Pradesh. Overview of the Previous Work, The Study Area, The People Setting, Occupation and Livelihood, The Methodology, Ethnobotany and Traditional Practices of Shiwalik Himalaya—Pragpur, Epilogue, Plates, and five useful appendices providing information on heritage status of Pragpur, index to families, index to local names, index to nutritional value of ethnobotanically important plants and index to glossary of words. The textural subject matter has been arranged according to the Bentham and Hooker's system of classification, and illustrated with the help of 17 Tables, 8 Figures and 235 Photographs.

We are indebted to Prof. T.N. Lakhanpal, Prof. D.C. Kalia, and Prof. V.K. Mattu for their valuable suggestions and encouragement. The authors heartily thank Mr. Bal Krishan for his help extended during the preparation of the manuscript. A word of appreciation is due to various authors whose works have been referred in the book. Our sincere thanks are also due to Anamika Publishers & Distributors (P) Ltd., New Delhi, for bringing out this book in an attractive form. Finally, it will not be out of place to acknowledge our deep sense of gratitude to all our family members for their moral support during this challenging and tedious endeavour.

S.K. Sood, Suresh Kumar
Sushil Kumar Bassi, J.C. Rana

ABBREVIATIONS

AD	*anno domini*
Alc.	Alcoholic
Alk.	Alkaline
Anon.	Anonymous
Aq.	Aqueous
Ass.	Assam
B.	Bengali
BC	Before Christ
B.P.	Blood Pressure
Bo.	Bombay
Bu-	Butyl
Ca	Calcium
C	Central
cm	Centimetre
CNS	Central Nervous System
Cu	Copper
E.	East
EBH	Ethnobotanical Herbarium
Eng.	English
et al.	(*et alia*) and other authors
etc.	(*et cetera*) and so on
Et	Ethyl
Extr.(s)	Extract(s)
Fig.(s)	Figure(s)
g	Gram
GIT	Gastro-intestinal tract
G.	Gujarat
H.	Hindi
H.P.	Himachal Pradesh
i.e.	(L. *id est*) that is
K	Potassium
Kan.	Kannada
Kash.	Kashmir
kg.	Kilogram
m	Metre
Mal.	Malyalam
Mar.	Marathi
Me-	Methyl
mg	Milligram
Mg	Magnesium
M.P.	Madhya Pradesh
ml	Millilitre
N.	North
(–)ve	Negative
N	Nitrogen
N.W.	North West
O	Oxygen
–OH	Hydroxy
P	Phosphorus
(+)ve	Positive
P.	Punjab
%	Per cent
Ph-	*Phenyl*
Pl.(s)	Plate(s)
S.	South
S.E.	South-East
Sans.	Sanskrit
Sp	Specie
Spp	Species
Sq km	Square Kilometre
Syn.	Synonym
Tam.	Tamil
U.P.	Uttar Pradesh
Var.	Variety
Vern.	Vernacular
viz.	*Vide licet* (namely)
Vit.	Vitamin
Vol. (s)	Volume (s)
W.	West
WHO	World Health Organisation
&	And

INTRODUCTION

Ethnobotany—A Multidisciplinary Science

Ethnobotany is the science which comprehends the relationship of a given society with its environment and, in particular, with the plant world. The term was coined for the first time by Harshberger in 1895 and the subject included mere identification and cataloguing of plants used by the primitive or aboriginal people. Subsequently, ethnobotany has been variously defined and interpreted by different workers as its discipline began to follow multidisciplinary approach combining a diversity of knowledge bases and methods through the use of anthropological methods (Conklin, 1980; Ford, 1978; Jain, 1986; Jones, 1941; Robbins *et al.*, 1916; Schultes and Reis, 1995; Wickens, 1990). According to Pushpangadan (1990), the term refers to the study of the knowledge system pertaining to the multidimensional perspective of life culture, traditions as well as interaction of traditional or less advanced human communities like tribals with their local flora (Ethnobotany) or fauna (Ethnozoology) whereas Heiser (1995) regarded ethnobotany as the study of wild and cultivated plants in relation to people. These definitions and concepts that have evolved during the last century give a broad dimension to the field of ethnobotany which, at present, is considered a rapidly expanding interdisciplinary science with its concept mainly including all aspects of direct relationship of plants with man, and with many workers throughout the world becoming involved in the practical application of its data in areas such as conservation biology, biodiversity prospecting and sustainable management (Álcorn, 1996; Arizpe, 1996; Rao, 1996; Redford and Stearman, 1993). Bioscientists consider ethnobotanical knowledge system as a first effective means for identifying as well as locating alternative food source, leads for drugs and pharmaceuticals, natural dyes, colours, gums, resins, etc. (Pushpangadan and Kumar, 2005).

Indigenous Societies, Traditional Knowledge and Tradomedicalism

Indigenous societies all over the world in different geographical regions have discovered multiple uses of natural resources around them in the form of traditional knowledge (Jain, 2004). Undeniably, traditional knowledge survives usually among the indigenous and local community as they maintain a balanced ecological rhythm in their surroundings. In this regard Chandra (1990) states: "These groups of people are not to be pitied for primitive existence; they rather deserve to be honoured and respected for their richness of human existence in harmony with nature". Traditionally, local communities worldwide are extremely knowledgeable about the local plant resources on which they are so intimately and immediately dependent. Indian region is also no exception as it has also been very rich in such traditional heritage and believed to evolve through sacred Vedas (Singh and Shankar, 1996). Unfortunately, much of the accumulated extraordinarily abundant knowledge on plants which have been acquired due to their long-term practices and handed down usually orally, from one generation to another, is dwindling because of the loss of their main culture and changes in sustenance economy (Hamilton, 1955). With the disappearance of indigenous culture all over the world the biodiversity is also disappearing, and the loss is beyond retrieval. Therefore, the need of hour is to give priorities to activities related to documenting of this knowledge before it is being lost forever to posterity.

Of the estimated 300 million indigenous people all over the world, about 67.7 million tribal people belonging to 573 tribal groups with various subsistence patterns have been reported from India, and it accounts for nearly 22% of world tribal population. Moreover, predominant tribal areas in India comprise of 15% of the total geographical area of the country. As many as 106 different languages and 227 subsidiary dialects are spoken by the tribals in India (Arora, 1995). So far, of the 2,50,000 to 3,00,000 plant species on earth only 7% of the vascular flora have been exploited for their medical potential (Iverson, 1988), whereas the number of plants utilized by man for food hardly exceeds 30 (Swaminathan, 1993; Khoshoo, 1995). The Indian scenario is also no different as with regard to the diversity of over 10,000 wild plant species used by tribals, 8,000 are used for medicinal purposes with over 1,75,000 specific preparations/applications, about 3,900 as edibles and 2,325 for miscellaneous purposes (Anonymous, 1998a, b); and the data on people inhabiting remote areas and ethnic groups in India still needs to be documented. By and large, tribal people throughout the world have their own system of medicine. They mostly use locally available plants and animals as ingredients to prepare their medicines for the treatment of various diseases. Mume (1976) referred to traditional medical system as 'Tradomedicalism' and according to him it is a system of treating diseases through the employment of the agencies and forces of nature. However, Gbile (1986) defined it as the use of plants by members of an indigenous culture for which there is no organized medical system. It is estimated that about 64% of the world's population rely on traditional medicines for their primary health-care (Wambene, 1990). Also, traditional medicine has given some of the valuable and prized herbal drugs to the modern medicine. Available data indicates that of the 119 plant based drugs today about 74% are from traditional medicine (Farnsworth and Soejarto, 1991). According to Dobson (1995): "Only 1,100 plants out of 3,65,000 known plant species have so far been thoroughly examined for their medicinal properties". Farnsworth (1985), Director of WHO Collaborating Centre for Traditional Medicine has strongly recommended that all those plants used by the primitive folk-healers and ethnic societies of world should be biologically investigated. Till todate, about 70% of Indian plant species are yet to be explored ethnobotanically and ethnopharmacology (Khan, 2005) and only 1,500 plant species are being used in ancient Indian system of medicine, i.e. Ayurveda (Bhatnagar, 1997) from which plant drugs worth Rs. 3,400 million per annum are being utilized for its various preparations. Today, ethnobotany plays a crucial role in the study of traditional medicine as it has an interfacial function linking nature with culture and traditional knowledge with modern technology contributing to the understanding of traditional medicinal knowledge (Pei, 2001).

Ethnic Communities of Himachal Pradesh

Himachal Pradesh, a hilly state in India situated in the North-West Himalaya between latitude 30°-3' to 33°-3' N and longitude 75°-3' to 79° E is considered a veritable emporium of medicinal and aromatic plants and diverse ethnic communities, viz., Gaddis, Gujjars, Kinners, Jads, Lahoulis, Spitians, Pangwalas, Swangalas. There are as many as 16,997 villages of different sizes covering around 55,678 sq. kms. area under its 12 districts, namely, Bilaspur, Chamba, Hamirpur, Kangra, Kinnaur, Kullu, Lahaul-Spiti, Mandi, Shimla, Sirmaur, Solan and Una having altitudinal range between 300-8,000 m. Since ages, the rural inhabitants of Himachal Pradesh are considered the inheritors and developers of rich traditions of plants and ecosystem. Another noticeable aspect of the state is that it also boasts of a wide diversity of plant forms of European, Tibetan and Chinese origin in its sub-tropical, temperate, alpine and cold desert regions, and the place is a very fertile area for ethnobotanical research for which the studies so far are a few and fragmentary and needs to be undertaken for ushering in economic prosperity to the people of the state.

Overview of the Previous Work

The scientific beginning of plant exploration work in India dates back to Garcia da Orta, a great pharmacist from Portugal who recorded this information in his book *Colloquious dos Simples* published in 1563. But, it was Heinrich Van Rheede—the Governor of Dutch possession in Malabar in 1667, who made the first real attempt to explore systematically the plant wealth of India. For the state of Himachal Pradesh, the pioneering floristic investigations were those of William Spencer in 1817, William Moorcroft in 1819 and Countess of Dalhousie in 1829, who collected plants from Sabathu; Kangra, Kullu, Lahaul and Spiti; and Shimla regions, respectively, falling under the erstwhile undivided Punjab or the territories of Himachal Pradesh before its reorganization in 1971. Since then, large number of investigators had dealt with the floristic studies of the region and notable contributions, viz., *Economic Botany of the Himalayan Region* (Atkinson, 1882), *The Flora of British India* (Hooker, 1872-1897), *Flora Simlensis* (Collett, 1902), *Flora of Bashahr Himalayas* (Nair, 1964), *Flora of Himachal Pradesh*, Vols. I-III (Chowdhury and Wadhawa, 1984), *Concise Flowers of the Himalaya* (Polunin and Stainton, 1984), *Flowers of the Himalaya—A Supplement* (Stainton, 1988), *Flora of Lahaul Spiti—A Cold Desert in North West Himalaya* (Aswal and Mehrotra, 1994), *Flora of Kullu District* (Dhaliwal and Sharma, 1999), *Flora of Great Himalayan National Park, Himachal Pradesh* (Singh and Rawat, 2000) and *Flora of Chamba District, Himachal Pradesh* (Singh and Sharma, 2006) and studies on diversity of medicinal and aromatic plants in different physiographic regions of the state, namely, **Kangra Valley** (Ahluwalia, 1952; Uniyal and Chauhan, 1971, 1973c), **Kullu** (Rastogi, 1960; Uniyal and Chauhan, 1972, 1973b; Dobriyal *et al.*, 1997), **Lahaul-Spiti** (Rau, 1961; Sarin, 1967; Uniyal *et al.*, 1973a; Arora *et al.*, 1980; Chauhan, 1989d; Chaurasia *et al.*, 2001); **Chamba** (Gupta, 1961, 1964, 1971; Shabnam, 1964), **Una** (Chauhan, 1974), **Mandi** (Singh, 1993, 1999b), and **Himachal Pradesh** (Sarin and Chopra, 1985; Chauhan, 1983, 1989a, 1990, 1999, 2003; Singh and Aswal, 1992; Sharma and Sood, 1997; Singh, 1999; Badola, 2001) have added to the wealth of information in this regard. Moreover, the information pertaining to the scope of phytochemical industry (Sharma and Misra, 1972); collection, cultivation, production potential, utilization and marketing strategies of important medicinal and aromatic plants (Sood *et al.*, 1978; Chauhan, 1987, 1988b, 1989b; Chauhan and Khosla, 1988; Chauhan and Negi, 1988; Negi and Bhalla, 2002); endangered Ayurvedic pharmacopoeia plant resources (Chauhan, 1989c); role of medicinal and aromatic plants in social forestry programme (Chauhan, 1989a); plant resources of economic use (Chauhan, 1996b); present status and future prospects for the cultivation and collection of medicinal plants (Gupta, 1988); and policy document on propagation and processing of herbal/medicinal plants in Himachal Pradesh is also available, and it has great value in the context of today's trade and patents scenario.

Even prior to the coining of the term Ethnobotany, the uses of plants by human beings found its mention in ancient Sanskrit, Greek, Chinese and Arabic literature. *An Introduction to Ethnobotany* by Faulks (1958) is the first book on ethnobotany. Since then uses of plants by the tribals are being recorded for a variety of purposes (Jain, 1981). From the Indian perspective, the real credit for initiating ethnobotanical researches goes to Dr. E. K. Janaki Ammal in 1955 who studied subsistence food plants of certain tribals of South India particularly to explore better prospects of Dioscoreas in India under an official programme in the Economic Botany Section of Botanical Survey of India (BSI). *Glimpses of Indian Ethnobotany* by Jain (1981) is the first book dealing with Indian Ethnobotany whereas the first attempt to publish *Bibliography of Ethnobotanical Works* goes to Jain *et al.* (1984) who compiled about 2000 Indian and foreign references on various aspects of ethnobotany, followed by that of Binu *et al.* (1992), Lalramnghinglova and Jha (1999) and Radhakrishnan *et al.* (2000). Thereafter, a graphic review of Indian ethnobotanical literature with analysis highlighting possible directions for future work (Jain and Srivastava, 2001) and also a recent detailed one listing 1,600 references of which over 1,250 relate strictly to ethnobotanical work or very closely allied themes on India and the others on concept, global evolution of the discipline, and methodology of new areas from work of some outstanding foreign ethnobotanists have been published (Jain, 2002). More importantly, Mudgal (1987) and Chandra (1990) compiled the diverse ethnobotanical literature existing in various languages of the world, and Manilal (1989) reported on the linkages of ethnobotany with other sciences and disciplines. In the last couple of decades, there has been an upsurge of activities for ethnobotanical studies in different phytogeographical areas and tribes of India as evident from the publications of valuable ethnobotanical data in a number of books, particular mention may be made of works of Maheshwari *et al.* (1981): *The Ethnobotany of Tharus of Kheri District, Uttar Pradesh*; Agarwal and Ghosh (1985): *Drug Plants of India (Root Drugs)*; Sensarma (1989): *Plants in the Indian Puranas—An Ethnobotanical Investigation*; Jain (1991): *Dictionary of Indian Folk Medicine and Ethnobotany*; Jain *et al.* (1991): *Notable Plants in Ethnomedicine of India*; Maheshwari (1992): *Ethnobotany in India*; Saklani and Jain (1994): *Cross-Cultural Ethnobotany of Northeast India*; Joshi (1995): *Ethnobotany of the Primitive Tribes in Rajasthan*; Arora and Pandey (1996): *Wild Edible Plants of India-Diversity, Conservation and Use*; Rama Rao and Henry (1996): *Ethnobotany of Eastern Ghats in Andhra Pradesh, India*; Sinha (1996): *Ethnobotany—The Renaissance of Traditional Herbal Medicines*; Varghese (1996): *Applied Ethnobotany—A Case Study Among the Kharia of C. India*; Vedanathy *et al.* (1997): *Tribal Medicine of Chittoor, District A.P.*; Kaushik (1998): *Indigenous Medicinal Plants (Including Microbes and Fungi)*; Pal and Jain (1998): *Tribal Medicine*; Pandey (1998): *Ethno-Forestry (Local Knowledge for Sustainable Forestry and Livelihood Security)*; Singh and Pandey (1998): *Ethnobotany of Rajasthan*; Dagar and Dagar (1999): *Ethnobotany of Aborigines of Andaman-Nicobar Islands*; Jain and Srivastava (1999): *Dictionary of Ethnoveterinary Plants of India*; Mudgal *et al.* (1999): *Ethnobotany of Totopora*; Parrotta (2000): *Healing Plants of Peninsular India*; Singh and Kumar (2000a): *Ethnobotanical Wisdom of Gaddi Tribe in Western Himalayas*; Varghese and Hembrom (2000): *Ethnomedicinal Lore of the Paharias*; Sood *et al.* (2001): *Ethnobotany of Cold Desert Tribes of Lahaul-Spiti (N.W.*

Himalaya); Singh *et al.* (2003a): *Herbal Medicine of Manipur: A Colour Encyclopedia*; Singh *et al.* (2003b): *Ethnomedicine of North-East India*; Sood *et al.* (2005a): *Ethnic Indian Plants in Cure of Diabetes*; Sood *et al.* (2005b): *Ethnic Plants of India Used in Cancer Cure—A Compendium*; Sood *et al.* (2005c): *Ethnic Aphrodisiac Plants*; Sood *et al.* (2005d): *Sacred and Magico-Religious Plants of India*; Kshirsagar and Singh (2007): *Ethnobotany of Mysore and Coorg Districts, Karnataka State*; and Sood and Prakash (2007): *Edible Roots and Underground Stems of Ethnic India*.

Ethnobotanical studies in different parts of Himachal Pradesh have been conducted by several workers. As many as 30 publications including 3 books have been published and the data is available only for the Gaddis, Lahaulas and Spitians. Of all the agro-climatic regions, there are a large number of publications on the tribes inhabiting cold desert areas of Lahaul-Spiti [Koelz (1979): Ethnobotany of Lahaul; Duhoon *et al.* (1996): Seabuckthorn (*Hippophae* spp): A less known wonder plant of ethno-medico-botanical importance in cold desert tribes of Lahaul-Spiti (N.W. Himalaya); Kala *et al.* (1999): Ethno-medico-botany of Indian Trans-Himalaya: A Case Study from Spiti; Singh *et al.* (2001): *Edible Wild Plants of Trans-Himalayan Cold Desert*; Sood *et al.* (2001): *Ethnobotany of Cold Desert Tribes of Lahaul Spiti (N.W. Himalaya)*; Chandra Sekar and Srivastava (2003): *Ethnomedicinal studies in Pin Valley National Park, Lahaul Spiti;* Srivastava and Chandra Sekar (2004): *Ethnomedicine of the Pin Valley National Park, H.P.: Plants used in Treating Dysentery;* Chandra Sekar and Srivastava (2005): New Reports on Aphrodisiac plants from Pin Valley National Park, H.P.; Singh and Chauhan (2005): Traditional practices of herbal medicines in Lahaul valleys], followed by that of district **Kangra** [Kapur (1993): Ethno-medico plants of Kangra Valley; Lal *et al.* (1996): Plants used as ethnomedicine and supplementary foods by Gaddis of H.P.; Singh and Kumar (2000a): *Ethnobotanical Wisdom of Gaddi Tribe of Western Himalaya*; Singh and Kumar (2000b): Observation on ethnoveterinary medicine among the Gaddi tribe of Kangra Valley; Sharma and Maheshwari (2005): Traditional medical practices of Gaddi tribes in Kangra district, H.P.], **Kullu** [Uniyal and Chauhan (1973b): Traditionally important medicinal plants of Kullu forest division, H.P.; Jain and Puri (1994): Ethno-medico-botanical survey of Parbati Valley; Singh *et al.* (1998a): Eco-cultural analysis of sacred species in Chhakinal watershed, H.P. in district Kullu; Singh (1999a): Ethnobotanical studies of useful plants of Kullu district in N.W. Himalaya, India; Uniyal (2003): Utilization of medicinal plants by the rural women of Kullu, H.P.; Sharma *et al.* (2004)—Observations on the traditional phytotherapy among the inhabitants of Parbati valley]; **Himachal Pradesh** [Chauhan (1988a): Ethnobotanical study of plants in Himachal Pradesh; Sharma and Rana (1999): Traditional medicinal uses of plants in Himachal hills; Rana (2004): Lesser known food plants of Himachal Pradesh]; **Mandi** [Singh (1993): Medicinal plants of Ayurvedic importance from Mandi district; Sood and Thakur (2004): *Ethnobotany of Rewalsar Himalaya*]; **Shimla** [Chauhan (1996a): Ethno-medico-botany of Pabbar valley in district Shimla]; **Solan** [Verma and Chauhan (2006): Ethno-medico-botany of Kunihar forest division; Verma and Chauhan (2007): Indigenous medicinal plants knowledge of Kunihar forest division]; **Chamba** [Uniyal *et al.* (2006): Traditional uses of medicinal plants among the tribal communities of Chhota Bhangal, Western Himalaya] and **Kinnaur** [Negi and Subramani (2002): Ethnobotanical study in the village Chhitkul of Sangla valley, district Kinnaur, Himachal Pradesh].

Inspite of the above literature, it is rather unfortunate that no attempt has been made so far to explore the empirical knowledge of the local inhabitants of the first heritage village of Himachal Pradesh—Pragpur, a region significantly rich in flora embodying an ancient heritage, endemism and cultural diversity. Against this backdrop, it was felt necessary to document information on this aspect available from this region and put it in proper perspective for the welfare of mankind.

THE STUDY AREA

The Land, People and Their Socio-Biology

Location and Extent

The Pragpur (Pl. 1C) nestles between 76°-20' East longitude and 31°-82' North latitude with an area of 464 sq. kms. in the Shiwalik ranges of North-West Himalaya. It is considered one of the blocks of district Kangra and situated at 568 m above sea level. The district Kangra is bounded on the South-West by Una, on the North-West by district Gurdaspur of Punjab, on the North by Lahaul-Spiti and Chamba districts, on the East by Kullu and Mandi and on South by Hamirpur (Fig. 1, 2). The entire population of Pragpur is rural and estimated to be around 3,500 in number according to 2001 census (Pl. 1A).

Etymology and Historical Perspectives

The word Pragpur is presumably composed of two different syllables, 'Prag' (pollen) and 'pur' (full of) which rightly describes the area when it is ablaze with blossoms in spring. The earliest references in recorded history suggest that the area of Pragpur came within the Jaswan kingdom whose rulers were the descendents of Katochs of the Kangra lineage. Around the 16-17th century, bands of marauders started invading the pretty and peaceful foothills of the Kangra valley. Prag Dei, a princess of the royal house of Jaswan, successfully organized resistance to these marauding bands. To commemorate this princess, Pragpur was founded and its layout carefully planned. Various clans and communities were allotted distinctive living spaces. One of the important communities that settled at Pragpur were some of the 52 clans of the hill Soods. Being enterprising, many made their fortunes at other destinations, especially Shimla. At Pragpur, they built elegant buildings that lie interspersed amongst lovely mud-plastered and slate-roofed houses alongside streets paved with dressed cobbled stones. They also invested in schools, dharamshalas and water systems. In course of time, they settled where their economic interests lay. However, the magnificent buildings constructed by them some decaying and crumbling have largely remained unspoiled, giving Pragpur a medieval ambience (Pl. 3B-D). Recognizing that this unspoiled traditional area could contribute to rural heritage tourism, the Government of Himachal Pradesh notified Pragpur as a heritage village in December 1997 (Pl. 1B; Appendix I).

Climate and Biogeography

The area nestles in the Shiwaliks (which in ancient times were known as 'Manak Parbat') or the outer Himalaya (Pl. 1C, D). Physiographically, it falls in humid sub-tropical zone.

The soil is loamy-clay with very mild acidic reaction. The area is suitable for cultivation of maize, wheat, ginger, sugarcane, paddy, potato and citrus fruits.

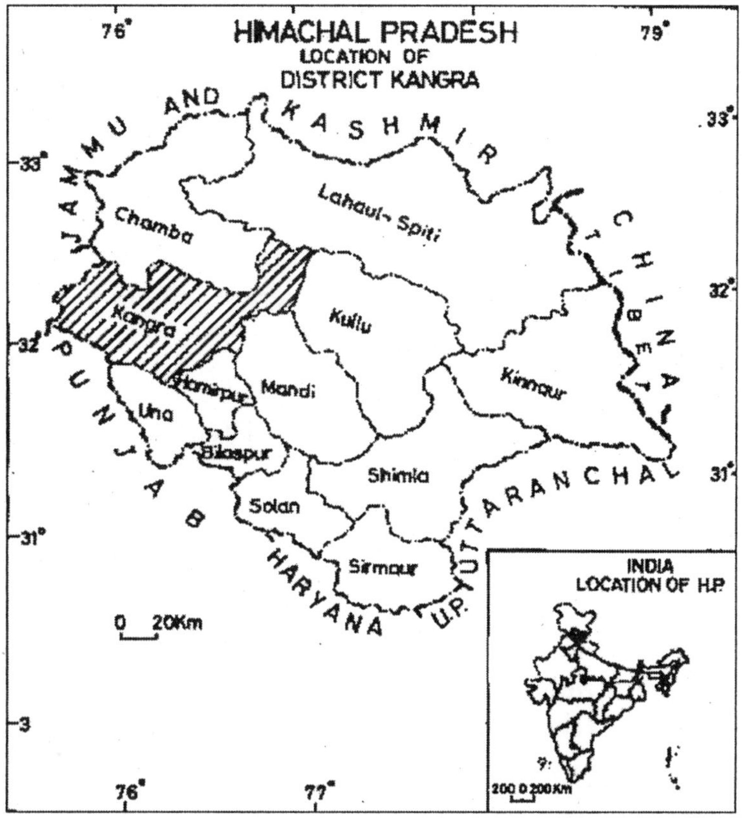

FIG. 1: *Map of Himachal Pradesh showing location of district Kangra.*

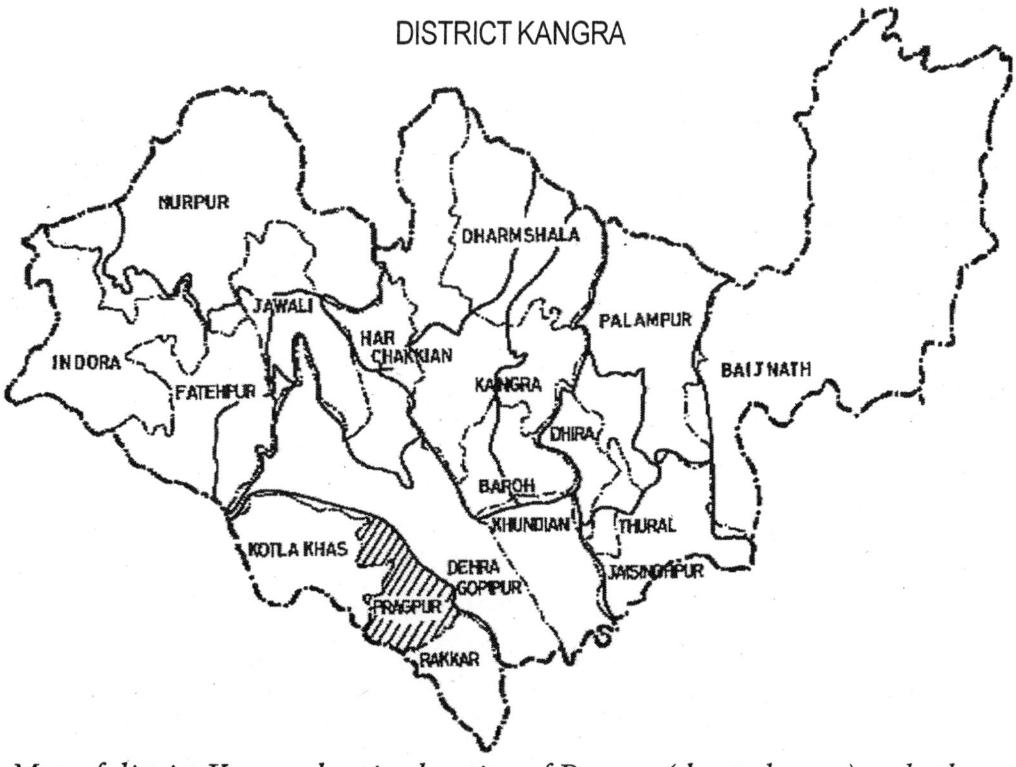

FIG. 2: *Map of district Kangra showing location of Pragpur (the study area) and other adjoining Blocks.*

There are three broad seasons in Pragpur, winter ('sayal', 'sardi', or 'thandi' in local language)—October to February; summer ('garmi')—March to June; and rainy season ('barsat')—July to September. The temperature during the winter months is cold due to lashing cold breeze of the nearby mountain ranges of Dhauladhar. The spring begins from mid-February to March, and the month of April is extremely delightful when the air is cool and fresh, and flowers of beautiful colour adorn the valleys, forests, slopes and grassy fields. The weather becomes extremely hot and dusty in the summer season. The monsoon arrives generally towards end of June. July and August are the rainiest months. The average rainfall is 1407.02-1500 mm. The average means temperature is 7.0°-40° C.

THE PEOPLE SETTING

Communities and Dialects

The population comprises of dominant Hindus followed by Muslims, Jain and Sikhs. Amongst Hindus, there is a predominance of Brahmins (priests and religious devotee) followed by 'Rajputs' (Thakurs—the landlords); 'Soods', 'Khatris', 'Mahajans' (businessmen); 'Chamar' (cobbler); 'Julahe' (weaver); 'Chureh' (sweeper); 'Dumane' (basket weaver); 'Teli' (oil miller); 'Lohar' (blacksmith); 'Tarkhan' (carpenter); 'Sunar' (goldsmith). The local populace by and large is simple, honest, god fearing and worship different deities (Pl. 8A, B). During princely days, the official business of the state was used to be conducted in Tankri and Urdu. At present, the dialects used are Pahari, Dogri and Kangri.

Dwellings

There exists magnificent havelis, mansions and Italianate buildings built in Anglo-Indian architecture (Pls. 2A, B; 3A-D; 4A, B). The houses are generally two-storeyed and well ventilated (Pls. 3B; 4B). The slates or ridged iron sheets (Pl. 3C, D) are used for roofing whereas bamboo wood is commonly employed for supporting the slates (Pl. 3D). The walls of the houses are built of unburnt bricks (Pl. 3C), stones or pressed clay (Pls. 4A, B). The doors and windows are made of 'tunni', 'ber' and/or 'simbal' wood (Pls. 2A; 3C, D; 4A, B; 7A-D). In single-storeyed houses the roofs are flat with slates or thatching grass generally used for roofing. Grass and 'khazoor' brooms (Pl. 6A) are used for sweeping the house; and tongs and brush used (Pl. 6A) for cleaning the earthen 'chullaha' (Pl. 6D). The various terms associated with dwellings in Pragpur are listed in Table I.

Due to simple living, 'charpai' (cot), mats of paddy straw and/or leaves of date palm trees and 'pattra' (low height wooden stool for sitting) and wooden or iron boxes are common traditionally available furniture in most of the houses. The household utensils are made of brass and bronze. The fuel wood (Pl. 6C) comprises dried sticks of 'beol' (*Grewia oppositifolia*), mango (*Mangifera indica*), 'garna' (*Carissa inermis*), and stalks of maize (*Zea mays*). To keep oneself warm during winter, burning 'gohttu' (cow dung cake) or charcoal placed in earthen pot-'kangri' is kept close to the body. The grains are stored in 'pehru'. For drying spices and other food items, 'changer' (bamboo basket) of different sizes are employed (Pl. 10D) whereas for rolling 'chappatis' wooden base and rolling pin are used (Pl. 10B).

TABLE I
ASSOCIATED TERMS OF TRADITIONAL DWELLINGS IN PRAGPUR

Associated Terms of Dwellings	*Name in English*
Baranda	Courtyard
Dwar	Door
Kar or Makan	House
Tabella	Stable
Owan	Ground floor living room
Kamra	Room
Bor	Room on first floor
Ladda	Roof top
Bethak	Sitting room
Kand	Wall
Dwari	Window
Sirrhi	Stairs

Food and Beverages

Maize and wheat are their staple food. People love to take rice at least once daily. Various pulses—'Kulthi', 'Moong', 'Masur', 'Arhar', 'Urad', 'Gram' and beans are eaten along with rice. During summer from April to September, potato (*Solanum tuberosum*), radish (*Raphanus sativus*), 'mungrae', 'sarson-ka-saag' (*Brassica campestris*), 'karyali' (*Bauhinia variegata*), 'lassure' (*Cordia dichotoma*), 'chaukah-ki-chutney' (*Rumex hastatus*), mango (*Mangifera indica*) ('chutney', 'amboa', 'chacha'), 'gia' (*Lagenaria siceraria*), 'gia-tori' (*Luffa aegyptica*), 'lanko' (*Sechium edule*), 'kaddo' (*Cucurbita pepo*) are grown for vegetables. Leaves of coriander (*Coriandrum sativum*) added for flavouring vegetables and curries. Mustard oil (Brassica campestris) and desi or hydrogenated ghee are used for the preparation of vegetables, curries and frying purposes. Earthen pots of different sizes are mostly used for storage and cooking (Pl. 10C). Milk and milk products are consumed in large quantity. Generally, men are fond of fish, egg, chicken, mutton and also hunt for pig ('sur'), 'kakkar', wild cock ('murga').

Food is mostly taken four times a day. The morning meal is called 'datyalu' (breakfast), mid-day as 'dupehri' (lunch), evening one as 'tukru 'or 'sanjyalu' and last one as dinner.

On the festival day and at the time of marriage and other functions special foods are prepared in large-sized metal vessels 'battoi' (Pl. 6B) and 'sagla'. The festival foods are 'khichri' on 'lohri', 'ankli'/'anklo', 'babroo' and 'kheer' on 'baishaki' (basaua), 'seviyan' on 'sair'. Dham (food served at the time of marriage ceremony or special occasions in the family) comprises rice, 'channa-ki-dal' (gram), 'mah', 'khatta', 'madra' and 'mitha' and enjoyed equally by one and all alike. The food taboos are also observed on certain occasions such as fasts, death in the family and festivals. Men are fond of country liquor.

Dress and Ornaments

Men wear 'kurta' (long shirt), trouser and 'safa' (a kind of turban) or shirt and trouser or pant whereas women wear 'kurti' (long shirt) with 'salwar' and 'chadru' ('duppata'). In winter, people generally wear cotton/woollen coat, 'basket' (half coat without arms) and sweaters. At the time of marriage, bride bears 'ghagri' (a type of long skirt or 'lahenga') made of coloured silken cloth.

TABLE II
ORNAMENTS WORN BY WOMENFOLK OF PRAGPUR

Name	*Material*	*Body part*
Anguthari	Silver or Gold	Toe
Balis	Silver or Gold	Ear
Bala	Silver or Gold	Ear
Balu	Silver or Gold	Nose
Baluk	Silver or Gold	Nose
Bang	Silver or Gold	Wrist
Churi	Silver or Gold	Arms
Chak	Silver or Gold	Head
Chiri (Dora)	Silver or Gold	Hair
Mang Tikka	Silver or Gold	Head
Clip	Silver or Gold	Hair
Dandi	Silver or Gold	Ear
Dhodun	Silver or Gold	Ear
Dhol	Silver or Gold	Neck
Jumkas (Kanta)	Silver or Gold	Ear
Jhanjar	Silver or Gold	Feet
Kanbali	Silver or Gold	Ear
Koka	Silver or Gold	Nose
Kantha (Kanthi)	Silver or Gold	Neck
Pher	Silver or Gold	Ear
Tili (Koka)	Silver or Gold	Nose
Pajeb	Silver or Gold	Anklet
Singi	Silver or Gold	Neck
Mundris (Anguthi)	Silver or Gold	Finger

Both men and women like to adorn themselves with jewellery made of silver or gold. The ornaments adorned by married women (Pl. 7 A-G) are different from those of unmarried ones. Details of different ornaments worn by womenfolk are given in Table II.

SOCIAL CUSTOMS AND BELIEFS

Birth Rituals

Birth ceremonies in Pragpur are very picturesque with quaint local touch. The delivery is attended by midwife and/or elderly women of the household. The umbilical cord is cut by midwife holding on to a silver coin or silver ornament. After delivery, the infant is bathed and wrapped in a clean cloth. The first thirteen days are called 'sootak' and the family observes abstention from all religious rituals. During this period no one accepts food and drink from newly become mother excepting lower caste people. The first feeding to the infant is known as 'ghutti' and it is believed that the child imbibes the temperament of the person who gives him/her its first dose. After 'sootak' is over, the cleansing up ceremony 'goontar' is performed by a Brahmin priest. A mixture of curd, sugar, cow's urine, milk and Gangajal (holy water from Ganges) is sprinkled all over the house and it is also taken by all the members of the family. Other associated ceremonies, namely 'namkaran' (naming of the infant), 'annaprashan' (first tasting of cereal by the baby), 'jamalu' or 'mundane' (first hair cut) are also performed with great joy and hoisting lunch for all the local people.

Marriage System

The first step in marriage ceremony is the betrothal ('sagai') in which boy's father sends a priest or relative to the girl's parents with a few ornaments and gifts and this offering is called 'roka'. Also, the date for proposed wedding is fixed in consultation with the priest. The marriage commences with 'saand' and 'samuhat' (associated religious rituals) followed by 'mehandi' (henna ceremony) and oil ceremony called 'tel' (bath rituals) for both the groom and the bride at their respective places. On the day of marriage, the groom covering his face with 'sehara' goes to bride's place in palanquin where the accompanying wedding party ('barat'), comprising his relatives and friends is welcomed by bride's kith and kin ('milani') and served with a special feast. At the bride's place, marriage rituals like 'lagan-bedi' and 'kanyadan' are performed and bride and groom circumambulate the sacred fire amongst the chanting of holy 'mantras' by the priest. At the bridegroom's place, both bride and groom worship 'pipal' (the holy fig) and 'dadan' (*Punica granatum*) trees for their long happy married life. 'Muklawa' (ritual visit of daughter after marriage to its parent's house) ceremony takes place on the next day after marriage.

Divorce

It is uncommon. When the relation between a wife and husband becomes unsustainable, the legal separation is granted by court.

Death Rituals

Death rituals are performed by a 'Acharj' as per Sanatan Dharma. As death approaches, the sick man's body is lowered on the ground with head pointing to the North and smeared with fresh cow-dung. After death of the person, conch shells are blown and mustard oil lamp is lit and placed near it. Before the funeral rites, the dead body is bathed and covered with a colourful cloth and taken in bier made of bamboos to cremation ground. Before cremating the body, a 'pind-dan' (ritual feeding of balls made of cooked rice, sesame seeds and curd to the birds in the name of the died one) ceremony is performed. On the third day, bones and ashes ('astu') are collected and immersed in Ganges. Mourning is observed for thirteen days. On the tenth day, clothes are washed ('kapad-dhulai') and a priest recites the holy 'Garud Purana'. At the end of the mourning period a ceremony 'kriya-karma' (a religious rite for the peace of the departed soul) is performed. The annual death rite is called 'barkhi' and the last death rite, 'chabarkh' is performed in the fourth year.

Faith and Worship

All different communities of the area are deeply religious in nature, and have their own 'Kul-Devta' and 'Kul-Devi'. On each happy occasion in the family, these are being worshipped with all reverence. Every year on the eve of 'bisowa', people from all walks of life gather at Kaleshwar Mahadev for taking a holy dip in the pond.

Fairs and Festivals

The colourful fairs and festivals of Pragpur provide healthy recreation and marketing place for its people. On the day of festival, people generally cook special food, wear neat and clean clothes

Table III
Festivals and Fairs of Pragpur

Fairs and Festivals	*Occasion*	*Remarks*
Festivals		
Chet (Dholru)	First month of the lunar year	Brings happiness and prosperity in family.
Navratras	Durga-Asthmi in April and October	Appeasing goddess Durga for her blessings.
Haryali	16th July	Signifies advent of the rainy season.
Rakhri	Full moon day in Shrawan	Sisters tie sacred thread on the wrist of their brothers and prey for long healthy life.
Sair	September	Worshipping of local deity for prosperous future.
Dashera	September/October	Signifies victory of good over evil and celebrated by burning effigies of Ravana, Kumbakarana and Meghnath on 'Dashmi', i.e. 10th day of celebration.
Diwali	November	Symbolising celebration of returning of Lord Rama's to his kingdom from exile after killing Ravana.
Shivratri	February	Occasion for keeping fast to appease Lord Shiva to invoke his blessings.
Fairs		
Bisowa	13th April	Agrarian fair when people celebrated it by taking holy dip at Kaleshwar Mahadev.
Gugga	August/September	Gugga worshipping constitutes a form of snake cult.
Lohri	1st Magh (mid January)	Celebrated by eating and dancing around bonhomie of fire.

and worship at the temple. Some of the important festivals of the region 'chet' ('dolru'), 'navratras', 'bisowa' ('baisakhi'), 'haryali', 'rakhi' ('rakhri'), 'sair', 'dashera', 'diwali', 'lohri' and 'shivratri' (Table III).

OCCUPATION AND LIVELIHOOD

Agriculture and Agricultural Implements

Agriculture is the mainstay of the people of Pragpur. Wheat, gram and sugarcane are the rabi crops whereas rice, maize, pulses and potatoes constitute kharif crops for this subtropical region. On an average, the area receives about 150 cm of annual precipitation of which 75% occurs during monsoon season. Due to inadequate and irregular rainfall, the crops usually face moisture stress during the remaining period of the year. The water channels called 'kuhls' and 'naun' (Pl. 9A-D) are the main source of irrigation. 'Kuhls' are constructed along the field borders carrying water from streams, springs, 'naun' (Pl. 9A-D), and nallahs to the terraced fields. 'Arbi' (*Colocasia esculenta*), 'zimikand' (*Amorphophallus campanulatus*), potato (*Solanum tuberosum*) and 'adarak' (*Zingiber officinale*) are grown as cash crops. Wooden plough with a pointed iron tip is the main agricultural implement, and it is got fabricated from the local artisan. Ploughing is usually done with the help of a pair of bulls. Some of the tools (Pl. 10A) and their uses in agricultural operations are listed in Table IV.

TABLE IV
AGRICULTURAL IMPLEMENTS AND HOUSEHOLD ACCESSORIES USED BY LOCAL FOLKS

Local Name/s	*Name in English*	*Used For*
Tools		
Kulhari	Axe	Cutting wood
Tokri	Basket	Carrying manure and cow dung
Kera		Dropping seeds in plough furrows
Chabbal	Digger	Digging
Thora	Hammer	Breaking big stones
Al	Plough	Ploughing
Drat or drati	Sickle	Cutting branches of trees
Jungla	Yoke	Yoking bullocks for ploughing
Mai		Levelling the fields
Pottery		
Chillam		For hookah
Jhawa (Chaun)		Scrubbing dead skin from feet
Karuae		Keeping 'suhagi' on 'Karva Chauth' fast
Ghara	Pitcher	Carrying water.
Handko	Small pitcher	Storing milk, ghee or lassi (sweetened milk)
Basketry		
Chatai	Mat	Woven from palm leaves and paddy husk-Pral
Chhabra/Chhabri	Basket	Keeping vegetables and fruits
Chhikri	Small basket	Covering mouth of cattle during ploughing
Pakhi	Fan	A hand held device for creating a current of cool air
Kardu or Karandu	Basket with handle and lid	Keeping vegetables, eatables or wool
Peru	Huge basket	Storing grains
Semi-circular basket	Tokri	Horticulture and agricultural operations
Utensil		
Belan		Rolling chappatis
Chakla		Base for rolling chappatis
Ghara		Storing water
Madani		Churning milk or curd
Tawa		Baking chappatis on fire
Furniture		
Baksa	Wooden box	Keeping clothes and valuables
Choki	Small square table	For worshipping
Patra	Small table	Stool for sitting on ground
Mattresses		
Khinde	Cushion	Bed cushion stuffed with small piece of clothes
Pand	Mat	Mat woven with palm leaves and bamboos for sitting purposes
Tapar	Mat	Mat stuffed with straw for sitting purposes

Horticulture and Livestock

The subtropical conditions are congenial for fruits like orange, mango, guava, litchi, citrus, lokat, figs, Jaman, peach, pear, plum, pomegranate and banana. It is impossible to think of improving agriculture without having good cattle. The common livestock in the region are cow, buffaloes, goats, sheep, horses, donkey, ass, mule and ox. Poultry, sericulture and apiculture are some important cottage industries, which provide employment to farmers to supplement their income. In addition, the people grow plums, pears, jack and citrus fruits and mangoes; rear domestic animals and birds, weave woollen shawls and make a variety of beautiful handicrafts from bamboo strips (Pl. 10D).

THE METHODOLOGY

Collection of Ethnobotanical Data

For collecting first hand data regarding various uses and local names of plants and anthropological aspects, intensive and extensive ethnobotanical surveys of all the localities of the heritage village Pragpur, district Kangra H.P. were conducted over a period of five years (2001-2006). Field tours to the study area were planned in such a way as to collect ethnobotanically interesting species either in flowering or fruiting stage. Participatory Rural Appraisal (Etkin, 1993; Lipp, 1989; Hedberg, 1993; Pei and Long, 1998), observations and key informant (family heads, healers, old experienced and knowledgeable informants, medicinemen) interviews were used to collect information for the field survey. A questionnaire based on the proforma designed by Jain and Goel (1995) was prepared to gather data for this purpose; and the resultant information was recorded in the ethnobotanical field notebook. Uses of plants narrated by a person were cross-checked and considered notable only if the author observed actual application, or similar application was reported by at least three informants. For the medicinal uses, information pertaining to method of preparation (powder, aqueous extract, infusion or decoction), dosage and combination with other plant, if any was also noted.

The Collection Sites, Flowering and Fruiting Season of Plants

For future reference and study, the information pertaining to flowering and fruiting season has been enumerated in Table V.

TABLE V
FLOWERING AND FRUITING SEASON OF ETHNOBOTANICALLY IMPORTANT PLANTS COLLECTED FROM PRAGPUR

Name of the Plant	*Flowering and Fruiting Season*	*Name of the Plant*	*Flowering and Fruiting Season*
Abrus precatorius L.	May-August	*Aloe vera* (L.) Webb. and Berth	March-April
Acacia catechu Willd.	May-July	*Amaranthus spinosus* Linn.	May-June
Acacia nilotica Linn.	June-August	*Amaranthus viridis* Linn	May-June
Achras zapota L.	December-January	*Amorphophallus campanulatus* (Roxb.) Blume *ex* Decne.	April-June
Achyranthes bidentata L.	May-October	*Annona squamosa* Linn.	May-August
Acorus calamus Linn.	June-July	*Argemone mexicana* Linn.	April-May
Adhatoda vasica Nees	April-May	*Artemisia japonica* Thunb.	September-December
Aegle marmelos (L.) Correa	April-July	*Artocarpus integrifolia* Linn.	May-June
Agave americana L.	March-May	*Artocarpus lakoocha* Roxb.	April-June
Ageratum conyzoides Linn.	July-August	*Asparagus adscendens* Roxb.	October-November
Albizia lebbeck Benth.	April-May	*Bambusa nutans* Wall.	Flowering sporadically
Allium cepa Linn.	March-April	*Bauhinia vahlii* Wight and Arn.	April-June
Allium sativum Linn.	March-April	*Bauhinia variegata* Linn.	March-August
Aloe maculata All.	June-January		

(*Contd.*)

Name of the Plant	*Flowering and Fruiting Season*
Boerhaavia diffusa Linn.	July-September
Bombax ceiba Linn.	February-May
Bougainvillea spectabilis Willd	January-June
Brassica campestris Linn.	March-September
Bryophyllum calycinum Salisb.	March-May
Butea monosperma (Lam.) Kuntze.	March-April
Callistemon lanceolatus DC.	March-April
Canna indica L.	January-February
Cannabis sativa Linn.	June-September
Capsicum annuum Linn.	Greater part of the year
Carica papaya Linn.	Greater part of the year
Carissa inermis Vahl.	April-July
Carissa spinarum Hook. f.	May-June
Cassia fistula L.	March-July
Cassia occidentalis Linn.	May-September
Catharanthus roseus G. Don	Throughout Year
Cedrela toona Roxb.	May-June
Celastrus paniculatus Willd.	April-June
Celosia argentea L.	May-September
Centella asiatica (Linn.) Urban	April-May
Cestrum nocturnum L.	August-October
Chenopodium album Linn.	January-March
Cicer arietinum Linn.	March-April
Cinnamomum camphora (L.) Nees and Ebesm	April-August
Cissampelos pareira Linn.	May-November
Citrus aurantium L.	October-March
Citrus grandis (L.) Osbeck.	November-March
Citrus jambhiri Lush.	October-March
Citrus limon (L.) Burm. f.	November-February
Citrus medica L.	April-September
Coccinia grandis (L.) Voigt.	June-September
Colebrookia oppositifolia Smith.	April-November
Colocasia esculenta (Linn.) Schott.	July-October
Cordia dichotoma Forst.	April-July
Coriandrum sativum Linn.	October-December
Cucumis sativus L.	May-September
Cucurbita hispida Thunb.	June-October
Cucurbita pepo L.	June-September
Curcuma domestica (Medik.) Valeton.	September-October
Cuscuta reflexa Roxb.	September-October
Cynodon dactylon (Linn.) Pers.	August-November
Dalbergia sissoo Roxb.	May-July
Datura innoxia Mill.	June-August
Dendrocalamus hamiltonii Nees and Arn.	May-August
Dioscorea bulbifera Linn.	August-September
Dioscorea deltoidea Wall.	August-September

Name of the Plant	*Flowering and Fruiting Season*
Dodonaea viscosa (Jacq.) Linn.	February-April
Duranta repens L.	April-August
Elettaria cardamomum (L.) Maton	April-August
Emblica officinalis Gaertn.	April-September
Eriobotrya japonica Lindl.	July-September
Eruca sativa Lam.	February-April
Eucalyptus umbellata Domin.	March-April
Eugenia cerasoides Roxb.	April-July
Eugenia jambolana Lam.	May-July
Euphorbia royleana Boiss.	June-July
Ficus carica Linn.	March-August
Ficus glomerata Roxb.	April-July
Ficus religiosa Linn.	March-October
Ficus roxburghii Wall. *ex* Miq.	April-August
Flacourtia ramontchi L. Herit	March-June
Foeniculum vulgare Mill.	August-September
Fragaria chiloensis (Linn.) Duchesne.	May-June
Fumaria indica (Haussk.) Pugsley	February-March
Gossypium arboreum Linn.	July-September
Grevillea robusta A. Cunn. *ex* R. Br.	March-July
Grewia oppositifolia Roxb.	May-June
Helianthus annuus L.	April-July
Hibiscus rosa-sinensis L.	February-October
Holarrhena antidysenterica (Buch.-Ham.) Wall.	June-February
Holoptelea integrifolia (Roxb.) Planch.	March-May
Impatiens balsamina Linn.	July-September
Ipomoea cairica (L.) Sweet	June-November
Ipomoea carnea Jacq.	May-July
Jasminum multiflorum Roth.	February-April
Jasminum sambac (L.) Ait	February-April
Jatropha curcus Linn.	February-May
Juglans regia L.	April-October
Lagenaria siceraria (Molina) Standl.	July-September
Lagerstroemia indica L.	June-October
Lannea coromandelica (Houtt.) Merrill.	July-September
Lantana camara Linn.	June-October
Lawsonia alba Lam.	June-September
Leucaena leucocephala (Lam.) de Wit.	September-November
Litchi chinensis Sonner.	April-June
Luffa aegyptica Mill.	May-September
Lycopersicum esculentum Mill.	May-June
Mallotus philippinensis (Lam.)	November-April

Name of the Plant	*Flowering and Fruiting Season*
Muell.-Arg.	
Malvaviscus conzatii Greenm.	Major part of the year
Mangifera indica Linn.	March-June
Martynia annua L.	June-October
Melia azedarach Linn.	March-May
Melia azadirachta Linn.	June-September
Mentha piperita Linn.	July-September
Mentha spicata L.	March-July
Mimosa pudica Linn.	July-September
Mirabilis jalapa Linn.	June-September
Momordica charantia L.	June-October
Momordica dioica Roxb. *ex* Wall.	July-October
Morus alba Linn.	March-April
Mucuna pruriens L.	July September
Murraya koenigii Spreng.	May-June
Murraya paniculata (Linn.) Jack.	June-December
Musa sapientum Linn.	Throughout year
Nerium indicum Mill.	April-November
Ocimum basilicum Linn.	June-September
Ocimum sanctum Linn.	July-November
Opuntia dillenii Haw.	April-May
Oroxylum indicum Vent.	May-August
Papaver rhoeus L.	May-October
Parthenium hysterophorus Linn.	August-September
Pedilanthes tithymaloides (L.) Poit	August-September
Pennisetum purpureum Schum.	September-November
Phoenix sylvestris Roxb.	May-June
Phyllanthus urinaria Linn.	August-September
Pinus roxburghii Sarg.	March-May
Pistacia integerrima Stewart *ex* Brand	March-May
Plumbago zeylanica Linn.	March-May
Pogostemon benghalensis Burn. f	February-April
Polyalthia longifolia (Sonn.) Thwaites	February-April
Populus nigra L.	April-July
Premna latifolia Roxb.	July-September
Prunus amygdalus Baill.	April-August
Prunus domestica L.	April-August
Prunus persica Benth. and Hook. f.	February-May
Psidium guajava Linn.	July-September
Punica granatum Linn.	May-September
Putranjiva roxburghii Wall.	March-April
Pyrus communis L.	June-August
Pyrus pashia Ham.	March-May
Randia dumetorum (Retz.) Lam.	April-June

Name of the Plant	*Flowering and Fruiting Season*
Raphanus sativus Linn.	November-March
Ricinus communis Linn.	March-August
Rosa damascena Mill.	July-September
Rumex hastatus Don.	November-April
Saccharum arundinaceum Reitz.	October-December
Saccharum officinarum L.	January-April
Saccharum spontaneum Linn.	Throughout year
Salvia coccinea Linn.	June-August
Sapindus mukorossi Gaertn.	May-August
Saraca asoca (Roxb.) de Wilde	March-September
Sechium edule Sw.	June-September
Sesamum indicum DC.	July-September
Solanum erianthum Don.	July-October
Solanum melongena L.	June-September
Solanum nigrum Linn.	January-July
Solanum surattense Burm. f.	April-October
Solanum tuberosum Linn.	May-September
Solanum viarum Dun.	May-August
Sonchus asper Hill.	March-April
Spilanthes acmella Murr.	May-August
Stephania glabra (Roxb.) Miers.	June-August
Tabernaemontana divaricata R. Br.	July-October
Tagetes erecta Linn.	October-January
Tamarindus indica Linn.	May-August
Taraxacum officinale Wigg.	March-November
Tecoma grandiflora Loisel.	Summer season
Terminalia arjuna Wt. and Arn.	May-September
Terminalia bellirica (Gaertn.) Roxb.	May-September
Terminalia chebula Retz.	April-June
Thevetia neriifolia Juss. *ex* Steud.	July-September
Thuja orientalis L.	July-October
Tinospora cordifolia (Willid.) Miers	May-July
Trifolium alxandrinum	March-June
Trigonella foenum-graecum Linn.	May-August
Triticum aestivum Linn.	November-April
Verbascum thapsus Linn.	January-April
Vitex negundo L.	March-June
Vitis jacquemontii Parker	April-May
Wendlandia heynei (Roem. and Schult.) Sant and Merch.	April-June
Woodfordia floribunda Salisb.	March-June
Xanthium strumarium Linn.	September-November
Zanthoxylum armatum DC.	April-June
Zea mays L.	July-August
Zingiber officinale Rose.	June-September
Zizyphus jujuba Lamk.	June-August

Methodology of Study

The herbarium sheets of all the ethnobotanically important species were dried, preserved and mounted as per standard herborizing practices outlined by Jain and Rao (1977). The specimens were identified using regional floras and various revisionary and monographic works (Chauhan, 1999; Chowdhery and Wadhwa, 1984; Collett, 1902; Dhiman, 1976; Hooker, 1872-1897; Nair, 1977; Polunin and Stainton, 1984; Stainton, 1988) and later got authenticated by carefully matching with the specimens at the herbaria of Botanical Survey of India (BSI), Northern Circle and Forest Research Institute (FRI), Dehradun. Nomenclature of these taxa was confirmed from Bennet (1986) and Wielgorskaya (1995). Total importance values (TIV) of all the presently studied ethnobotanically important species were worked out as per the methodology given by Belal and Springuel (1996). Festivals and other ceremonial occasions on which plants are used were also attended and relevant information was documented. The novelty of the claims has been confirmed from The Wealth of India (Anonymous, 1948-1976) and the publications of the earlier workers (Agarwal, 1986; Ambasta, 1986; Jain, 1991; Jain and De Filipps, 1991; Kirtikar and Basu, 1935; Maheshwari, 1996; Nayar et al., 1989). Likewise, data on nutritional value of ethnobotanically important plants of the region have also been collected (Bennet et al., 1991; Dwivedi, 1993; Pruthi, 1976). Photographs of ethnically important plants in natural habitats were also clicked. The voucher specimens of all the presently studied taxa have been deposited in the herbarium section of Ethnobotanical Laboratory, Department of Biosciences, Himachal Pradesh University, Shimla.

Enumeration of Data

The ethnobotanical observations on uses of plants are presented with correct botanical names of the plants arranged in alphabetical order, followed by synonyms (wherever available), family, vernaculars, English, Sanskrit and Regional Names, Distribution, a brief description, Flowering and Fruiting period, Habitat Ecology, Voucher specimen number and Date of Collection, Plant Part/s Used, Folk Uses, Active Constituents, Biological Activity, Uses in Literature, Major Ayurvedic Formulations, Adulterant or Substitute, Warning (if any) and Trade. The new/interesting uses have been marked with an asterisk. Besides, appendices include information regarding the government notification pertaining to heritage status for Pragpur, indices to families, local names, nutritional value and glossary of words.

Ethnobotany and Traditional Practices of Shiwalik Himalaya

1. *Abrus precatorius* L. (Pl. 11A)

Syn. : *A. minor* Desv., *A. pauciflorus* Desv.
Family : Fabaceae.
Vern. : Rati.
English, Sanskrit and Regional Names:

Eng. : Indian-liquorice, Jerquity, Papernoster pea, Rosary pea, Weather plant;
Sans. : Angaravallari, Aruna, Bhilabhushana, Chatak, Gunja, Kanchi, Rakta, Raktala, Raktika;
Ass. : Gunja, Krishanala, Kakachinchi;
B. : Ghunhati, Gunj, Kawet;
Bo. : Ghungchi Gunja;
G. : Chanoti, Gunja;
H. : Chirmiti, Ghumghachi;
Kan. : Guengungi;
Kash. : Shangir;
Mal. : Irattimadhuram, Shekkunni;
Mar. : Chanoti, Kunch;
Oriya : Kainsho, Kotibepolo;
P. : Labri, Rattak;
Tam. : Adisamiyai, Attighurie-ghenza;
Tel. : Kukkutamu, Sinnaguruginja;
U.P. : Ghumchi, Labri, Rattak, Ujarki.

Distribution : Throughout India.

Description : A copiously branched woody climber having pinnate leaves and reddish or white flowers. Seeds bright scarlet with a black spot on the hilum.

Flowering and Fruiting : May-August.

Habitat Ecology : Wastelands. Open dry slopes and shrubberies.

Material Examined : EBH-HVP-855, 20.05.2005.

Parts Used : Roots. Leaves. Seed.

Folk Uses : ½ teaspoon of powdered roots given with honey once daily for 5 days to cure **whooping cough**. Seeds used as measure of **weight** by goldsmiths. Its paste mixed with root paste of *Plumbago zeylanica* Linn. good for applying locally on **skin diseases**.

Active Constituents : These are precol, abrol precasin, isoflavoquinones I, II, III (roots), abrine A and B, abricin, abridin, abric acid, hypaphorine, alanine, serine, abralin, abrin, abramine (seeds), xyloglucosyldelphinidin and p-coumaryl galloyl-glucosldephnidin (seed coat).

Biological Activity : Antitumorous, anthelmintic, antispermatogenic and spasmogenic activities +ve.

Uses in Literature : Known to be used as an abortifacient, alexiteric, antidote to poison, antiperiodic in preventing malaria, antiseptic, antitumorous, aphrodisiac, blood purifier, CNS depressant, colic, contraceptive, diuretic, emetic, febrifuge, laxative (vet.), masticatory, measure of weight for goldsmiths, purgative, refrigerant, symbolic (to ward off evil spirits), tonic; and for affections of hair and nervous system, anthrax (vet.), boils, bronchitis, convulsions, cough, cold, coma, congestion, decoration of baskets and bracelets, cuts, diarrhoea, dropsy, expulsion of placenta (vet.), eye complaints, facilitating delivery, flatulence, gall neck (vet.), gonorrhoea, haemorrhoids, headache, hemicrania, indigestion, inflammation, insect bites (vet.), liver disorders, menorrhoea, menstrual pains, muscular pains, nervous disorders (vet.), paralysis, parturition, peptic ulcers, pneumonia, retained placenta (vet.), retention of urine, rheumatism, ringworms, scabies, sciatica, scorpion bite, skin eruptions, snake bite (vet.), sore throat, stiffness of the shoulder joints, swelling, syphilis, ulcer (vet.), and wounds (vet.) **(Ambasta, 1986; Bhatt *et al.*, 2002; Chauhan, 1999; Dager & Dager, 1996; Datt, 1996; Gill & Nyawuame, 1994; Girach *et al.*, 1998; Goud & Pullaiah, 1996a, b; Jain, 1991, 1996; Katewa *et al.*, 2001; Khanna, 2002; Kothary & Moorthy, 1996; Kshirsagar *et al.*, 2003; Kumar & Chauhan, 2005; Kumar & Pullaiah, 1998; Kumar *et al.*, 2004; Kurian, 1995; Malik, 1996; Mandal & Basu, 1996; Mishra *et al.*, 1996b; Nazarudeen *et al.*, 1996; Painuli & Maheswari, 1996; Pawar & Patil, 2004; Prasad *et al.*, 1996; Pullaiah & Kumar, 1996; Ram *et al.*, 2002; Rama Rao & Henry, 1996; Rao & Pullaiah, 2001; Ravisankar & Henry, 1992; Reddy *et al.*, 1998; Retnam & Martin, 2006; Saini, 1996b; Satapathy & Brahmam, 1996; Sen & Behera, 2003; Sharma, 1996; Singh & Pandey, 1996; Singh & Prakash, 1994; Singh *et al.*, 1996b; Siwakoti & Verma, 1996; Verma & Chauhan, 2006).**

Major Ayurvedic Formulations : Gunjadi-taila, Gunja-pralepa, Mritasanjivani-gutika, Nilibhringadi-taila, Uchatta.

Trade : Sold in market as 'Gunchi' and 'Rati'.

Warning : Seeds highly poisonous as these contain aborin—a highly toxic protein similar to ricin and should not be taken orally.

Substitute or Adulterant : Roots used as adulterant or substitute for 'Mulhati' (*Glycyrrhiza glabra* Linn.).

2. *Acacia catechu* Willd. (Pl. 11B)

Syn. : *A. polyacantha* Willd., *A. wallichiana* DC, *A. catechuoides* DC, *A. sundra* Bedd., *M. catechu* Roxb., *M. catechuoides* Roxb.

Family : Fabaceae.

Vern. : Khair.

English, Sanskrit and Regional Names:

Eng. : Black catechu, Catechu, Nigraum, Cutuch, Pegu, Catechu;
Sans. : Dantadhavana, Jivumangalaya, Vakrakantaka;
Ass. : Kat, Koir;
B. : Kahyer, Kuth;
Bo. : Khaderi, Lalkhair;
G. : Kher, Kheribayal;
H. : Babul, Katha, Khair;
Mal. : Kadaram, Senkainnali;
Mar. : Khaderi, Lalkheir;
Oriya : Bimbu, Khodira;
Tam. : Karangalli, Kasukkalti;
Tel. : Khadiramu, Mallasandra.

Distribution : Throughout Peninsula and from Indus eastwards to Assam and N. India: 500-1,350 m.

Description : Trees with recurved prickles, rough dark coloured bark and bipinnate leaves. Flowers pale-yellow, sessile. Pods stalked, glabrous-oblong, flat shining with a triangular beak.

Flowering and Fruiting : May-July.

Habitat Ecology : Roadsides. Drier areas. Wastelands. Hilly slopes.

Material Examined : EBH-HVP-859, 10.06.2005.

Part Used : Heartwood ('Katha').

Folk Uses : 'Katha' obtained from its heartwood used for curing ulcerated or **spongy gums**, **tonsils**, ulceration of **mouth** and **throat diseases**.

Active Constituents : These are catechin, catechu, tannic acid, tannin (bark), α-β-γ catechin and 1-epicatechin (Wood).

Biological Activity : Antiviral, anti-inflammatory, hepatoprotective, hypoglycaemic and spasmolytic activities confirmed.

Uses in Literature : So far reported in India as an abortifacient, anticancerous, astringent, aphrodisiac, dentrifice, digestive, dye, expectorant, fibre, gum arabic, host for lac, insect flies, adhesive for wood, tonic; and for affection of gums, asthma, bleeding from nose, blood dysentery, bronchial affection, burns, curing impotency in excess, chest pain, child birth, colic pain, conjection of faces, conjuctivitis, constipation (vet.), cough, diarrhoea with pyrosis, eczema, gonorrhoea, gravel pthisis, haemoptysis, hoarseness, increasing the secretion of milk after delivery, inflammation, intermittent fevers, itching, leprosy, leucorrhoea, menorrhagia, menstrual complaints, paintings and muscular growth (vet.), otorrhoea, tonsils, pain in chest, piles, printing, rheumatism, scurvy, skin disease (vet.), suppressing sexual desire, swelling (vet.), syphilis, tanning, toothache, ulcer and wounds (vet.) **(Ahuja, 1965; Ambasta, 1986; Asolkar *et al.*, 1992; Chauhan, 1999; Jain, 1991; Jain *et al.*, 1991; Jha & Basak, 1994; Katewa *et al.*, 2001; Maheshwari & Singh, 1987; Masih, 2003; Mishra *et al.*, 1996a; Retnam & Martin, 2006; Saini, 1996a; Shukla &**

Verma, 1996; Siwakoti & Verma, 1996; Watt, 1972).

Major Ayurvedic Formulations : Arimedadi-taila, Chandanadi-vati, Dasnar-churna, Khadiraadi-kwatha, Khadira-astaka, Khadiradi-gutika, Khadiradi-vati, Khadirishta, Khadirsar, Kutajarishta, Lavangadi-vati.

Trade : Kathaa, Katthaa, Paakhraa.

3. *Acacia nilotica* Linn. (Pl. 11C)

Syn. : *A. arabica* (Lam.) Willd.

Family : Fabaceae.

Vern. : Babul, Kikkar.

English, Sanskrit and Regional Names:

- **Eng.** : Arabic Tree, Black Babool, Indian Babul;
- **Sans.** : Babbuka;
- **H.** : Babul, Kikkar;
- **P.** : Kikkar;
- **Tam.** : Karuvelei;
- **Tel.** : Nallatumma.

Distribution : Throughout Warmer regions of India.

Description : A tree having young pubescent branches and straight stipular spines. Flowers yellow, sessile; pod stalked, pubescent and moniliform.

Flowering and Fruiting : July-September.

Habitat Ecology : Along Roadsides.

Material Examined : EBH-HVP-782, 05.07.2003.

Parts Used : Leaves. Wood.

Folk Uses : Fresh leaves chewed twice a day for five days to cure **mouth sores.** Twigs used for **scouring teeth.** Wood used for **agricultural implements**.

Active Constituents : Steroids, glycosides, flavonoids, tannins, gums, amino acids, fatty acids, organic acids, polyphenols; tannin (av.c. 12%), + catechin-5-gallate, gallic acid its methyl ester naringenin, catechin, catechol, pyrogallol (bark); stearic acid, kaemferol glucoside, isoquercetin, leucocynidin (flowers) have been isolated.

Biological Activity : Stem bark hypoglycaemic, antiprotozoal, spasmolytic, CVC active and analgesic.

Uses in Literature : Known in India for agricultural implements, body swelling, burning sensation in eyes, cholera, cold, cooling effect (vet.), cough, dental caries, diarrhoea, dysentery, ear discharge, earache, fencing, fodder, indigestion (vet.), leprosy, leucorrhoea, mouth sores, musical instruments, premature ejaculation, pyorrhoea, rinderpest, syphilitic ulcer, toothache, whooping cough; and as an astringent, galactagogue, spermatorrhoea

(Ambasta, 1986; Bhatt *et al.*, 2002; Kalpesh *et al.*, 2004; Khanna, 2002; Khanna & Kumar 2000; Kshirsagar *et al.*, 2003; Kumar & Chauhan, 2005, 2006; Kumar *et al.*, 2004; Mohanty *et al.*, 1998; Pandey & Verma, 2002; Patel *et al.*, 2003; Punjani, 1998; Ram *et al.*, 2002; Sen *et al.*, 2002; Singh, 2003; Sen & Batra, 1997; Sen & Behera, 2003).

Major Ayurvedic Formulations : Khadiradi-vati, Khadirishta, Khadie-sar, Larangadi-vati, Chandaradi-vati, Kutajarista.

Adulterant or Substitute : Seeds used as adulterant or substitute for 'Kaladana' *Ipomoea hederacea* (Linn.) Jacq. and for bark *Quercus* spp.

4. *Achras zapota* Linn. (Pl. 11D)

Syn. : *A. sapota* Linn., *Mimusops manilkara* G. Don.

Family : Sapotaceae.

Vern. : Chiku.

English and Regional Names:

Eng. : Sapodilla Plum, Bully Tree;

H. : Chiku.

Distribution : Throughout India.

Description : An evergreen small tree with pubescent branches; shining, dark green, oblong-elliptic leaves crowded at apex and solitary—axillary white flowers. Fruits brown and globose berry. Seeds black, polished and embedded in a light-brown pulp of fruit.

Flowering and Fruiting : December-January.

Habitat Ecology : Cultivated.

Material Examined : EBH-HVP-781, 15.12.2003

Part Used : Fruit.

Folk Uses : Fruit **edible**.

Active Constituents : Bark contains latex 20-25% (gutta-percha), tannin (11.8%), α- and β-amyrinacetate, taraxerol, ursolic acid.

Uses in Literature : Known in India as edible, base for chewing gum, dental surgery, colouring sails and fishing tackles **(Ambasta, 1986)**.

5. *Achyranthes bidentata* Blume (Pl. 12A)

Syn. : *A. wightiana* Wall., *A. lanceolata* Wall., *A. aspera* Wall.

Family : Amaranthaceae.

Vern. : Puthkanda.

Regional Names:

Ass. : Apamarga, Bankhat;

H. : Adhajhara, Oga;

Kash. : Puthkandiari;
U.P. : Latzeera.

Distribution : Temperate and subtropical Himalaya.

Description : An undershrub with stiff stem and thin membranous ovate-lanceolate leaves. Flowers greenish-white, numerous, stiffy deflexed and having spinous bracteoles. Seeds subcylindric, inverted, truncate at apex and rounded at base.

Flowering and Fruiting : June-September.

Habitat Ecology : Wastelands. Grasslands. Drier places.

Material Examined : EBH-HVP-724, 28.06.2002.

Parts Used : Roots. Seeds.

Folk Uses : ½ teaspoon of powdered roots prescribed twice a day for 5 days with black pepper (*Piper nigrum* Linn.) and honey to cure **cough**. Gargles of its juice good for **throat infection***. Also, its amulet tied around waist of pregnant women to facilitate **delivery**. Seeds considered good for **suppressing hunger***.

Active Constituents : These are ecdysterone, ecdysone, emodins and physcion (roots).

Biological Activity : Antimicrobial (root extract) and abortifacient (stem bark) activities confirmed.

Uses in Literature : Reported so far as an astringent, diuretic, galactagogue (vet.), tooth brush, tooth powder; and for asthma, blisters in mouth, cholera, dental caries, diarrhoea, edible purposes (leaves, seeds), facilitating delivery, hemicrania, indigestion, jaundice, pyorrhoea, scorpion stings, scrofula, skin diseases, swelling, toothache and whooping cough **(Chhetri, 2005; Chowdhery, 1996; Jain, 1991; Jain *et al.*, 1991; Kapur & Singh, 1996; Lalramnghinglova, 1996; Manandhar, 1990, 1996; Molla & Roy, 1996; Pawar & Patil, 2004; Retnam & Martin, 2006; Singh *et al.*, 1980; Verma & Chauhan, 2006)**.

Major Ayurvedic Formulations : Abhaya, Agastiharitaki, Agnimukha, Apamargak-sara, Lavana, Gudapippali, Jyotismati-taila.

6. *Acorus calamus* Linn. (Pl. 12B)

Family : Araceae.

Vern. : Barae, Barre, Barya.

English, Sanskrit and Regional Names:

Eng. : Calamus, Sweet flag;
Sans. : Bhadra, Galani, Golomi, Jalaja, Jatila, Kunga, Shadagrantha, Ugra, Ugragandhaha, Vacha;
B. : Bach;
G. : Gandhilovaj, Godavaj, Vekhand;
H. : Bach, Ghor, Gorbach, Ghorbach;

P. : Bariboj, Wach;
Tam. : Vadaja, Vasa, Vashambutel, Wasa.

Distribution : Throughout India.

Description : Aromatic marshy herbs. Rootstock branched and creeping. Leaves bright green, thickened in middle and wavy margin. Spadix sessile, dense flowered and possesses yellow-green flowers. Berries turbinate, few-seeded and oblong seeds.

Flowering and Fruiting : June-July.

Habitat Ecology : Frequent along water banks and marshy places.

Material Examined : EBH-HVP-790, 28.06.2003.

Parts Used : Leaves. Rhizome. Whole Plant.

Folk Uses : Roasted rhizome in ghee given to check **flatulence***. Small piece of rhizome chewed for cure of **cough**. Whole plant considered good for **stomachache**. Also, paste of its rhizome and leaves in mustard oil makes good lotion for **joint pains** and **body massage*** of children.

Active Constituents : Main ingredients are vit. C, luteolin-6, 7-diglucoside (leaves, rhizome), oxalic acid 0.078 and calcium 0.006% (leaves), asarone, α pinene, d-camphene, calamene, calamenol and sesquiterpenes (oil extracted from rhizome), acorin, calamene, tannin, mucilage starch, vit. C, fatty acid, and calcium oxalate (roots).

Biological Activities: Rhizome exhibitis CNS depressant and sapsmogenic activities.

Uses in Literature : Known earlier for the treatment of abdominal pains, aching joints, arthritis, asthma, bodyache, bronchitis, cold, cough, cuts, diarrhoea, dysentery, dyspepsia, epilepsy, fever, headache, heart and lung trouble, hysteria, indigestion, inflammation, insect bites, leucorrhoea, malaria, pain in neck, pthisis, rheumatism, skin disease, snake bites, teething in children, throat pain, wounds (vet.); and as analgesic, emetic, sedative, insecticidal and tonic (**Ambasta, 1986; Borthakur *et al.*, 2004; Chak & Sharma, 1965; Chhetri, 2005; Dandiya & Sharma, 1961, 1962; Gogoi & Borthakur, 2001; Jain, 1991; Kumar & Jain, 1998; Mohanty *et al.*, 1998; Nath & Borodoloi, 1989; Retnam & Martin, 2006; Upadhye & Kulkarni, 2004; Verma & Chauhan, 2006**).

Major Ayurvedic Formulations : Abana tablets, Ashwagandharista, Bacha-churna, Chandra-prabha-vati, Laghu, Mahashankha-vati, Sanjivani-vati, Saraswat-churna.

Trade : Decorated and transversely sliced rhizomes sold in market.

Remarks : Rhizome emetic in large doses. Also it may cause disturbed digestion and gastroenteritis.

Adulterant or Substitute : Rhizome used as adulterant or substitute for that of 'Galangal' (*Alpinia galangal* Linn.) and 'Bish' (*Aconitum* spp.).

7. *Adhatoda vasica* Nees (Pl. 12C)

Syn. : *A. zeylanica* Medic., *Justicia adhatoda* L.
Family : Acanthaceae.
Vern. : Safed basuti.
English, Sanskrit and Regional Names:

Eng. : Malabarnut leaf, Nut tree;
Sans. : Arus, Atarusha, Bhisangmata, Raktappittaghni, Vajidantakaha, Vasaka;
Ass. : Herb uksha;
B. : Bakas, Basaka patra, Vasaka;
Bo. : Adalsa, Vasuka;
G. : Adalsa, Adsoge, Adulaso, Arduosa, Arusa, Arusha, Paandru;
H. : Adalsa, Adarsa, Arusa, Banasa, Vasaka;
Kan. : Adusoge, Attalotakam;
Mal. : Atalotakam;
Mar. : Adulsa, Baksha;
Oriya : Rottoomuli;
P. : Bhekar;
Tam. : Adadoli, Kattumurunghai;
Tel. : Adasara, Addasaramu;
U.P. : Adusa, Arusa, Rusa, Roos.

Distribution : Distributed in India from Panjab-Assam: 500-1,600 m.

Description : An evergreen glabrous dense shrub with dark green, elliptic-lanceolate leaves and white flowers in dense axillary pedunculate spikes. Seeds orbicular, oblong, tubercular-verrucose and glabrous.

Flowering and Fruiting : March-May.

Habitat Ecology : Open slopes. Damp places. Wastelands.

Material Examined : EBH-HVP-805, 28.03.2004.

Parts Used : Root. Leaves. Flowers.

Folk Uses : 10 ml. decoction of roots given twice a day for 2-4 days to cure **high fever**. 1 teaspoon decoction of leaves (twice daily for 5 days) checks **cough**. Tender leaves given to cattle for 4-5 days for **loss of appetite*** and also as **fodder***.

Active Constituents : Active principles are due to vascinin, vascinine, vascicone vasinone, vascine, β-sitosterol, tritriacontane, vasicinine, vascicinol, vascicoline, adnatodine, vasciolinone, antistione (leaves), vasicinol, 6-hydroxyvasicine (root), vasicoline, adhatodine, anisotine, betaine, 6-hydroxypaganine (plant).

Biological Activities : Alkaloid vasicine acts as uterine stimulant, uterotonic, abortifacient, bronchodilator, respiratory stimulant and hypotensive. Root extr. exhibits positive hypoglycaemic activities.

Uses in Literature : Known to be used as an anthelmintic, antipyretic, antiseptic, blood purifier, edible, expectorant, febrifuge, insect repellent for stored grains, psychoactive plant, insecticidal, fuel for brick burning, green manure; and considered useful for abortion, asthma, boils, bronchitis, child birth, cold, cough, dislocation of joints, dysentery, eczema, eye complaints, fencing, gastric problems, gonorrhoea, headache, heart troubles, jaundice, loss of memory, malaria, pthisis, pulmonary affections, rheumatism, swelling and making beads of rosaries, ingredient of 'Chyawanaprasha', toothache, wounds, wrapping paddy and preparation of rice bear (**Ambasta, 1986; Aswal, 1996; Borthakur, 1993; Borthakur & Sarma, 1996; Borthakur *et al.*, 2004; Chopra *et al.*, 1956; Chowdhury *et al.*, 1996; Dixit & Pal, 1996; Girach, 1992; Girach *et al.*, 1996; Gogoi *et al.*, 2003; Huidrom, 1996; Jain *et al.*, 1968, 1991; Kapur & Singh, 1996; Khanna & Kumar, 2000; Kurian, 1995; Lalramnghinglova, 1996; Mehrotra *et al.*, 1995; Mokat & Deokule, 2004; Negi *et al.*, 1993; Prasad *et al.*, 1996; Punjani, 1998; Raghupathy & Mahadevan, 1991; Rama Rao & Henry, 1996; Retnam & Martin, 2006; Sharma & Borthakur, 2003; Sharma & Sood, 1997; Sikarwar, 1996; Srivastava *et al.*, 2003; Swami & Gupta, 1996; Tirkey, 2004; Tirkey *et al.*, 2003a, b; Varma, 1997; Watt, 1972**).

Major Ayurvedic Formulations : Asokarista, Chandanadi-taila, Chyawanaprasha, Kanakasava, Kasamitra, Kasni, Punarnavarishta, Pradarai-rasa, Vasaaca-churna, Vasakamadhu, Vasakasava, Vasarista, Vasaputaphaka, Vasavaleha, Viaspanaka.

Trade : Adusa, Baansapatti, Vasakapatti.

Adulterant or Substitute : Roots used as an adulterant or substitute for Snake root or 'senega' (*Polygala senega* Linn.).

8. *Aegle marmelos* (L.) Correa (Pl. 12D)

Syn. : *Feronia pellucida* Roth., *Crateava marmelos* L., *C. religosa* Ainstie.

Family : Rutaceae.

Vern. : Bel.

English, Sanskrit and Regional Names:

Eng. : Bael fruit tree, Bengal Quince, Holy fruit, Indian Quince, Stone Apple;
Sans. : Asholam, Adhavaruha, Atimangaliya, Bilva, Duraruha, Mahakapithakya, Malura, Shandilya, Satyadharma, Tripatra;
Ass. : Bel;
B. : Bela, Vilva;
Bo. : Bila;
G. : Bil, Billy;
H. : Bel, Bili, Siriphal, Sirphal;
Kan. : Bilvapatra;
M.P. : Bael;
Mal. : Kuvulam, Mavilamu;
Mar. : Bael, Bel;
Oriya : Biliva, Sripholo;
Tam. : Alvuvigam, Iyalbudi;
Tel. : Bilvamu, Sailushamu.

Distribution : Throughout India.

Description : A small or medium-sized glabrous tree armed with sharp straight axillary thorns and 3 glabrous, ovate-lanceolate leaflets. Flowers greenish-white, sweet scented and in short axillary panicles. Berries globose, yellowish and having woody rind.

Flowering and Fruiting : April-July.

Habitat Ecology : Wastelands. Open slopes. Drier areas. Temple courtyards.

Material Examined : EBH-HVP-713, 28.04.2001.

Parts Used : Leaves. Fruits. Fruit Gum.

Folk Uses : Ripe fruits **edible** and considered **refrigerant**. 10g fruit gum in combination with 10g 'gur' (jaggery) taken twice a day to cure **dysentery**. Its leaves ('Bilpatri') offered to appease Lord Shiva. Also these are used in **death rituals***. The gummy substance obtained from fruits used as an **adhesive**.

Active Constituents : Main constituents are aegeline, γ-sitosterol (leaves), umbelliferone, dictamnine, xanthotoxol, xanthotoxin, scoparone, isopimpinellin, isoimperation, N-2-methoxy 2 (4 methoxy phenyl), ethyl cinnamamide, marmelin, bergapten, marmesin, osthol and auraptin (fruit pericarp).

Biological Activity : Hypoglycaemic activity +ve.

Uses in Literature : Known to be used in India as abortifacient, adhesive, appetiser, astringent, beads, candy, cardiac depressant, cementing mixture, fish poison, ingredient of 'Chyawanaprasha', laxative, opacity of cornea, psychoactive plant, refrigerant, sacred, scorbutic, soaps, soothing agent, sweet drinks, fuel, tree motif of Shiva temples, substitute for tea, tannin and varnish, wish granting tree; and for abdominal accumulation of cellular tissue accompanied with constipation and jaundice, abdominal pain for abscess, allaying urinary troubles, anthrax (vet.), boils (vet.), bowel complaints, breast pains, bronchitis, burning sensation of soles of the feet, catarrhs, chest pains, cholera, cold, convulsions, diabetes, diarrhoea (vet.) dyeing, dysentery, edible purposes, eye complaints, gastric trouble, headache, heart palpitation, hyperacidity, indigestion, injuries, irritation of alimentary canal, jaundice, joint pains (vet.), keeping evil spirits away, menstrual disorders, mouth sores, nausea, night fever, paintings (Mithila), skin infections, spermatorrhoea, sunstroke, toys, tympany, vomiting and wounds (**Ambasta, 1986; Aminuddin & Girach, 1993; Bajpayee & Dixit, 1996; Bhatt *et al.*, 2001; Chowdhury *et al.*, 1996; Datt, 1996; Dwarakan & Ansari, 1996; Goud & Pullaiah, 1996b; Huidrom, 1996; Jain *et al.*, 1994; Jha & Basak, 1994; Kadel & Jain, 2006; Katewa & Arora, 1997; Kothari & Moorthy, 1996; Kshirsagar & Singh, 2000; Kurian, 1995; Lalramnghinglova, 1996; Manandhar, 1990; Masih,**

2003, Mehrotra *et al.*, 1995; Mishra *et al.*, 1996a; Mohanty *et al.*, 1998; Negi & Gaur, 1991; Painuli & Maheswari, 1996; Patole & Jain, 2002; Pullaiah & Kumar, 1996; Rama Rao & Henry, 1996; Ravishankar & Maheshwari, 1992; Reddy *et al.*, 1998; Retnam & Martin, 2006; Sen & Behera, 2003; Sharma & Singh, 2001; Singh, 2003; Singh & Rao, 2003; Singh *et al.*, 1997; Srivastava *et al.*, 2003; Tirkey *et al.*, 2003a, b; Venkataratnam & Venkata Raju, 2004; Vihari, 1995; Watt, 1972).

Major Ayurvedic Formulations : Amritarista, Bilva-churna, Bilvadi-kwatha, Bilvadi-leha, Bilvadi-majadi-yog, Bilwamooltwak, Brihatgangadhara-churna, Chyawanaprasha, Dasmula-taila, Dhanyapanchka-kwatha, Gangadhara-churna, Mahanarayan-taila, Pushyanuga-churna, Vatsakadi-kwatha.

Trade : Dried pulp roots and 'Belgiri' fruits sold in the drug markets of India.

9. *Agave americana* L. (Pl. 13A)

Syn. : *A. cantula* Roxb.

Family : Agavaceae.

Vern. : Koaed, Kuer Chatyan.

English, Sanskrit and Regional Names:

- **Eng.** : American Aloe, Carata, Century plant;
- **Sans.** : Kantala;
- **B.** : Banskeora, Bilati-ananash, Bilatipat, Carata, Koyam;
- **G.** : Fangli-kunvara;
- **H.** : Banskeora, Barakanwar, Hatisengar, Kantala, Rakas-pattah, Ramkanta;
- **Kan.** : Bhuttale, Budukattalenaru;
- **Mal.** : Panam, Katrazha;
- **Mar.** : Parkand;
- **Tam.** : Anaik-kat-razhai, Pithakalabuntha;
- **Tel.** : Rakshimatalu.

Distribution : Cultivated as hedge plant throughout India.

Description : Stout shrubby rhizomatous plants. Leaves stout, variegated, narrow lanceolate, spine tipped, spongy margins and in lux rosette. Flowers in large branched cluster.

Flowering and Fruiting : March-May.

Habitat Ecology : Wastelands. Roadsides. Drier areas.

Material Examined : EBH-HVP-706, 19.03.2001.

Parts Used : Leaves. Shoots.

Folk Uses : Tender shoots ('Mannu') **edible**. Leaves used to **tie** bundle of grasses and also as **fodder** for animals. Fibre extracted by thatching leaves kept in water for 7 days are used for making **ropes** and **cordage**.

Active Constituents : Active constituents of leaves are hecogenin, cholrogenin, 10 steroidal saponins, agavoside A, B, C, D, E, F, G, H, I, glucose, galactose, xylose and rhamnose.

Biological Activity : Antibacterial and anticancerous activities found to be positive.

Uses in Literature : Known so far as anthelmintic, antisyphilitic, fire wood, fish poison, fodder, diuretic, vegetable; and for abortion, boils, burns, constipation, cuts, dropsy, dysentery, earache, expelling worms, fencing, fever, goitre, paralysis, skin diseases, syphilis, venereal sores, veterinary diseases, waist pain, mats and cordage, thatching warts (**Ambasta, 1986; Chopra *et al.*, 1956; Goel *et al.*, 1984; Jain, 1991; Jain *et al.*, 1991; Kapur, 1996; Kumar & Jain, 1998; Lalramnghinglova, 1996; Maliya & Singh, 2003; Mittre, 1981; Punjani, 1998; Rana *et al.*, 1996; Rao, 1996; Reddy *et al.*, 1998; Saxena & Vyas, 1981; Singh, 1996b; Watt, 1972**).

Adulterant or Substitute : Extr. of roots and leaves used as an adulterant or substitute for Sarsaparilla (*Smilax* spp.).

10. *Ageratum conyzoides* Linn. (Pl. 13B)

Syn. : *A. cordifolium* Roxb.

Family : Asteraceae.

Vern. : Phulanu.

English and Regional Names:

- **Eng.** : Goat weed, White weed;
- **B.** : Dochunty;
- **Bo.** : Osari, Sahadevi;
- **G.** : Ajgandha, Gandharisedardi, Mankdamari;
- **M.P.** : Koobhi;
- **Mal.** : Appa, Muryanpacha;
- **Mar.** : Ghaneraosadi;
- **Oriya** : Boksunga, Poksunga.

Distribution : Throughout India.

Description : An annual erect hairy herb; leaves opposite, petaloid, ovate and flowers in small purple heads. Achenes black; pappus awned.

Flowering and Fruiting : July-August.

Habitat Ecology : Edges of Cultivated Areas. Wasteplaces.

Material Examined : EBH-HVP-860, 21.07.2005.

Part Used : Whole Plant.

Folk Uses : Plant extr. used for **repelling insects**.

Active Constituents : These are phenols (eugenol) 5% and phenol ester (leaves, flowers), friedelin, β-sitosterol, stigmasterol, α-sphinasterol, coumarin, flavonoids and anti-gonadotropic hormone (plants).

Biological Activity : Antibacterial, anthelmintic and anticancerous activities confirmed.

Uses in Literature : Reported earlier as antilithic, antiseptic, haemostatic, insecticidal, nerve tonic, shampoo, symbolic; and for accelerating delivery, body swellings, boils, burns, cancer, cuts, diarrhoea, dysentery, fever, flatulence, headache, healing, kidney stone, leprosy, muscular pain, piles, prolapsed anus, ringworm, scabies, sores, skin diseases, tetanus, uterine disorders, wounds and edible purposes **(Ambasta, 1986; Augustine & Sivadasan, 2004; Baruah & Sarma, 1984; Chetri, 2005; Jain, 1991; Kumar & Jain, 1998; Sikarwar, 2002; Sinha, 1986; Upadhyay & Singh, 2005)**.

Adulterant or Substitute : Leaves used as substitute for leaves of 'Ayapana' (*Eupatorium triplinerve* Vahl.).

11. *Albizia lebbeck* Benth. (Pl. 13C)

Syn. : *A. latifolia* Boivin., *Acacia speciosa* Willd., *A. sirisa* Ham., *A. lebbeck* Willd., *Mimosa speciosa* Jacq., *M. sirisa* Roxb., *M. lebbeck* L.

Family : Fabaceae.

Vern. : Sirin.

English, Sanskrit and Regional Names:

Eng. : Parrot tree, Siris tree, Sizzling tree;
Sans. : Barhapuspha, Bhandi, Bhandika, Kapitaka, Kapitana, Krishnasiris, Lomashapushpaka, Murdhapushpa, Plavaga, Shankiniphala, Vishavghati, Vishahanta;
B. : Siris, Sirisha;
Bo. : Garso, Harreri, Siris;
G. : Kaliosavas, Kalokansakio, Kalosadasadu, Kaloshirish, Pilosarshio;
H. : Garso, Kalshish, Kalsir, Lasrin, Mathirshi, Shiris, Shirish, Sirai, Sirin,Tantia;
Kan. : Baggera;
M. : Vagu;
Mal. : Kattuvaka, Nenmani, Vaku;
Mar. : Chichola, Kalashiras, Mothasiras, Shirish;
Oriya : Vaghe, Kotvaghe;
Tam. : Adukkavajai, Kalindi, Kattuvagai, Pandil, Perunvagai, Piram, Undugam;
Tel. : Dirisan.-amu, Girisamu, Kapitanamu, Paddadirisan-amu, Sirishamu, Siriso.

Distribution : Throughout India ascending up to 1,650 m.

Description : An unarmed deciduous tree having bipinnate leaves and white, fragrant flowers in globose, umbellate heads. Pods linear-oblong, pale-yellow, smooth shining with ellipsoid-oblong, pale brown seeds.

Flowering and Fruiting : April-May.

Habitat Ecology : Wasteplaces. Along sides of cultivated fields; frequently planted.

Material Examined : EBH-HVP-743, 16.04.2002

Parts Used : Leaves. Wood.

Folk Uses : Wood used for making **furniture** and agricultural implements

(plough, 'jungla'). Leaves used as **fodder** and twigs as **fire wood**.

Active Constituents : Flavandiols, melacacidin, lebbeacacidin (heart wood), d-catechin, leucocyanidin, condensed tannins (bark), vicenin-2, γ-cynoutrin, rutin, myricitin, robibin and albegenin (leaves) are the main ingredients.

Biological Activity : Antiasthmatic and antianaphylactic activities found to be +ve.

Uses in Literature : Known as an alternative, antidote to poison, aphrodisiac, astringent, camel fodder, detergent, expectorant, fire wood, psychoactive plant, restorative, symbolic (god of love uses its flowers on arrow of desires), tannin, tonic; and for agricultural implements, arthritis, asthma, boats, boils, carriages, cataract, cold, carbuncle, cooling, cough, diarrhoea, dyeing, dysentery, edible purposes, eruptions, eye troubles, fever (vet.), furniture, gonorrhoea, gum ulcers, house building, infected teeth, interior decoration, leather, leprosy, leucoderma, night blindness, oil mills, picture frame, piles, railway carriage works, reducing enlargement of cervical glands, scorpion sting, snake bite, sugar cane crushers, swelling, toy manufacturing, tympany (vet.) and ulcer **(Ambasta, 1986; Chauhan, 1999; Gill & Nyawwame, 1994; Goud & Pullaiah, 1996b; Jain, 1991; Jain & Patole, 2001; Khanna *et al.*, 1996b; Kurian, 1995; Raghupathy & Mahadevan, 1991; Ram *et al.*, 2002; Rama Rao & Henry, 1996; Siwakoti & Verma, 1996; Srivastava *et al.*, 2003; Watt, 1972)**.

Major Ayurvedic Formulations : Deshanglepa, Mahasirisagad, Sirisarishata, Sirisavlena.

12. *Allium cepa* Linn. (Pl. 13D)

Family : Liliaceae.

Vern. : Piaz.

English, Sanskrit and Regional Names:

Eng. : Onion;
Sans. : Basl, Dirghapatra, Mahakanda, Palandu, Rajapriya;
Ass. : Piaz;
B. : Piyas;
Bo. : Kanda;
H. : Piaj, Palandu;
Kan. : Bawang;
Mal. : Lund;
Mar. : Kanda;
Tam. : Vulli-gaddalu, Nirulli;
Tel. : Vengayam, Nirulli, Kunballi.

Distribution : Cultivated Throughout India.

Description : Perennials with thick globular bulb having membranous tunic.

Scape tall, hollow, inflated and leafy near base. Leaves in 2 rows, shorter than the scape. Flowers white, many.

Flowering and Fruiting : March-April.
Habitat Ecology : Cultivated.
Material Examined : EBH-HVP-707, 18.03.2001.
Parts Used : Bulb (underground).
Folk Uses : 5-7 drops of onion extr. help to cure **earache**. One teaspoon infusion prepared from its bulbs given thrice daily for fifteen days to expel **intestinal worms** and for **piles**. Also considered **diuretic**; eaten raw and used as **condiments** for **flavouring** food stuff. Roasted onions applied as poultice on **boils**.
Active Constituents : Diosgenin (0.18%), (+) S-methyl-L-cystein-sulphoxide and (+) S-propyl-L-cystein sulphoxide, propenyl sulphonic acid and lipooxygenase metabolites of linoleic acid, isomeric 9, 10, 13 trihydroxy-11-octadecenoic acid (bulb) have been isolated.
Biological Activity : Antiperiodic, antibacterial, aphrodisiac, and antitumour activities exhibited.
Uses in Literature : So far known as anthelmintic, diuretic, expectorant, heart stimulant, bactericidal, appetiser, emetic, pungent, spice, tonic; and for asthma, bad taste of mouth, body pain, boils, bleeding piles, cold, diarrhoea, foot sores, impotency, indigestion, urinary discharge, pimples, promoting bile production, pyorrhoea, opthalmia, scabies, soup, earache, flavouring, flatulence, dysentery, unconsciousness, vegetable and pickling (**Ambasta, 1986; Kadel & Jain, 2006; Katewa *et al.*, 2001; Khanna & Kumar 2000; Mohanty *et al.*, 1998; Pandey & Verma, 2002; Retnam & Martin, 2006; Samati, 2004; Sharma & Singh, 2001; Viswanathan *et al.*, 2001**).

13. *Allium sativum* Linn., Roxb. (Pl. 14A)

Family : Liliaceae.
Vern. : Lasun.
English, Sanskrit and Regional Names:

Eng. : Garlic;
Sans. : Sum, Sir;
Ass. : Rasun;
B. : Naharu;
G. : Shunam;
H. : Rasun, Lasun, Lashan;
Kan. : Gokpas;
Mar. : Lasan;
Tam. : Velluli-fella-gadda;
Tel. : Belluli.

Distribution : Cultivated all over India.

Description : Perennials. Leaves flat; scape slender. Spathes long-beaked. Sepals lanceolate and acuminate.

Flowering and Fruiting : March-April.

Habitat Ecology : Cultivated.

Material Examined : EBH-HVP-708, 20.03.2001.

Parts Used : Bulb (underground). Bulbils (aerial).

Folk Uses : Mustard oil fried in garlic cloves used as drops to cure **ear-ache**. Its cloves consumed raw in morning to cure **stomach disorder***. ½ spoon of its juice diluted with 4 spoons of water and a pinch of salt given to cure **stomachache***. Gargles of 1/2 spoon of 'lassun' powder in hot water check **tonsillitis***. Paste of cloves applied locally for **joint pains***.

Active Constituents : These are allicin (2-propenyl propane thiosulphate), allin, allinase, scorclinins, selenium, vitamin A, B, C and E (fresh garlic oil), allitin, sativin I and II, seordinine, scordine, anthocyanins, glycosides of kaempferol and querectin and essential oil.

Biological Activity : Aq. garlic extr. (2ml)/100mg, intra gastric) significantly increases level of blood serum urea (6.27 vs 4.1 m mol/I) and activity of blood serum aspartate amino transferase (130 vs 120 units) in rats as compared to control oil.

Uses in Literature : Used earlier for arthritis, ascites, brain diseases, duodenal ulcer, earache, gangrene of lung, hepatitis, laryngeal tuberculosis, lumps, malaria, pthisis, ringworms, respiratory disorder, scorpion bite, sexual potency, skin trouble, earache, whooping cough, wounds; and used as anthelmintic, aphrodisiac, colic, digestive, diuretic, fattening, tonic, spice and condiments **(Ambasta, 1986; Borthakur *et al.*, 2004; Gogoi & Borthakur, 2001; Kadel & Jain, 2006; Retnam & Martin, 2006; Samati, 2004; Sen *et al.*, 2002; Sen & Behera, 2003; Sharma & Singh, 2001; Sharma *et al.*, 2001)**.

Major Ayurvedic Formulations : Rasan-vati, Rason-pind, Rasonashtka, Lahsunadya-ghrita, Garlic capsule, Garlic pearls.

14. *Aloe maculata* All. (Pl. 14B)

Syn. : *Aloe latifolia* Haw., *Aloe perfoliata* L. var. *saponaria* Aiton, *Aloe saponaria* (Aiton) Haw.

Family : Liliaceae.

Vern. : Ghi-kawer, Kuarian.

English Names : Broad-leaved aloe, Soap aloe.

Distribution : Kerala, Tamilnadu and other warm regions of India below 1,500 m.

Description : Erect succulent herbs with lanceolate-ovate leaves and upper surface pale to darker green with white blotches; lower surface pale green without blotches. Capsule oblong. Seeds blackish brown.

Flowering and Fruiting : June-January.
Habitat Ecology : Escape from cultivation.
Material Examined : EBH-HVP-742, 30.06.2003.
Part Used : Leaves.
Folk Uses : Leaf pulp or juice applied externally in **burns***. 10 ml. extr. of leaves given once a day for 3 days in **pain*** during menstruation.
Uses in Literature : Reported to be used as an emollient and for pickles **(Watt, 1972)**.

15. *Aloe vera* (L.) Webb. and Benth. (Pl. 14C)

Syn. : *A. barbadensis* Mill., *A. perfoliata* var. *vera* L.
Family : Liliaceae.
Vern. : Ghi-kawer, Kuarian.
English, Sanskrit and Regional Names:

Eng. : Aloe, Barbado, Alo, Common Indian, Aloe, Curcago Aloe, Jaffarabad Aloe;
Sans. : Adala, Ajara, Amara, Ambudhisrava, Aphala, Atipichhila, Bahupatri, Brahmaghni, Grihakanya, Kantakapravrita; Sthuladala, Sukantaka, Tarani, Vipulsrava;
Ass. : Chaikuwari;
B. : Ghirtakumari;
G. : Kadvikunwar, Kunvar, Kuvara;
H. : Ghiguvara, Ghikumari, Gvarpatha, Kumari, Kuvarapatha;
Kan. : Loli-Sara;
Mal. : Cheninayakam, Cherukattala, Ghirtakumari, Kaltala, Kumari;
Mar. : Koraphada, Koraphanta, Korkand;
Oriya : Ghikumari, Ghritokumari, Kumari;
Tam. : Angani, Angini, Ettakalavanta, Kalavandu, Musamvaramu, Sottukkattala.

Distribution : Warmer regions of India below 1,500 m. Planted in Indian garden.
Description : Perennial, stoloniferous shrubs having rosette of leaves arising from the ground. Flowers vermillion coloured.
Flowering and Fruiting : March-April.
Habitat Ecology : Wasteplaces. Cultivated along path of field borders.
Material Examined : EBH-HVP-804, 20.03.2004.
Parts Used : Leaves. Latex. Plant.
Folk Uses : Decoction prepared from Aloe leaves, 'ajwain' (*Trachyspermum ammi* Linn.) and black pepper (*Piper nigrum* Linn.) given to cattle for 7 days to cure **stomach disorders***. Massage of leaves considered good for **joint pains**; and its pulp applied to heal **boils** and **burns**. Also, the pulp roasted in desi ghee given to cure **'pitta' disease** (liver ailments) and considered **refrigerant**. Plant regarded

sacred* and its garland hung at the entrance of newly constructed house to ward off evil spirits.

Active Constituents : Chief constituents are barbaloin, aloesin, aloesone, β-sitosterol (leaves) and 4 partially acetylated glucomannans, B-barboline and aloeanodin (jelly).

Biological Activity : Barbaloin inhibits the growth of *Mycobacterium tuberculosis*. Plant extr. (100mg/kg) prevents implantation in 50 and 100% rats at oral dose of 100 and 200mg/kg. Supernatant of gel cytotoxic against C3 HIOTI/2 cells at intiation stage of transformation.

Uses in Literature : Reported to be used as an aperient in apoplexy, aromatic, carminative, cathartic, coolant in fever, diaphoretic, diuretic, emmenagogue, fibre, flavouring agent, mastitis, refrigerant, stimulant, stomachic, tonic, colic, vegetable, vermicidal and for acute indigestion; and for arthritis, asthma, blisters, blunt injuries, boils (vet.), breast tissue hardening, bronchitis, burns (vet.), burnt part, checking growth of tumour, chest pain, cold, cough, bone sprain, ostisis in horses, dermatitis, digestion (vet.), dyeing, easy shedding of placenta during delivery (vet.), enlarged lymphatic glands, eye flu, falling hairs, fencing, flatulence, giddiness, preparation of gin, gonorrhoea, headache, healing wounds, fresh cuts and injuries, itching (vets.), jaundice, joint diseases, leprosy, leucorrhoea, liver disorder, menstrual complaints, miscarriage (vet.), constipation, obesity, piles, relieving muscle pains, skin diseases, spleen ailments, sexual vigour and vitality, foot swelling, ulcer, washing hairs (vet.), wounds and X-ray burns **(Ambasta, 1986; Bhatt *et al.*, 2001, 2002; Borthakur & Sarma, 1996; Borthakur *et al.*, 2004; Chauhan, 1999; Datt, 1996; Dwivedi *et al.*, 2005; Geetha *et al.*, 1996a, b; Girach & Aminuddin, 1995; Gogoi & Borthakur, 2001; Gogoi *et al.*, 2003; Hosagoudar & Henry, 1996a; Jain, 1991; Jain & Patole, 2001; Jadhav, 2006; Kadel & Jain, 2006; Katewa *et al.*, 2001; Khanna & Kumar, 2000; Kurian, 1995; Mishra *et al.*, 1996b; Punjani, 1998; Rama Rao & Henry, 1996; Retnam & Martin, 2006; Saini, 1996b; Samwatsar & Diwanji, 1996; Satapathy & Brahmam, 1996; Sen *et al.*, 2002; Sharma & Borthakur, 2003; Shome *et al.*, 1996; Silori & Rana, 2000; Singh & Pandey, 1996; Verma & Chauhan, 2006; Watt, 1972).**

Major Ayurvedic Formulations : Aloes compound, Bringamaladi, Cukkumtipalyadi-gutika, Kumaryadi-vati, Kumaryasava, Loharasayana, Rajarpravartini-vati, Rajpravartain-vati.

Trade : Fresh Leaves, Aloe Vera Gel, Aloe Vera Powder.

16. *Amaranthus spinosus* Linn. (Pl. 14D)

Family : Amaranthaceae.
Vern. : Chelri, Cholyai.
English, Sanskrit and Regional Names:

Eng. : Prikly Amaranth;
Sans. : Alpamarisha, Bahuvirya, Bhandira, Granthila, Kandakamarisha, Kandera, Meghananda, Sushaka, Tandula;
B. : Kanta notya;
H. : Kataili chauli;
Guj. : antanu-dant;
Kan. : Mulla harive soppu;
Mal. : Kattu-mullen-keera;
Mar. : Kante math;
Tam. : Mulluk-kirai;
Tel. : Mullatota-kura.

Distribution : Throughout India.
Description : A robust spinous herb. Leaves ovate-lanceolate. Flowers greenish in axillary, panicled slender spikes. Utricles rugose, acute, indehiscent. Seeds shining black.
Flowering and Fruiting : May-June.
Habitat Ecology : Wastelands. Cultivated fields. Roadsides.
Material Examined : EBH-HVP-725, 11.05.2002.
Parts Used : Whole Plant. Leaves.
Folk Uses : The plant used as **pot herb***. Also considered **good appetiser*** and **source of iron***. Leaves used as **fodder***.
Active Constituents : Reported to contain saponins (plant), hentriacontane and a-spinasterol (leaves, stem).
Biological Activity : Petroleum ether extr. juvenomimetic positive.
Uses in Literature : Considered useful in fever, eczema, colic pain, diarrhoea, gonorrhoea, leucorrhoea, scorpion and snake bites, warmth; and as antiseptic, diuretic, emollient, lactagogue, laxative, spermatorrhoea, febrifuge and vermifuge (**Ambasta, 1986; Bondya *et al.*, 2006; Chandra Sekar & Srivastava, 2003; Gogoi & Borthakur, 2001; Kirtikar & Basu, 1935; Kshirsagar *et al.*, 2003; Kumar & Chauhan, 2005; Kumar & Goel, 1998; Maliya, 2004; Maya *et al.*, 2003; Mohanty *et al.*, 1998; Patil, 2004; Retnam & Martin, 2006; Upadhyay & Singh, 2005**).

17. *Amaranthus viridis* Linn. (Pl. 15A)

Syn. : *A. caudatus* L., *Chenopodium caudatum* Jacq., *Euxolus viridis* Moquin.
Family : Amaranthaceae.
Vern. : Chelri, Cholyai.

English, Sanskrit and Regional Names:

Eng. : Green or Wild Amaranth, Green Pigweed;
Sans. : Tanduliya,Vishaghna;
B. : Ban note;
Guj. : Dhindo;
Mal. : Bayam munyit;
Mar. : Lhanamat;
Tam. : Kuppai-keerai, Sinna-keerai;
Tel. : Chailaka thotakura.

Distribution : Throughout India.

Description : An erect glabrous herb possessing grooved stem; long, petiolated, ovate leaves and pale green flowers in axillary and panicled slender spikes. Utricles rugose, acute, indehiscent. Seeds shining black.

Flowering and Fruiting : May-June.

Habitat Ecology : Wastelands. Cultivated fields. Roadsides.

Material Examined : EBH-HVP-809, 15.05.2004.

Parts Used : The Whole Plant. Leaves. Root.

Folk Uses : Plant considered as a good **source of iron, appetiser*** and **pot herb**. Leaves used as **fodder**. 10 ml root juice taken with 'amla' (*Emblica officinalis* Gaertn.) juice twice a day for 3-4 days to control excessive **bleeding*** during **menstruation**.

Active Constituents : Plant contains saponins.

Biological Activity : Petroleum ether extr. juvenomimetic positive.

Uses in Literature : Considered useful for acne, edible purposes, scorpion sting, snake bites, swelling; and as laxative, pot herb, cattle fodder and vermifuge (**Ambasta, 1986; Chopra *et al.*, 1956; Girach *et al.*, 1998; Kirtikar & Basu, 1935; Jain, 1991; Patole & Jain, 2002; Retnam & Martin, 2006**).

Adulterant or Substitute : Young shoots are used as an adulterant or substitute for *Asparagus* and *Spinach*.

18. *Amorphophallus campanulatus* (Roxb.) Blume *ex* Decne. (Pl. 15B)

Syn. : *Amorphophallus paeoniifolius* (Densst.) Nicolson, *Arum campanulatum* Roxb.

Family : Araceae.

Vern. : Zamikand.

English, Sanskrit and Regional Names:

Eng. : Elephant Food Yam;
Sans. : Arsaghana;
B. : Ban note;
Guj. : Suran;
H. : Arsaghana, Balukand, Zamikand;
Kan. : Surarna gerdde;

	Mal.	: Chenakarona-kizhana;
	Mar.	: Jangli suran;
	Tam.	: Kurruma kalungh;
	Tel.	: Kunda-muncha-kunda.

Distribution : Humid parts of E. and W. India. Also cultivated in some parts of India.

Description : A cormous herb with large lobed leaves and tuberous roots. Flowers unisexual. Fruit clustered, obovoid, red.

Flowering and Fruiting : April-June.

Habitat Ecology : Cultivated.

Material Examined : EBH-HVP-861, 20.06.2005.

Parts Used : Tuber. Leaves.

Folk Uses : Tuber used as **vegetable*** and also **pickled***. Leaf juice used as **antiseptic*** to clean wounds.

Active Constituents : Reported to contain tyrosinase, laccase betulinic acid, β-sitosterol, stigmasterol, lupeol, triacontane, glucose, galactose and xylose (corm).

Biological Activity : Tyrosinase exhibits mono- and di-phenolase activities specific for L-DOPA as substrate.

Uses in Literature : Known to be used for bruises, dysentery, liver and spleen enlargement, piles, stomach disorders, toothache and glandular enlargement; and also as anthelmintic, carminative and expectorant **(Ambasta, 1986; Retnam & Martin, 2006)**.

19. *Annona squamosa* Linn. (Pl. 15C)

Family : Annonaceae.

Vern. : Sitaphal.

English, Sanskrit and Regional Names:

	Eng.	: Custard Apple, Sweet Sop, Sugar Apple;
	Sans.	: Agrimakhya, Atripya, Bahubijika, Gandagatra, Sitaphala, Subha, Suda;
	Ass.	: Mandargon;
	B.	: Ata, Katal;
	H.	: At, Atasitaphal, Sitaphal;
	Mal.	: Sita-Palam, Sita Pasham;
	Oriya	: Ato, Sitapholo;
	Tam.	: Sitapandu;
	Tel.	: Atta.

Distribution : Naturalised in Western Region.

Description : A small tree possessing oblong, lanceolate leaves and solitary, greenish-yellow flowers. Fruits fleshy, areolate and yellow green.

Flowering and Fruiting : May-August.

Habitat Ecology : Cultivated.

Material Examined : EBH-HVP-726, 15.06.2001.

Parts Used : Fruits. Seeds.

Folk Uses : Fruit **edible** and considered good in **anaemia** and **burning sensation**. Seed paste applied over head to kill **lice**.

Active Constituents : Reported to contain folic acid (fruit); polyphenol (fruit pulp); a kaurane, diterpenoid, aporphine, alkaloids-anonain, glaucine, camphor and new monoterpenoid (root bark).

Biological Activity : Anthelminitc activity against *Caenorhabditis elegans* +ve.

Uses in Literature : Reported as abortifacient, astringent, antiscorbutic, cooling, emetic, febrifugal, pectoral, purgative, tonic, tooth brush, wormicide; and for carbuncle, diabetes, diarrhoea, edible purposes, excess hairfall, indigestion, dysentery, killing lice, licks, pyorrhoea, tumours and worms (**Ambasta, 1986; Asolkar *et al.*, 1992; Bhatt *et al.*, 2002; Chatterjee & Pakrashi, 1997; Chopra *et al.*, 1956, 1983; Girach *et al.*, 1997; Kadel & Jain, 2006; Katewa *et al.*, 2001; Kirtikar *et al.*, 1935; Kumar & Goel, 1998; Maliya & Singh, 2003; Mohanty *et al.*, 1998; Mokat & Deokule, 2004; Patole & Jain, 2002; Rao & Pullaiah, 2001; Retnam & Martin, 2006; Sen & Behera, 2003**).

Remarks : Seeds strong irritant to conjuctiva and cause blindness.

Trade : Annona seeds. Annona leaves. Seed oil.

20. *Argemone mexicana* Linn. (Pl. 15D)

Family : Papaveraceae.

Vern. : Bhindi.

English, Sanskrit and Regional Names:

- **Eng.** : Mexican Poppy, Prickly poppy;
- **Sans.** : Brahmadandi, Hemashikha, Kshirini, Srigalakanta;
- **B.** : Baroshial-kanta, Sialkanta;
- **H.** : Bharbhand, Satiyanashi;
- **Kan.** : Datturi;
- **Mal.** : Brahmadanti;
- **Mar.** : Daruri, Kontedhotra;
- **P.** : Bhatkateya, Bherband, Kandiari, Sialkanta;
- **Tam.** : Bramadandu, Kurukkum;
- **Tel.** : Brahmadandi, Chettu;
- **Urdu** : Baramdandi.

Distribution : Naturalised Throughout India.

Description : Prickly annual herbs with yellow latex and thistle-like leaves. Stem clasping, oblong, minutely pinnatifid; spinous veins white. Flowers yellow, poppy-like prickly, ovate. Capsule oblong-ovoid, prickly, opening by 4-6 valves. Seeds numerous, globose.

Flowering and Fruiting : April-May.

Habitat Ecology : Rocky slopes. Drier areas, along nallahs.

Material Examined : EBH-HVP-746, 18.04.2002.

Parts Used : Seeds. Stem.

Folk Uses : 1 teaspoon of pulverized twigs and seeds given twice a day with honey for 5 days to cure **cough**. Smoke of seeds and 'ajwain' (*Trachyspermum ammi* Linn.) considered effective for **tooth problems** and **spongy gums**.

Active Constituents : Protopine (0.36%), berberine (0.74%), resin and toxic principle (roots, stems), norsanguinarine, norchelerythrine, cryptopine, berberine, benzo-phenanthridine, coptisine, sanguinarine and chelerythrine (plant), sanguinarine and allocryptopine (seed oil) have been isolated.

Biological Activity : Antifungal, anthelmintic, antileprotic, antidiarrhoeal, antiviral activities confirmed. Berberine exhibits significant myocardial depressant activity and total alkaloid fractions reported to cause fall in blood pressure.

Uses in Literature : Known earlier as an anodyne, antiallergic, antidote against snake bite, blood purifier, diuretic, edible, mouth wash; and for arthritis, asthma, boil, cataract, constipation, chronic cough, dog bites, dropsy, eczema, eye complaints, fertility, gum trouble, headache, indigestion, inflammation, itching, jaundice, joint diseases, leucorrhoea, leprosy, ophthalmia, piles, psoriasis, pyorrhoea, regulate fertility, ringworm, scabies, scorpion bites, skin diseases, sun stroke, syphilis, toothache, typhoid, ulcers, and whooping cough (**Ambasta, 1986; Bhatt & Sabnis, 1987; Chopra *et al.*, 1956; Dey, 1973; Dwivedi *et al.*, 2005; Jain, 1991; Jain & Patole, 2001; Jain *et al.*, 1991; Khanna & Kumar, 2000; Kumar & Chauhan, 2006; Kumar & Jain, 1998; Kumar & Pullaiah, 1998; Maliya, 2004; Patil, 2004; Retnam & Martin, 2006; Sarangi & Sahu, 2004; Sen *et al.*, 2002; Sikarwar, 2002; Silori & Rana, 2000; Upadhyay & Singh, 2005; Verma & Chauhan, 2006**).

21. *Artemisia japonica* Thunb. (Pl. 16A)

Syn. : *A. parviflora* Don.

Family : Asteraceae.

Vern. : Bhangru.

Regional Names:

H. : Pati;

P. : Kanyurts.

Distribution : Temperate Himalaya.

Description : A tall aromatic undershrub with lower leaves wedge-shaped. Heads small, greenish-yellow. Pappus none.

Flowering and Fruiting : September-December.
Habitat Ecology : Meadows. Open slopes. Wastelands.
Material Examined : EBH-HVP-796, 20.09.2003.
Part Used : Leaves.
Folk Uses : Leaf paste along with that of *Vitex negundo* applied externally for **sprains*** and **swellings***. Leaf paste applied in **skin diseases**.
Active Constituents : Active constituents are coumarins 6-7-di-methyl-aesculetin, aescule in and scopoletin.
Biological Activity : Antiviral and diuretic activity +ve.
Uses in Literature : Used so far for convulsion in children, skin diseases and throat infections (**Jain, 1991; Megoneitso & Rao, 1983**).

22. *Artocarpus integrifolia* Linn. (Pl. 16B)

Syn. : *Artocarpus heterophyllus* Lam.
Family : Moraceae.
Vern. : Katal.
English and Regional Names:
Eng. : Jack Fruit, Jack tree, Kathal;
Ass. : Dawa;
B. : Dahu;
H. : Lovi, Kathal;
Kan. : Daho;
Kol. : Dewa, Chama, Chamba;
Mar. : Vonte;
Mal. : Chakka;
Tel. : Lakucha.
Distribution : Cultivated throughout India.
Description : Evergreen trees. Leaves entire, coriaceous, base rounded and obtuse apex. Solitary flowers in axillary receptacles. Fruits tuberculated.
Flowering and Fruiting : May-June.
Habitat Ecology : Cultivated.
Material Examined : EBH-HVP-750, 13.05.2002.
Part Used : Fruit.
Folk Uses : Sorosis used as **vegetable** and also **pickled**. Ripe fruits **edible** and good for **dysentery***.
Active Constituents : Ataocarpetin, artocarpanone, artocarpin, isoartocarpin (heart wood) isolated.
Biological Activity : Antiameobic, aphrodisiac (unripened fruits), acholine (seeds) and esterogenic (leaves) activity confirmed.
Uses in Literature : Known to be used as vegetable, gum, pickles and for carpentry, furniture, turnery and musical instruments **(Ambasta, 1986; Samati, 2004)**.

23. *Artocarpus lakoocha* Roxb. (Pl. 16C)

Syn. : *Artocarpus lacucha* Buch.-Ham.
Family : Moraceae.
Vern. : Dheu.
English, Sanskrit and Regional Names:

Eng. : The Jack Fruit tree, Monkey Jack;
Sans. : Peingnai, Pienne;
Ass. : Kanthar;
B. : Knathal;
H. : Kanthal, Kathal;
Kan. : Teprong;
Kol. : Panasa;
Mar. : Pila,Pilapasham;
Tam. : Panasa-pandu, Pansa, Veru-panasa;
Tel. : Halsu, Heb-helsu, Halsina.

Distribution : Sub-Himalayan track and Outer Hills.
Description : A large deciduous tree. Bark rough, grey. Leaves coriaceous, oblong, elliptic, entire, obtuse, glabrous and shining above. Flowers in axillary globose pedunculate heads. Fruits lobulate, smooth, yellow. Seeds oblong, few broad.
Flowering and Fruiting : March-June.
Habitat Ecology : Cultivated.
Material Examined : EBH-HVP-715, 26.03.2001.
Parts Used : Fruit. Leaves.
Folk Uses : Fruit milk applied on **boils***. Sorosis **pickled**. Heated leaves applied to heal **wounds***.
Active Constituents : Some water soluble bases (fruits), cycloartenyl acetate, cycloartenol and cycloartnone (bark).
Uses in Literature : Known to be used as edible; and for dental caries, diarrhoea, toothache, beams, posts, vegetable, furniture, gum, rafter and boat building **(Ambasta, 1986; Kumar & Goel, 1998; Mohanty *et al.*, 1998; Samati, 2004)**.

24. *Asparagus adscendens* Roxb. (Pl. 16D)

Syn. : *A. volubilis* Wall., *Asparagus adscendens* Kunth.
Family : Liliaceae.
Vern. : Sansban.
English, Sanskrit and Regional Names:

Eng. : Asparagus;
Sans. : Bon, Musli;
Bo. : Dholimusali, Saphedmusali;
G. : Dholimusali, Saphedmusali, Ujlimusli;
H. : Satavar, Safedmusali;
Mar. : Dholimusali, Safedmusali.

Distribution : W. Himalaya. Punjab to Kumaon.

Description : Branched creeping herbs having tall, stout, suberect, terete grooved stem and long stout spines. Cladodes densely crowded in tufts. Flowers pure white, fragrant. Berries globose, green, small, 1-seeded.

Flowering and Fruiting : October-November.

Habitat Ecology : Damp places. Along edges of cultivation. Wastelands.

Material Examined : EBH-HVP-737, 13.10.2001.

Parts Used : Root. Leaves. Whole Plant.

Folk Uses : Poultice of roots, twigs and leaves applied on stomach to cure **menorrhagia***. Mixture of its root with 'gur' (jaggery) given to animals for **stomach disorders***. Root considered **emetic***. Plant regarded as **sacred*** and worshipped during marriages.

Active Constituents : Asparagine and saponin steroidal (roots), β-sitosterol, disogenin and sarsasapogenin (fruits) are the principle constituents.

Biological Activity : Antimicrobial activity found to be +ve.

Uses in Literature : So far used as an aphrodisiac, appetiser, astringent, diaphoretic, diuretic, demulcent in veterinary medicine, edible, galactagogue, spermatorrhoea and tonic; and for anthrax, debility, diarrhoea, dysentery, inducing heat, joint diseases, increasing lactation in animals, kibe, leucorrhoea, liver disorder, miscarriage, premature ejaculation, throat complaints, snake bites, stomach pain, and as substitute for 'Salep misree' **(Ambasta, 1986; Borthakur *et al.*, 2004; Dhiman, 1976; Dwivedi *et al.*, 2005; Girach *et al.*, 1998; Hembrom & Goel, 2005; Kadel & Jain, 2006; Kirtikar & Basu, 1935; Kumar & Pullaiah, 1998; Manandhar, 1980; Mohanty *et al.*, 1998; Mokat & Deokule, 2004; Rao & Pullaiah, 2001; Retnam & Martin, 2006; Reddy *et al.*, 1998; Sarangi & Sahu, 2004; Sen & Behera, 2003; Singh, 2003)**.

Major Ayurvedic Formulations : Musali-Pak, Aswagandharista, Geriforte tablets, Maharishi Amrit Kalash-5, Shatavari-ghrita, Narayana-taila, Vishnu-taila.

Trade : Dried decorticated harvested roots sold in the market.

25. *Bambusa nutans* Wall. (Pl. 17A)

Syn. : *Bambusa falconeri* Munro.

Family : Bambusaceae.

Vern. : Bans.

English, Sanskrit and Regional Names:

Eng. : Thorny Bamboo;
Sans. : Vansh;
B. : Mahlbans;
H. : Bans.

Distribution : H.P., Sikkim; Commonly cultivated.
Description : A tall solitary stem arising from creeping rhizome and without spiny branches. Culms not crowded, nodes not much raised. Leaves linear-lanceolate, pubescent beneath.
Flowering and Fruiting : Flowering sporadically.
Habitat Ecology : Roadsides. Forests. Dry areas.
Material Examined : EBH-HVP-870, 15.07. 2005.
Parts Used : Stem. Leaves. Plants.
Folk Uses : Bamboos used for making **ladders**, **baskets**, '**chicks**' (window curtains)*. Twigs used for '**fencing**' agricultural fields and as **fuel***. Plant considered **sacred*** and used for performing religious rituals at the time of marriage and to make **bier** '(**arthi**)' for carrying dead body for final rites. Young stems ('Manu') **pickled***. Leaves used as **fodder***.
Uses in Literature : Reported to be used for fencing, house building, and as a source of pulp **(Ambasta, 1986; Jain, 1991; Punjani, 1998)**.
Major Ayurvedic Formulations : Shitopladi-churna, Banslochan.
Trade : Bans.

26. *Bauhinia vahlii* Wight and Arn. (Pl. 17B)

Syn. : *B. racemosa* Vahl., *Phanera vahlii* Benth.
Family : Fabaceae.
Vern. : Tor.
English and Regional Names:
Eng. : Camel's foot climber;
B. : Chehur;
G. : Chambelli;
H. : Jallur, Malijhan, Malghan, Mahul, Malo, Malu;
Mal. : Motanvally;
Mar. : Chambal, Chambura;
Oriya : Shiali;
P. : Taur;
Tam. : Mandarai;
Tel. : Adattige, Madapur.
Distribution : Foot of Himalaya. Kashmir to Sikkim up to 1,500 m.
Description : A large climber with wide spreading stem, densely pubescent branchlets and circinate tendrils. Bark rough, dark reddish brown. Leaves glabrescent above when mature. Flowers white changing into cream yellow. Pods sublignose, rusty-brown with velvet hairs. Seeds suborbicular, flat, dark brown.
Flowering and Fruiting : April-June.
Habitat Ecology : Forests. Rocky slopes. Dry areas.
Material Examined : EBH-HVP-716, 20.04.2001.
Parts Used : Stem. Leaves. Wood.

Folk Uses : Leaves used for making meal plates ('**pattal**') and bowl ('**duna**'). Leaves also used as **fodder***. Wood used as **fuel***. Powdered leaves (2-3 g) taken with water to check **dysentery**.

Active Constituents : Therapeutic properties attributed to glycosides of quercitol, kaempferol, quercitroside, isoquercitroside, rutoside, taxifoline rhamnoside, agathisflavone and its mono and di-O-methyl derivatives, quercetin, betulinic acid, 3 glucoside tannins and proteins (plant).

Biological Activity : Stem antiacetylcholine and hypotensive.

Uses in Literature : Known till date as an aphrodisiac, demulcent, dysentery, edible (seed pulp and stem eaten raw), adhesive for fixing hair, stomachic, tonic, vegetable and vermifuges; and for amenorrhoea, bone fracture (vet.), antifertility, diarrhoea, dysentery, facial paralysis, gonorrhoea, making meal plates, malaria, mucilaginous remedies, rain hats and caps, ropes and cordage, syphilis, thatching and for making bidis **(Ambasta, 1986; Banerjee, 1996; Chhetri, 2005; Goud & Pullaiah, 1996a; Jain, 1991; Kirtikar & Basu, 1935; Pandey & Rout, 2006; Pawar & Patil, 2004; Rama Rao & Henry, 1996; Sahu, 1996; Retnam & Martin, 2006; Saini, 1996b; Singh & Prakash, 1994; Tosh, 1996; Tirkey, 2004; Watt, 1972)**.

27. *Bauhinia variegata* L. (Pl. 17C)

Syn. : *B. purpurea* Wall., *B. candida* Roxb., *Phanera variegata* Benth.

Family : Fabaceae.

Vern. : Kachnar, Karale, Karayalae.

English, Sanskrit and Regional Names:

Eng. : Buddhist bauhinia, Mountain ebony, Orchid tree, Variegated bauhinia;
Sans. : Ashmantaka, Chamrika, Gandari, Girija, Kanchanara, Karbadra, Kaddara, Mahapushpa, Pakari, Tamrapushpa, Uddalaka, Yugmaptra;
B. : Raktakanchan;
Bo. : Kanaraj, Kanchan, Segapa-Munthari;
G. : Kovindara;
H. : Barial, Gurial, Gwiar, Kachnar, Kaniar, Padrian;
Kan. : Kanchavala;
Mal. : Chuvannamandaram;
Mar. : Kanchan, Raktakanchan;
Oriya : Boroda;
Tam. : Mandarai, Sagapumanchori, Vellaippuvatt;
Tel. : Devakanchanamo; Urdu-Kachnal.

Distribution : Throughout India.

Description : Deciduous trees. Leaves simple, suborbicular, cordate. Flowers white having red and blue strips. Pods hard, flat, glabrous.

Flowering and Fruiting : March-August.

Habitat Ecology : Forests. Roadsides. Planted in villages and in agricultural fields.

Material Examined : EBH-HVP-755, 18.03.2002.

Parts Used : Leaves. Buds. Flowers.

Folk Uses : Flower buds used as **vegetable** and for making **pickles**, '**pakoras**'*, '**chutney**'*, '**rayata**'* (culinary preparations). Leaves employed as fodder for increasing **lactation**. One teaspoon of powdered dried buds thrice a day checks **diarrhoea** and **dysentery**.

Active Constituents : Isoquercetin, astragalin, quercetin, β-p-glucopyranoside, hentriacontene, octacosanol, reducing sugar, stigmasterol, β-sitosterol (flower), and kaempferol, p-glucopyranoside, glycoside, quercitroside, isoquercitroside and rutoside (plants) have been characterised.

Biological Activity : Seed extr. exhibits marked agglutinating activity for human blood. Antibacterial and CNS depressant activity also found to be +ve.

Uses in Literature : Considered as an astringent, alternative, anti-inflammatory, anthelmintic, antidote to snake poison, astringent, blood purifier, tonic, vegetable, symbolic offering to Lord Buddha, purgative, carminative (roots), culminative, edible, fodder, gentle laxative, pot herbs (flowers and buds), sacred, tonic, vermifuge (vet.); and for agricultural implements, beedi manufacturing (leaves), bleeding gum, cold, cough, cuts (vets.), diarrhoea, dysentery, enhancing lactation, expelling placenta (vet.), fatness, goitre, gripping pains from stomach and bowels, indigestion, mouth ulcers, paintings (Mithila), curd preparation (petals), pickling, piles, skin diseases, tanning, toothache, tuberculosis, uterine bleeding, ulcers and wounds (vet.) (**Ambasta, 1986; Banerjee, 1996; Bondya *et al.*, 2006; Chettri, 2005; Chauhan, 1999; Gill & Nyawuame, 1994; Hosagoudar & Henry, 1996a ; Jain, 1991; Jain *et al.*, 1991; Jha & Basak, 1994; Jha *et al.*, 1996; Khanna *et al.*, 1996a, b; Kurian, 1995; Lal *et al.*, 1996; Lalramnghinglova, 1996; Masih, 2003; Mohanty *et al.*, 1998; Negi *et al.*, 1985; Pandey & Verma, 2002; Pandey & Rout, 2006; Patole & Jain, 2002; Rao, 1996; Retnam & Martin, 2006; Saini, 1996b; Singh, 1996, 1999b; Sinha, 1996; Verma & Chauhan, 2006; Watt, 1972**).

Major Ayurvedic Formulations : Chandanasava, Kanchanaaradi-kwatha, Kanchanara-guggulu, Kanchanara-gutika, Kanchanar-kwatha, Ushirasava, Vidangasava.

Trade : Stem bark sold as 'Kachnar Chal'.

Adulterant or Substitute : Bark used as an adulterant or substitute for bark of 'Ashoka' (*Saraca asoca* Roxb.).

28. *Boerhaavia diffusa* Linn. (Pl. 17D)

Syn. : *B. pricumbens* Roxb., *B. erecta* Gartn., *B. repens* L., *B. r.* var. *diffusa* (L) Hook., *B. r.* var. *procumbens* (Roxb.) Hook.

Family : Nyctaginaceae.

Vern. : Itsat.

English, Sanskrit and Regional Names:

- **Eng.** : Hogweed, Pigweed, Spreading Hog weed;
- **Sans.** : Nilini, Punarnava, Rakthkanda, Raktapatrịka, Raktapunarnava;
- **H.** : Gadapurana, Sant, Sarikinjngi;
- **Mal.** : Thumizhazhma, Thazhuthama;
- **Tam.** : Mukkurattai;
- **Tel.** : Atikamamidi.

Distribution : Weed in drier parts of India.

Description : A prostrate herb having reddish branches. Leaves ovate, margin undulate. Flowers crimson-red, minute and in small bracteate umbels. Fruits calvate, penta-ribbed.

Flowering and Fruiting : July-September.

Habitat Ecology : Drier Areas.

Material Examined : EBH-HVP-732, 20.07.2001.

Parts Used : Root. Leaves. Whole Plant.

Folk Uses : Paste of the plant prepared in 'til' oil (*Sesamum indicum* Linn.) good for **joint pains**. Plant paste also applied locally to cure **insect bite***. Root and leaves considered **diuretic** and given for curing **jaundice**.

Active Constituents : Hentriacontane, β-sitosterol, ursolic acid, a phytoecdysome-β-ecdysterone and a new (methyal flavone, 5, 7-dihydroxy-3', 4'-dimethoxy-6-8-dimethyl flavone, triacontanol, a purine nucleoside-hypoxanthine-9-arabinofuranoside, 2 rotenoside, boeravinones A, B and C, punarvine, C-methyl flavone, punarnavoside (root) and crystaline amino acid (leaves) are the active principles.

Biological Activity : Antispasmolytic, antiarrhythmic, antiviral, anti-inflammatory activities confirmed.

Uses in Literature : Reported to be used for asthma, blood dysentery, cold, cough, colitis, cuts, deworming, diarrhoea, dislocated joints, epilepsy, elephantiasis, eye disease, fever, gall bladder stones, jaundice, piles, pimples, leprosy, leucorrhoea, jaundice, joint diseases, odema, paralysis, spermatorrhoea, urinary troubles, wounds; and as anti-arrhythmic, anthelmintic, anti-inflammatory anti-viral, analgesic, diuretic, febrifuge, purgative and spasmolytic **(Ambasta, 1986; Bondya *et al.*, 2006; Borthakur *et al.*, 2004; Dwivedi *et al.*, 2005; Gogoi & Borthakur, 2001; Khanna & Kumar, 2000; Kshirsagar *et al.*, 2003; Kumar & Chauhan, 2005; Mohanty *et al.*, 1998; Maya *et al.*, 2003; Retnam & Martin, 2006; Sen & Behera, 2003)**.

Major Ayurvedic Formulations : Punarnava, Punarnavasana, Punarnavadi, Mandoor, Guggulu,

Abana Tablets.

Trade : Roots sold as 'Punarnava' or 'Itsit'.

Adulterant or Substitute : Leaves and shoots used as an adulterant or substitute for 'Sweta Punarnava' or 'Varshabhu' (*Trianthema portul-castrum* Linn.).

29. *Bombax ceiba* Linn. (Pl. 18A)

Syn. : *B. heptaphylla* Cav., *B. heptaphyllum* Roxb., *B. malabaricum* DC, *Salmalia malabarica* (DC) Schott. and Endl., *Gossampinus malabarica* (DC) Merill., *Gossampinus rubra* Ham.

Family : Bombacaceae.

Vern. : Simal, Simbal.

English, Sanskrit and Regional Names:

Eng. : Red silk cotton tree, Semal tree, Silk Cotton Tree;
Sans. : Apurari, Bahuvirya, Nirgandhapuspi, Mocha, Rakta-pushpa, Salmali;
B. : Roktosimul, Simul, Tula;
Bo. : Saer, Somr, Shembal;
G. : Ratoshemalo, Shimlo, Shimul;
H. : Kantisembal, Pagun, Semal, Simal, Somr;
Mal. : Illavu, Mocha, Mullilavu, Unnamurrika;
Mar. : Kanteri samar, Savara, Shevari, Simlo, Tamari;
Oriya : Buro, Mochorso;
Tam. : Agigi, Ilavam, Muli-ilava-maram, Pougar, Selavagu;
Tel. : Kondaburaga, Mundla-buraga-chettu, Pinnaburaga.

Distribution : Throughout Hotter Parts of India.

Description : A large deciduous tree with prickly stem and wide spreading branches. Leaves glabrous, lanceolate. Flowers red, appearing before the leaves. Capsule 5-valved. Seeds enveloped in dense silky hairs.

Flowering and Fruiting : February-May.

Habitat Ecology : Drier Areas. Wastelands. Roadsides.

Material Examined : EBH-HVP-845, 20.02.2005.

Parts Used : Bark. Spine. Wood. Seeds.

Folk Uses : One teaspoon of bark decoction recommended twice a day for 15 days to cure **sex diseases**. Crushed seeds mixed with wheat flour fed to livestock for **stomach disorders**. Poultice made from its spine good for **boils**. Wood employed for **shuttering purposes*** (house construction). Silky hairs used for **stuffing cushions** and **pillows**.

Active Constituents : Lupeol, β-sitosterol (stem), hexaacosanol, tocopherol, terpenes (seeds), arabinose, galactose, galacturonic acid, catechu-tannic acid and traces of rhamnose (gums) have been isolated.

Biological Activity : Antiviral, hypoglycaemic (flowers) and oxytocic (seeds) activities found +ve.

Uses in Literature : Reported so far as an alternative, astringent, aphrodisiac, coolant, contraceptive, filling material for jackets, cushion and pillows, diuretic, fibre, firewood, fodder, laxative, cordage, spermatorrhoea, tonic and vegetable; and for abdominal pains, acne, anaemia, antifertility, asthma, atrophy, blood dysentery, boils, bone fracture, burns, carbuncle, chicken pox, cholera, cough, debility, diarrhoea, fever, gonorrhoea, gum trouble, hematuria, inflammation, leucorrhoea, liver enlargement, menorrhagia, pain in legs, pimples, pneumonia, prolapses of uterus, rheumatism, scabies, scorpion-bites, sex weakness, toothache, ulcers, urinary complaints, vitality, mattresses, ulceration of bladders, indigestion (vet.), vitality, wounds, water troughs and window shutters (**Ambasta, 1986; Aminuddin & Girach, 1991; Asolkar *et al.*, 1992; Aswal, 1996; Bajpayee & Dixit, 1996; Banerjee, 1996; Chauhan, 1999; Girach, 1992; Girach & Aminuddin, 1995; Gogoi & Borthakur, 1991, 2001; Kadel & Jain, 2006; Katewa & Arora, 1997; Lalramnghinglova, 1996; Maliya, 2004; Manandhar, 1989a, b; Mishra *et al.*, 1996b; Mukherjee & Namhata, 1990; Patil, 2004; Rao, 1996; Retnam & Martin, 2006; Sarangi & Sahu, 2004; Sen & Behera, 2003; Shome *et al.*, 1996; Singh, 2003; Singh & Maheshwari, 1992; Singh & Zaheer, 1996; Watt, 1972**).

Major Ayurvedic Formulations : Abhaya-dirista, Salmali-ghrita, Laghugangadhar-churna, and several aphrodisiac, restorative and expectorant preparations.

Trade : Mochavasa, Suparikas Phool.

Adulterant or Substitute : Seed oil used as an adulterant or substitute for cotton seed oil.

30. *Bougainvillea spectabilis* Willd (Pl. 18B)

Family : Nyctaginaceae.

Vern. : Garden-glory.

English and Regional Names:

Eng. : Garden glory;

H. : Bogenvillea.

Distribution : Often cultivated.

Description : A woody straggling thorny shrub. Leaves orvicular-ovate. Flowers small yellow having red coloured showy bracts with number exceeding the flowers.

Flowering and Fruiting : January-June.

Habitat Ecology : Often cultivated.

Material Examined : EBH-HVP-741, 21.01.2002.

Parts Used : Whole Plant.

Folk Uses : Grown as an **ornamental** and for **fencing***.

Active Constituents : Glycosides of isorhamnetin and quercetin (flowers), pinitol (leaves) have been isolated.

Biological Activity : Pinitol (0.01 g/kg p.o.) exhibits significant hypoglycaemia in normal fasted mice.

Uses in Literature : Known to be used as an ornamental, and for paintings (Mithila), ulcers and diabetes **(Gogoi & Borthakur, 1991; Jha & Basak, 1994; Kshirsagar *et al.*, 2003; Lalramnghinglova, 1996)**.

31. *Brassica campestris* Linn. (Pl. 18C)

Syn. : *Brassica rapa* Linn.
Family : Brassicaceae.
Vern. : Sarson.
English, Sanskrit and Regional Names:

- **Eng.** : Yellow sarson;
- **Sans.** : Bhutaghana, Bimbata, Duradharsha, Grahagna, Kadamba, Kadambaka, Rajika, Sarishapa, Sarahapa, Sidhartha, Tantuka, Ugragandha;
- **B.** : Kalisarson, Sadrai, Sanshi, Shurshi, Sursi;
- **H.** : Bangasarson, Baralai, Dain, Jariya, Kalisarson, Lahota, Pilasarson, Sarson, Sursi, Tori, Toriya;
- **Mal.** : Karupakatuka;
- **Mar.** : Kalamohare, Sherasa;
- **Tam.** : Karuppukkadugu;
- **Tel.** : Nallaavalu.

Distribution : Throughout India.
Description : An annual erect herb with yellow flowers in corymbose heads.
Flowering and Fruiting : March-September.
Habitat Ecology : Cultivated. As an escape.
Material Examined : EBH-HVP-753, 11.04.2001.
Parts Used : Leaves. Tender shoots. Seeds.
Folk Uses : The young leaves and tender shoots used as **vegetable***, 'sarson-ka-saag'. Oil cake used as **cattle feed***. The oil combined with camphor applied externally for **muscular rheumatism***.
Active Constituents : Active constituents are glycerides of erucic acid.
Uses in Literature : Known to be used for bronchitis, epilepsy, fever, inflammation, itching, leucoderma, piles, skin eruptions, toothache, ulcers **(Kirtikar & Basu, 1935)**.

32. *Bryophyllum calycinum* Salisb. (Pl. 18D)

Syn. : *B. pinnatum* (Lam.) Kurz., *Kalanchoe pinnata* Pers., *Cotyledon rhizophylla* Roxb.
Family : Crassulaceae.
Vern. : Gillar pattar, Patharchat.

English and Regional Names:

Eng.	:	Ahiravana-mahiravana, Ghayamari;
Ben.	:	Koppata;
Guj.	:	Ghayamari;
H.	:	Zakhm-Haiyat-Ka-Patta;
Kan.	:	Yo-kiya-pin-ba;
Mal.	:	Lonnuhadakana-gidd;
Tam.	:	Simajamunda, Runakalli;
Tel.	:	Ela-marunna, Elamarunga, Murikuti, Simaja-muda.

Distribution : Throughout tropical parts of India.

Description : Tall erect herbs. Flowers reddish green or purple. Fruits enclosed in persistent peppery corolla and calyx.

Flowering and Fruiting : March-May.

Habitat Ecology : Drier areas.

Material Examined : EBH-HVP-744, 27.03.2002.

Part Used : Leaves.

Folk Uses : Toasted leaves applied to cure **wounds** and **boils**. Leaves **eaten** raw and considered **antilithic***.

Active Constituents : Active principles are due to bryophyllol, bryophollone, bryophollenone, bryophynol and two homogenous phenanthren derivatives-2 (9-decenyl) phenanthrene (I) and 2 (undecenyl), malic, isocitric and citric acid (leaves).

Biological Activity : Bersaldegenin-1, 3, 5, orthoacetate inhibited growth of several cancer cell lines.

Uses in Literature : Toasted leaves applied to heal wounds, bruises, boils, bite of venomous insects, dysentery, leprosy, swelling, discoloration of affected part, kidney and liver troubles and sloughing ulcers; and also as antiseptic, astringent, carminative, emollient, (**Ambasta, 1986; Jain & Patole, 2001; Retnam & Martin, 2006; Sharma *et al.*, 2001**).

33. *Butea monosperma* (Lam.) Kuntze. (Pl. 19A)

Syn. : *B. frondosa* Koen *ex* Roxb., *Erythrina monosperma* Lamk., *Plaso monosperma* Kuntze.

Family : Fabaceae.

Vern. : Palah, Palash, Plah.

English, Sanskrit and Regional Names:

Eng.	:	Bastard Teak, Bengal Kino, Butea gum, Flame of the forest;
Sans.	:	Bijasneha, Kinsuka, Ksharashreshtha, Palasa, Palasham, Vatapotha;
B.	:	Palas, Palasa;
Bo.	:	Khakara, Palasi;
G.	:	Khakara, Palasso;
H.	:	Chichra, Kankrej, Palas;
Kan.	:	Muttuga, Muttugamara,Thoras;
Mal.	:	Brahmavrikshan, Palasinsamatha;
Mar.	:	Kakradha, Phalas, Paras, Palas;

	Oriya :	Kinjuko, Polaso, Porasu;
	Tam. :	Kutumurukku, Murukkam, Parasa, Porasan;
	Tel. :	Modugachettu, Mohtu, Palas, Tellamodugu; Tellamoduga, Vatapodhamu;
	U.P. :	Dhak.
Distribution	:	Throughout India.
Description	:	A medium-sized deciduous tree possessing pinnately trifoliate leaves. Leaflets coriaceous, silky tomentose beneath. Bright orange red flowers in terminal racemes appear before the leaves. Pods 1-seeded, grey downy.
Flowering and Fruiting	:	March-April.
Habitat Ecology	:	Drier parts. Along the borders of agricultural fields.
Material Examined	:	EBH-HVP-807, 27.03.2004.
Parts Used	:	Leaves. Flowers. Wood.
Folk Uses	:	One teaspoon decoction of leaves prescribed twice daily for seven days to cure **amenorrhoea**. Flowers tied as an amulet on stomach for relief from **pain**. Decoction of flowers (thrice daily for fifteen days) good against **jaundice**. Wood considered **sacred** and used for performing '**havan**'.
Active Constituents	:	Principle constituents are butrin, coreopsin, isobutrin-5, monospermoside isocoreopsi and three iso-derivatives, sulphurein, chalcones (flowers); butin, butrin, isobutrin, palasonin, palastrin (plant).
Biological Activity	:	Antifungal, antibacterial, antiimplantation, antiovula-tory, spasmolytic and anthelmintic (due to palasonin) activities +ve.
Uses in Literature	:	Used so far as an abortifacient, antiseptic, anthelmintic, aphrodisiac, astringent, contraceptive, coolant, diuretic, febrfuge, haemorrhagic, spermatorrhoea, stomachic, tonic, vegetable, fodder, fish poison, beedi wrapper, colouring material, fibre, fuel, lac host, laxative, sacred ornamental, and steriliser; and for bad breath, boil, bone fracture, cooling, cuts, delirium, diabetes, diarrhoea, dog bite, dysentery, eye complaints, expelling worms, fever, foot and mouth diseases, gastric problems, heart burns, inflammation, irregular menstruation, jaundice, ophthalmia, pains, piles, pimples, ringworms, skin diseases, sun stroke, tooth and gum disorder, leucorrhoea, traditional dwelling, tuberculosis, urinary complaints, wounds, plates, bowls, dummy corpse, wounds, water scoops, well curb and worshipping (**Ambasta, 1986; Bajpayee & Dixit, 1996; Banerjee, 1996; Bhandary** ***et al.*****, 1996; Bhatt** ***et al.*****, 2001, 2002; Chopra** ***et al.*****, 1956; Gill & Nyawuame, 1994; Goel** ***et al.*****, 1984; Jain, 1968, 1991, 1992, 1996; Jain** ***et al.*****, 1994; Jadhav, 2006; Jha & Varma, 1996; Kadel & Jain, 2006; Katewa** ***et al.*****, 2001; Khanna & Kumar, 2000; Kothari & Moorthy, 1996; Kurian, 1995; Lalramnghinglova, 1996; Masih, 2003; Mohanty** ***et al.*****, 1998; Pandey & Rout, 2006; Pandey & Verma, 2002; Patel** ***et al.*****, 2003; Pawar & Patil, 2004; Rama Rao & Henry,**

1996; Ravishankar & Henry, 1992; Reddy *et al.*, 1996; Retnam & Martin, 2006; Saini, 1996b; Sarangi & Sahu, 2004; Satapathy & Brahmam, 1996; Silori & Rana, 2000; Singh, 2003; Singh & Prakash, 1994; Singh & Zaheer, 1996; Varma, 1997; Venkataratnam & Venkata Raju, 2004; Watt, 1972).

Major Ayurvedic Formulations : Ayaskrti, Janmagautti, Krimighna-churna, Krimikuthara-rasa, Krimimudgar-rasa, Mahanarayan-taila, Nyagrodhadi-kwatha-churna, Palasabhijadi-churna, Palsakshara-ghirta, Plasapuspasava, Plash-ghrita.

Trade : Traded in Indian market as 'Plashapapra' (seeds), 'Dhaak phool' or 'Tesh' flowers and 'Bengal kino' (gum).

Adulterant or Substitute : Dried trunk juice used as an adulterant or substitute to East Indian Kino (*Pterocarpus marsupium* Roxb.). Seeds used as substitute for Sanotin, Kirmala (*Artemisia maritime* Linn. var *thomsoniana* Clarke).

34. *Callistemon lanceolatus* DC. (Pl. 19B)

Syn. : *C. citrinus* Skeels.

Family : Myrtaceae.

Vern. : Bottle Brush.

English Name : Bootle Brush Tree.

Distribution : Cultivated in Himalaya up to 1,500 m.

Description : A small evergreen tree. Leaves glossy, lanceolate. Flowers with very long red stamens.

Flowering and Fruiting : March-April.

Habitat Ecology : Planted.

Material Examined : EBH-HVP-774, 20.04.2003.

Part Used : Flowers spike.

Folk Uses* : Planted as an **ornamental**. Spike used for **decorative purposes***.

Uses in Literature : There is no mention of its use in literature.

35. *Canna indica* L. (Pl. 19C)

Syn. : C. *chinensis* Willd., C. *indica* var. *orientalis* Rosc., C. *orientalis* Rosc.

Family : Cannaceae.

Vern. : Aeran, Hardarshan, Sudarshan.

English, Sanskrit and Regional Names:

Eng. : Indian Bead, Indian Reed, Indian Shot;

Sans. : Devkili, Kamkshi, Krishnamantara, Sarvajaya, Shilarambha, Vanakadali, Kamakshi, Lalsarbojaya, Sarbajaya;

B. : Akalabera;
G. : Sabbajaya, Sarvaliaya;
H. : Hudigana, Kalahu;
Kan. : Kattuvala;
Mal. : Devakali;
Mar. : Sorobojoya;
Oriya : Hakik;
Tam. : Kalvalai, Kalvalaimani, Kundimani, Puvalai, Siramalai;
Tel. : Guruginja.

Distribution : Mostly throughout India.

Description : Erect undershrubs with a creeping, subterranean tuberous rhizome and fleshy thick roots. Leaves sub-sessile, oblong, acuminate. Corolla bright red. Capsule red.

Reproductive Cycle : January-February.

Habitat Ecology : Oftenly cultivated in gardens as an ornamental.

Material Examined : EBH-HVP-702, 15.02.2001.

Parts Used : Flower. Leaves.

Folk Uses : Flowers used in **worshipping**, and poultice of leaves heated in oil used against **stomachache** and **internal swelling***.

Uses in Literature : So far used in India as an antidote to poison, febrifuge, fodder (vet.), demulcent, diaphoretic, diuretic, antidote to snake bite; and for burns, chest pain, cough and cold, edible, hounding dogs (vet.), offerings and adornments, ulcers, pin-worms; and for dropsy, dyeing, headache, meal plates, necklace, ornaments, stomachache, swelling of abdomen in cattle, worshipping and thatching houses **(Ambasta, 1986; Bajpayee & Dixit, 1996; Dagar, 2003; Datt, 1996; Gogoi & Borthakur, 1991; Jain, 1991; Kapur & Nanda, 1996; Rama Rao & Henry, 1996; Watt, 1972)**.

Major Ayurvedic Formulations : Aragvahadi-kwatha-churna.

36. *Cannabis sativa* Linn. (Pl. 19D)

Syn. : *C. indica* var. *indica* (Lamk.) Wehmer.

Family : Cannabinaceae.

Vern. : Bhang.

English, Sanskrit and Regional Names:

Eng. : Hemp, Indian hemp;
Sans. : Bhanga, Chapola, Ganja, Ganjika, Hursini, Indrasana, Jaya, Ujaya, Viyaya, Vrijpatta;
B. : Bhang, Ganja, Sidhi;
H. : Bhang, Charas, Ghanja, Gur, Kinnab, Siddhi;
Kan. : Bhangi-gida;
Mal. : Ginjilachilachi, Kanchavachetto;
Mar. : Bhangachajhada;
Tam. : Ganja-chedi, Kalpam, Korkkarmuli;
Tel. : Ganjari-chettu.

Distribution : Throughout India.
Description : An erect aromatic annual herb with palmately divided leaves. Flowers yellowish-green. Achenes small, glandular hairy.
Flowering and Fruiting : June-September.
Habitat Ecology : Wastelands. Edges of fields.
Material Examined : EBH-HVP-730, 15.06.2001.
Parts Used : Leaves. Seed.
Folk Uses : Poultice of leaves with mustard oil applied on joints for relieving **inflammation*** and **rheumatic pains***; and its juice applied on insect bites as an **antidote***. Leaves also used for making '**pakora**'* and '**kota**'* (hallucinagic drink); and as an **offering*** to appease Lord Shiva. Roasted seeds fed to cattle in **diarrhoea**.
Active Constituents : These are glucocapparins (seeds), pthalic acid, 1-stachydrine (flower and fruit husk), cannabidiolic acid, canabidiol, cannabinol, tetrahydrocannabinol, eugenol, guaiacol, cannabigerol.
Biological Activity : Antibacterial, hypotensive and spasmolytic activity and effect on CNS found +ve.
Uses in Literature : So far reported as an anthelmintic, appetiser, diuretic, laxative, narcotic, nervine, stimulant, edible (seeds), and sleep pills; and for asthma, bowel complaints, bronchitis, cold, cough, convulsion, cramps, delirium, epilepsy, cuts, diarrhoea, dyspepsia, ear complaints, eye diseases, gonorrhoea, hydrocele, labour pains, paralysis of tongue, piles, skin diseases, sores, tetanus and making rough clothes **(Chopra *et al.*, 1956; Jain, 1991; Khanna & Kumar, 2000; Retnam & Martin, 2006; Sharma, 2003)**.
Major Ayurvedic Formulations : Jatiphaladi-churna, Madnanand-modaka, Kameshwar-modaka, Vedantak-rasa, Vijaya-vatika.
Adulterant or Substitute : Seed oil used as substitute for Linseed oil in paints, varnish and making soft soap.

37. *Capsicum annuum* Linn. (Pl. 20A)

Family : Solanaceae.
Vern. : Mirach.
English and Regional Names:
Eng. : Red chillies, Red pepper;
H. : Mirch.
Distribution : Cultivated.
Description : A shrub with lanceolate-ovate leaves and white flowers. Fruits elongated berry which are red on maturity.
Flowering and Fruiting : Greater part of the year.

Habitat Ecology : Cultivated.
Material Examined : EBH-HVP-742, 16.07.2002.
Parts Used : Fruit. Plant.
Folk Uses : Grounded fruits used for preparing pepper **sauce*** and various types of **pickles***. Dried powder of fruits applied on parts bitten by **rabid dog**. 200 ml decoction of fruits along with *Trigonella foenum-graecum* L., *Cuminum cyminum* L., *Allium cepa* L. and jaggery given twice daily for treating **skin eruptions*** in cattle. Plant considered **magico-religious***.
Active Constituents : Lutein, capsorubin, cryptoxanthin, capasnthin, rubixanthin phytofluene, capasaicin, capsicin, capsaicin and solanin (fruits) are the main constituents.
Biological Activity : Capsaicin in conc. of 0.1 to 1.0 μ m induces relaxation in pre-contracted longitudinal muscle strips of gastric fundus potentiated by hexamethonium whereas it activates sensory afferent neurons at conc. of (0.1 to 1.0 μ m).
Uses in Literature : Used as analgesic, carminative, expectorant, fruit, spice, stimulant, externally as rubefacient; and for common cold, dysentery, sore throat, scarlatina, dyspepsia, dog bite, yellow fever, diarrhoea, foot diseases, piles, rheumatism, and snake bite (**Bhatt *et al.*, 2001; Gogoi & Borthakur, 2001; Kirtikar & Basu, 1935; Kumar *et al.*, 2004; Retnam & Martin, 2006; Samati, 2004; Sen *et al.*, 2002; Sharma, 2003**).
Major Ayurvedic Formulations : Visham-jwaraghini-vati, various liniments and embrocations.

38. *Carica papaya* Linn. (Pl. 20B)

Family : Caricaceae.
Vern. : Papita.
English, Sanskrit and Regional Names:

Eng. : Melon tree, Papaw, Papaya, Paw-Paw;
Sans. : Chirbita, Eranda, Naikadala;
B. : Papai;
Bo. : Chibhado, Katha, Paputa;
G. : Chibda, Eranda kakdi, Thadchibadi;
H. : Anda Kharbuja, Papita;
Mal. : Kappalam, Nalikadala;
Mar. : Papaya;
Oriya : Omrytobhondo, Popoya;
P. : Popai;
Tam. : Pappli, Parangiyananakki;
Tel. : Madananaba, Madhurnakamu.

Distribution : Cultivated throughout India.

Description : A small soft wooded tree having succulent trunk with milk sap and large glabrous, palmatifid and palminerved leaves. Flowers pale-yellow, fragrant. Fruits succulent, indehiscent, 1-celled. Seeds numerous, black, enclosed in sweet mucus pulp.

Flowering and Fruiting : Greater part of the year.

Habitat Ecology : Planted.

Material Examined : EBH-HVP-806, 13.06.2004.

Part Used : Fruit.

Folk Uses : Unripe fruits cooked as **vegetable** and stuffed in 'Prantha'. Ripe fruits **edible** and considered good for **liver ailments**, **diabetes** and as **cardiac tonic**. Fruits also taken with milk to cure **constipation**; and its latex applied externally for treating **eczema***.

Active Constituents : Hentricontane, β-sitosterol, benzoyl thiourea, carpaine (seeds), myristoleic, palmitoleic, arachidic acid, pectin, free sugar, papain, vit. C, nicotinic acid (fruits) have been characterized.

Biological Activity : Carpaine exhibits antitumour activity *in vitro* against mouse lymphoid leukaemia (1:210), lymphocytic leukaemia (P:338) and ehlrich ascites tumour cells, and also as antimicrobial against gram +ve and –ve bacteria.

Uses in Literature : Reported earlier to be used as an abortifacient, adulterant, anthelmintic, antidote for cancer, blood anticoagulant, carminative, emmenagogue, diuretic, face creams, febrifuge, galactagogue, hair shampoo, purgative, stomachic, source of vitamin and papain, substitute for soap, tonic, vegetable; and for amoebic dysentery, asthma, beri-beri, bleeding piles, boils, bone fracture, candies, cardiac diseases, chest problems, cooling, cough, cuts, dental caries, diabetes, diptheria, dyspepsia, eczema, excessive bleeding of the kidney, flatulence, flavouring meat, gastric, gingivitis, gonorrhoea, haemoptysis, heart depression, itching, jaundice, kidney stone, liver disorder, malaria, marmalades, pain in gums, pickling, piles, promoting menstrual flow, reducing elephantoid growth, rheumatis, ringworm, skin diseases, soft drinks, spleenic and hepatic enlargement, tonsils, tooth and gumache, tuberculosis, ulcers, urinary trouble, warts and wounds (**Ambasta, 1986; Bhardwaj & Gakhar, 2003; Borthakur, 1993; Borthakur *et al.*, 2004; Dagar & Dagar, 1999; Ghosh *et al.*, 1996; Gill *et al.*, 1993, 1997; Gogoi *et al.*, 2003; Hosagoudar & Henry, 1996a; Jain, 1991; Jain *et al.*, 1991; Kulkarni & Kumbhojkar, 1993; Kurian, 1995; Lalramnghinglova, 1996; Maliya & Singh, 2003; Pandey & Verma, 2002; Rama Rao & Henry, 1996; Retnam & Martin, 2006; Sharma *et al.*, 2002; Sen & Behera, 2003; Singh & Zaheer, 1996; Siwakoti & Verma, 1996; Watt, 1972**).

39. *Carissa inermis* Vahl. (Pl. 20C)

Syn. : *C. macrophylla* Wall.
Family : Apocynaceae.
Vern. : Badda Garnae.
Distribution : Throughout India
Description : Erect spinous shrubs possessing leathery leaves. Flowers white, sweet-scented and in corymbose cymes. Thorns long and sharp. Berries ovoid, red to dull-dark purple.
Flowering and Fruiting : April-July.
Habitat Ecology : Open slopes. Wastelands in drier areas.
Material Examined : EBH-HVP-889, 15.07.2006.
Parts Used : Stem. Leaves. Fruit.
Folk Uses : Fruits edible and good for **anaemia**. Fruits **pickled***. Leaves used as **fodder*** for goats and sheep.
Active Constituents : Carrissonecarissol glucose, galactose, serine, glutamine, alamine, valine, phenylalanine, glycine and carissic acid (fruits) are chief active constituents.
Biological Activity : Hypotensive effect and cardiotonic activity confirmed.
Uses in Literature : Fruits edible **(Ambasta, 1986)**.

40. *Carissa spinarum* Hook. f. (Pl. 20D)

Syn. : *C. opaca* Stapf. *ex* Haines.
Family : Apocynaceae.
Vern. : Garnae.
English, Sanskrit and Regional Names:

Eng. : Bengal currents;
Sans. : Bokkarambuka, Jalipushpa, Krishna, Pakapha-la,Vanalaya;
B. : Karamcha, Kurumira;
Bo. : Hartundi, Karinda;
G. : Karamadan, Karamarda;
H. : Karaunda, Karrona;
Kan. : Karekayi;
Mal. : Kalavu, Karakka;
Mar. : Boronda, Haradundi, Karavandi;
Oriya : Khyirokoli, Korondo;
Tam. : Kalakkay, Perungala;
Tel. : Paddavaka, Peddakalive, Vaka.

Distribution : Throughout India.
Description : An erect, spinous shrub possessing leathery leaves and white, sweet-scented flowers in corymbose cymes. Thorns long and sharp. Berries ovoid, red to dark purple.
Flowering and Fruiting : May-June.
Habitat Ecology : Open slopes. Wastelands in drier areas.
Material Examined : EBH-HVP-727, 16.06.2001.

Parts Used : Stem. Leaves. Fruit.

Folk Uses : Gargles of stem decoction cure **mouth ulcer***. Fruits **edible** and good for **anaemia**. Wood, twigs and dry leaves used as **fuel***. Leaves used as **fodder*** for cattle and stem for **fencing***.

Active Constituents : A new triterpine carissol (fruits) along with glucose, galactose, serine, glutamine, alamine, valine, phenylalanine, glycine and carissic acid have been isolated.

Biological Activity : CNS depressant and hypotensive activity found (+)ve.

Uses in Literature : So far known for constipation, cough, dysentery, fever, edible fruits, veterinary ailments, wounds; and as laxative **(Jain, 1991; Pal, 1981; Retnam & Martin, 2006)**.

Major Ayurvedic Formulation : Kusthara.

41. *Cassia fistula* L. (Pl. 21A)

Syn. : *C. rhombifolia* Roxb., *Cathartocarpus fistula* Pers., *Cassia fistula* Willd.

Family : Fabaceae.

Vern. : Aahali, Allia, Kanyar.

English, Sanskrit and Regional Names:

Eng. : Drum stick, Golden shower, Indian Laburnum, Purging Cassia;
Sans. : Aragvadha, Chatuangula, Kritamalluka;
Ass. : Sonaru;
B. : Amultas, Sonali;
G. : Balla, Garmala;
H. : Amaltas, Bandarlauri;
Kan. : Kakke;
Mal. : Saturangulam, Svuarnnakam;
Mar. : Bhawabaya, Chimkani;
Oriya : Soturongulo, Sunari;
P. : Alash, Karangal;
Tam. : Arogaram, Irayavinnadagam;
Tel. : Aragvadhamu, Sampakamu.

Distribution : Throughout India, Common in Himalaya up to 1,400 m.

Description : Medium-sized trees having straight trunk and long, pinnately compound leaves. Main rachis pubescent. Leaflets ovate, margin entire. Flowers yellow and in hanging raceme. Pods cylindric, straight, smooth, shining, brownish black, indehiscent with numerous broadly-ovate seeds immersed in dark coloured sweet pulp.

Flowering and Fruiting : March-July.

Habitat Ecology : Forest. Open fields. Roadsides.

Material Examined : EBH-HVP-803, 20.05.2004.

Parts Used : Leaf. Seed. Fruit. Wood.

Folk Uses : One tablespoon decoction of fruit pulp taken twice daily for a week to relieve **gastric ailments*** and **constipation** both in man and livestock. Poultice of tender twig applied for

toe infection*. Seeds prescribed as **laxative** in animals. Wood used as **fuel wood*** and for making **agricultural implements**.

Active Constituents : Fistucacidin identified as racemic or meso 3, 4, 4, 7, 8 pentahydroxyflavan (bark), leucoanthocyanidin-5-4-dihydroxy flavon-3-4-diol (I) (sap wood), rhein, glucose, sucrose and fructose (fruit pulp), galactomannan (seed), seven new biflavinoides β-sitosterol (I-VII) and two triflavinoids (VIII-IX), (–) epicatechin, procyanidin B-2, (–) epiafzelechin and its 3-o-glucosides (leaves), steroids, glycisides, flavonoids, tannins, quinones, disaccharides, phlobaphenes, oxyanthroquinone have been isolated.

Biological Activity : An aq. extr. of fruit pulp exhibits antibacterial activity.

Uses in Literature : Known so far as antibacterial (vet.), antinarcotics, antiseptic, blood purifier, cooling agent, edible (vet.), emetic, febrifuge, insecticidal (vet.), laxative (vet.), mild cathartic, ornamental, psychoactive, symbolic abode to god, tonic (roots), vegetable vermicides; and for abdominal pains, adjust bile, agricultural implements, anthrax (vet.), black water fever, blindness, body swelling, bloated stomach, body heat, boils, burns, cancer, carts, cataract, chest infection, chest pain, cold (vet.), diabetes, digestion, dog bite, dyeing, dysentery, eczema, epilepsy, expelling hook worms, gout, house construction, indigestion, jaundice, liver complaints, malarial fever, pimples, ploughs, scorpion bite, swelling in throat (vet.), tool handles and wheels, toothache, whooping cough and wounds (vet.) **(Ambasta, 1986; Balasubramaniam & Prasad, 1996; Bhandary *et al.*, 1996; Bondya & Sharma, 2003; Chauhan, 1999; Dixit & Pal, 1996; Gill & Nayawuame, 1994; Girach & Amin-uddin, 1995; Gogoi & Borthakur, 1991; Goud & Pullaiah, 1996b; Jain, 1968, 1991, 1996; Jain & Patole, 2001; Jain *et al.*, 1994; Kothari & Moorthy, 1996; Kumar, 1996; Kumar & Jain, 1998; Kurian, 1995; Lalramnghinglova, 1996; Manandhar, 1989a; Masih, 2003; Molla & Roy, 1996; Patil & Merat, 2003; Patole & Jain, 2002; Pullaiah & Kumar, 1996; Rama Rao & Henry, 1996; Retnam & Martin, 2006; Sahoo & Mudgal, 1995; Saini, 1996a; Sharma, 1996; Sharma & Sood, 1997; Sikarwar, 1996; Singh, 1999b; Siwakoti & Verma, 1996; Srivastava *et al.*, 2003; Tiwari & Tiwari, 1996; Tosh, 1996; Venkataratnam & Venkata Raju, 2004; Watt, 1972)**.

Major Ayurvedic Formulations : Aargavadhaadi-arishta, Argavadhaadi-taila, Ragvadhadi-kwatha, Aragvadhadi-leha, Aragvadharistaz, Agnikumar-churna, Arishta, Kankayan-gutika, Leha-taila, Mahamanjisth-Yadyarista, Rasnadi-kwatha-yoga.

Trade : Traded for its dry fruits and pulp.

42. *Cassia occidentalis* Linn. (Pl. 21B)

Syn. : *Senna occidentalis* Roxb., *C. foetida* Pers., *C. sophera* Wall.

Family : Fabaceae.

Vern. : Alliuna, Bada-rellu.

English, Sanskrit and Regional Names:

Eng.	:	Fetid Cassia, Nergo coffee, Rubbish Cassia, Stinking weed;
Sans.	:	Arimarda, Kasari, Kasamara;
B.	:	Kalkashunda;
Bo.	:	Hikal;
H.	:	Barikasondi, Chakunda, Kasunda;
Mal.	:	Karintakara;
Oriya	:	Kasundri;
Tam.	:	Nattandagarai, Peyavirai;
Tel.	:	Kasindha;
Urdu	:	Kasonji.

Distribution : Common throughout India; naturalised.

Description : An annual undershrub. Branches subglabrous, purplish. Leaves fetid. Flowers yellow, in short peduncled racemes. Pods recurved, glabrous, tortulose. Seeds ovoid-compressed at one end and rounded at other, hard, dark olive green.

Flowering and Fruiting : May-September.

Habitat Ecology : Wastelands. Edges of cultivated areas.

Material Examined : EBH-HVP-728, 04.06.2001.

Parts Used : Leaves. Seeds.

Folk Uses : Tender pods used for making **pickles**. **'Chutney'** comprising of its seeds, **'methi'** (*Trigonella foenum-graecum* Linn.), **'ajwain'** (*Trachyspermum ammi* Linn.) and **'gur'** (jaggery) good for **gastric disorders.** Leaf paste applied as poultice to check swelling due to **clotting***; and its juice with honey good against **mouth ulcers***.

Active Constituents : These are aloe-emodin, a bianthraquinone, chrysophenol, dianthronic heterosides, jaceidin-7-rhamnoside, matteucino 1-7-rhamnoside rhein and potassium chelidonate (leaves), chrysolein, linoleic, linoleic and oleic acid, campesterol, 8-carbomethoxy-1-7-dihydroxy-3 methylxanthane, 1-8-dihydroxy-2 methyl anthraquinone, 1, 4, 5, trihydroxy anthroquinone derivatives, tannic acid and toxalbumin (seeds), steroid, glycosides, flavones, dianthronic, heteroside, sennoside A and B anthraquinone.

Biological Activity : Seed, root and leaf exhibit antibacterial properties.

Uses in Literature : Known in India as antiperiodic, diuretic, purgative, substitute for coffee, stimulant drink, stomachic, pot herb, tonic, tooth brush; and for bone fracture, dog bite, dropsy, diarrhoea, dysentery, edible purposes, eczema, fever, gastric complaints, lactation, rheumatism, ring worm, skin diseases, snake bites, throat infections, whooping cough, wounds and veterinary ailments (**Ambasta, 1986; Chopra *et al.*, 1956; Jain, 1991; Khanna, 2000; Kshirsagar *et al.*, 2003; Mohanty *et al.*, 1998; Patole & Jain, 2002; Retnam & Martin, 2006; Sharma & Sood, 1997; Uphof, 1968**).

Major Ayurvedic Formulation : Liv.-52.

43. *Catharanthus roseus* G. Don (Pl. 21C)

Syn. : *Vinca rosea* L., *Lochnera rosea* (L.) Reichb.

Family : Apocynaceae.

Vern. : Sadabahar.

English and Regional Names:

- **Eng.** : The Periwinkle;
- **B.** : Nayantara;
- **H.** : Sadabahar;
- **Mal.** : Ushamalari;
- **Mar.** : Sadphul;
- **Oriya** : Ainskati;
- **Tel.** : Billaganneru.

Distribution : Cultivated throughout India.

Description : Erect annual herbs. Leaves opposite, obovate, glossy. Flowers white or pink and in cymose axillary clusters. Fruit a cylindrical follicle, many seeded.

Flowering and Fruiting : Throughout Year.

Habitat Ecology : Wastelands.

Material Examined : EBH-HVP-824, 20.08.2004.

Parts Used : Roots. Leaves.

Folk Uses : Leaf juice applied as **antidote** for wasp stings. Roots **stomachic** and good for **diabetes** (25g powder thrice daily till cure). Plant grown as an **ornamental***. Flowers offered to appease **deities***.

Active Constituents : Flowers rich in vinblastine, coronaridine, vindorosine, vindoline, catharanthine, mitraphlline, ajmalicine, tetrahydroal stonnie, 11-methoxytabersonineb and reserpine.

Biological Activity : Antitumorous (catharanthamine), anticancerous (vinblastine-i.p. or i.v.), (γ ohimbine-1mg/kg of rats), cytotoxic hypotensive and hypoglycaemic activities (aq. extr. of fruits and seeds) positive.

Uses in Literature : Known earlier for anthrax (vet.), boils, cold (vet.), cut (vet.), diabetes, diarrhoea, dysentery, fever, high blood pressure, madness (vet.), mastitis (vet.), scabies, skin infections, swelling of body (vet.), toothache, wasp sting, wounds; and as antileukaemic, anticancerous, haemostat (vet.), hypotensive, sedative, mosquito repellent (vet.), purgative, stomachic

(Ambasta, 1986; Bondya & Sharma, 2003; Hembrom & Goel, 2005; Hosagoudar & Henry, 1996b; Jain, 1991; Kurian, 1995; Lalramnghinglova, 1996; Prasad *et al.*, 1996; Retnam & Martin, 2006; Sharma & Sood, 1997; Siwakoti & Verma, 1996; Watt, 1972).

44. *Cedrela toona* Roxb. *ex* Rottl. (Pl. 21D)

Syn. : *C. febrifuga* Blume, *C. hexandra* Wall., *C. longifolia* Blume, *C. serrata* Royle, *C. australis* F. Muell., *Toona ciliata* Roem.

Family : Meliaceae.

Vern. : Tooni.

English, Sanskrit and Regional Names:

- **Eng.** : Happy tree, Indian Mahagany Tree, Moulmein Cedar, Red Cedar, San.-dal, Toon;
- **Sans.** : Kubheraka, Nandi vriksha, Apina, Apita, Kachla, Kanta, Nanda, Tooni;
- **Ass.** : Henduripoma, Jia;
- **B.** : Lud-Tunna, Tuni;
- **Bo.** : Deodari, Tundu, Tupa, Tuni, Tunna, Kudaka, Limb, Mahanim;
- **H.** : Kudaka, Limb, Mahanim, Tun, Tuna, Tupa, Tuni, Tunna;
- **Mal.** : Akil, Arana, Devabaram, Malarveppu, Kacham;
- **Mar.** : Deodari, Kuruk;
- **Oriya** : Mona Limba;
- **Tam.** : Madagirvempu, San.-davanembu;
- **Tel.** : Gali, Nandi.

Distribution : Common throughout the greater parts of India.

Description : Deciduous trees. Leaves abruptly pinnate. Panicles drooping, terminal. Flowers small, white, fragrant. Capsule 5-celled, ovoid, dehiscent.

Flowering and Fruiting : May-June.

Habitat Ecology : Roadsides. Open slopes.

Material Examined : EBH-HVP-835, 07.05.2004.

Parts Used : Leaves. Wood.

Folk Uses : Wood used for making doors, windows, **furniture** and **agricultural implements***. Leaves used as **fodder.**

Active Constituents : Isolated constituents are cedrelone, β-sitosterol (seed, wood), (–)-alloaromadendrene, cadalene, (+) α calacorene, (–)-calamenene, (+) Δ cadinene, (–) Δ-cadinol, (–)-copaene, cubenol, farnesene, geranylgeraniol, (–)-ledol, leucocyanidin, T-murolol and siderin (wood).

Biological Activity : Stem bark anticancer, spasmolytic whereas leaf exhibits antiameobic, CNS depressant, hypoglycaemic and spasmolytic activities.

Uses in Literature : Used earlier as an astringent, antiperiodic, antipyretic, antiseptic, cattle feed, dye (yellow), tonic; and for bridges, doors,

floors, furniture, insect repellent, dysentery, fever, gastric troubles, mouth ulcers, plywood, tonsillitis, window, wood carving and wounds **(Ambasta 1986; Chopra *et al.*, 1956; Jain, 1991; Kapur & Singh, 1996; Pandey & Verma, 2002; Siwakoti & Verma, 1996; Watt, 1972).**

45. *Celastrus paniculatus* Willd. (Pl. 22A)

Syn. : *Celastrus alnifolia* Don., *C. dependens* Wall., *C. nutans* Roxb.
Family : Celastraceae.
Vern. : Sankheere, Sankhiryan.
English, Sanskrit and Regional Names:

Eng. : Black Oil, The Oleum Nigrum Plant;
Sans. : Myinkhoung-na-young;
B. : Kujari, Kujri;
Bo. : Malkangana;
G. : Kangani, Malkangoni, Malkanganitela, Pigavi;
H. : San.-khu, San.-khii;
Kan. : Jiotishmati;
Mal. : Kariganne;
Mar. : Ruglim;
P. : Malkakni;
Tam. : Ruglim;
Tel. : Valushuva.

Distribution : Tropical and Subtropical Himalaya.
Description : A large woody scandent climber with terete branches and oval-oblong, serrate, membranous, coriaceous and glabrous leaves. Flowers yellowish-green and in pendulous panicles. Fruits globose, small, green-yellow when ripe. Seeds 1-2 in each cell, enveloped in red and fleshy coat.
Flowering and Fruiting : April-June.
Habitat Ecology : Forests. Along Roadsides.
Material Examined : EBH-HVP-776, 15.05.2002.
Part Used : Seeds.
Folk Uses : One teaspoon of powdered seed given twice a day till cure for **cough** and **bronchitis***. Seed oil mixed with that of mustard in equal ratio applied externally for **skin affections**.
Active Constituents : Main ingredients are celastrin, β-dihydrogaroturan sesquiterpenoids (VI, VII, VIII, IX), sesquiterpenoids, 1 β, 6-diacetoxy-9-β-benzoyloxy-8 β-connamoyloxy-β-dihydro-agarofuran (V), malkanguinol, sesquiterpene, ester-malkangunin, celapanine and celapanigine, paniculatadiol along with b-amyrin; b-sitosterol and fatty acids (seed oil).
Biological Activity : Antiviral, hypotensive and spasmolytic activities found to be positive.

Uses in Literature : Reported as abortifacient, aphrodisiac, brain tonic, laxative, mosquito repellent, stimulant; and for arthritis, blood clotting, body pains, cough, cold, fencing, fever, headache, hiccups, impotency, leucoderma, psoriasis, piles, urinary inflammation, rheumatism, skin diseases, spermatorrhoea, stomach pain, strengthening memory, syphilis, tuberculosis, leucorrhoea, amenorrhoea, pneumonia, paralysis and kidney troubles (**Ambasta, 1986; Jain & Patole, 2001; Khanna & Kumar, 2000; Kirtikar & Basu, 1935; Pandey & Rout, 2006; Punjani, 1998; Rao & Pullaiah, 2001; Sarangi & Sahu, 2004; Silori & Rana, 2000; Tirkey, 2004**).

Major Ayurvedic Formulations : Geriforte tablets, Himcolin cream, Joytismati-tel, Remem tablets and syrups and Seotex.

Trade : Seeds sold in the drug markets under the name of 'Malkangani' or 'Jyotismati beej'.

46. *Celosia argentea* L. (Pl. 22B)

Syn. : *Celosia cristata* Linn.

Family : Amaranthaceae.

Vern. : Kukkar phul.

English, Sanskrit and Regional Names:

Eng. : Cox comb plant, Quail Grass;
Sans. : Pothura;
B. : Swetmurga;
G. : Lapadi;
H. : Safed murgha;
P. : Sarwali sil;
Tel. : Gurugu.

Distribution : Throughout India.

Description : Erect herbs. Leaves elliptic, ovate below and linear lanceolate above. Flowers crowded in terminal pink shining heads.

Flowering and Fruiting : August-December.

Habitat Ecology : Cultivated. As an escape.

Material Examined : EBH-HVP-880, 15.06.2005.

Part Used : Seeds.

Folk Uses : One teaspoon of powdered seed prescribed twice daily for **kidney stones***.

Active Constituents : Celosianin and isocelasianin.

Uses in Literature : Used so far as antipyretic, antiscorbutic, aphrodisiac, galactagogue; and for blood dysentery, diarrhoea, eye diseases, inflammation, wasp and scorpion stings, uterine bleeding (**Chopra *et al.*, 1956; Chatterjee & Pakrashi, 1997; Kirtikar & Basu, 1935**).

47. *Centella asiatica* (Linn.) Urban (Pl. 22C)

Syn. : *Hydrocotyle asiatica* Linn.
Family : Apiaceae.
Vern. : Brahmi.
English, Sanskrit and Regional Names:

Eng. : Indian Pennywort;
Sans. : Bhekaparni, Divya, Manduki;
Ass. : Manimini;
B. : Brahmamanduki;
Bo. : Karinga, Karivana;
H. : Brahmamanduki, Khulakhudi;
Kan. : Von-de-laga;
Mal. : Kodagam, Muthal;
Mar. : Brahmi;
Tam. : Babassa,Vallari;
Tel. : Bokkudu.

Distribution : Tropical and subtropical regions, common throughout India.

Description : A prostrate mat forming herb rooting at nodes. Leaves reniform, margin toothed. Flowers purplish-white, and in axillary umbels. Fruit minute, orbicular, reticulated, laterally compressed.

Flowering and Fruiting : April-May.

Habitat Ecology : Along water channels in fields.

Material Examined : EBH-HVP-717, 13.04.2001.

Part Used : Leaves.

Folk Uses : One teaspoon of powdered leaves with cow's milk given in the morning to improve **memory** and as **astringent***. 100 ml leaf juice with sugar candy given thrice a day for **cooling**.

Active Constituents : Therapeutic properties attributed to centoic, centellic, centic acid, brahmoside, brahminoside, two triterpenic acid-brahmic and iso-brahmic acid, betulic acid, stigmasterol, 2 saponins-brahmoside and brahminoside.

Biological Activity : EtOH (50%) extr. of plant antiprotozoal and spasmolytic.

Uses in Literature : Known so far for curing blood disorders, blood purifier, bodyache, boils, cholera, cough, diarrhoea, dysentery, eczema, eye complaints, fever, gastric disorders, headache, insanity, jaundice, leprosy, liver complaints, mouth sores, memory, nervine disorder, kidney stone, respiratory disorders, scabies, syphilis, skin diseases, tubercles, tumour, urinary complaints, wounds; and as anthelmintic, brain tonic, coolant, diuretic, insecticidal, postnatal tonic, spermatorrhoea, stomachic, vegetable and local drink **(Ambasta, 1986; Augustine & Sivadasan, 2004; Bondya *et al.*, 2006; Borthakur *et al.*, 2004; Chettri, 2005; Gogoi & Borthakur, 2001; Jain, 1991; Jain & Patole, 2001; Khanna & Kumar 2000; Kshirsagar & Singh, 2000; Kumar & Pullaiah, 1998; Rao & Pullaiah,**

2001; Sharma *et al.*, 2002; Sharma & Sood, 1997; Singh, 2003; Retnam & Martin, 2006; Upadhyay & Singh, 2005).

Major Ayurvedic Formulations : Brahmi-vati, Brahmi-ghrita, Brahmi-taila, Brahmi-amala-taila, Brahmi-rasayana, Geriforte tablets and Saraswatarista.

Trade : Marketed as 'Brahmi'.

Adulterant or Substitute : Plant used as an adulterant or substitute to Herpestris, 'Brahmi' (*Bacopa monniera* Wettst.)

48. *Cestrum nocturnum* L. (Pl. 22D)

Family : Solanaceae.

Vern. : Rat-ki-Rani.

English and Regional Names:

- **Eng.** : Lady of Night, Night Blooming Jasmine;
- **B.** : Hasna;
- **H.** : Rat ki Rani.

Distribution : Throughout India. Commonly cultivated up to 1,500 m.

Description : Evergreen shrubs. Leaves long, pointed, ovate-lanceolate, shining. Flowers yellowish-green in axillary branched cluster. Fruits a berry.

Flowering and Fruiting : August-October.

Habitat Ecology : Agricultural Fields. Wastelands.

Material Examined : EBH-HVP-823, 16.07.2004.

Parts Used : Leaves. Flowers.

Folk Uses : Flowers considered **sacred*** for offering to deities and burnt as **incense***. Decoction of leaves used as a good wash for **mouth sores***.

Active Constituents : Active properties attributed to benzaldehyde, benzyl alcohol, benzyl acetate, methyl benzoate, phenyl acetaldehyde, phenyl ethyl alcohol, linalool and α farnesene (flowers); tigogenin, yuccagenin, nornicotine, cotinine and myosmine (leaves).

Biological Activity : Aerial parts spasmolytic, diuretic and hypotensive.

Uses in Literature : Recorded in India for epilepsy; and as perfumes, cosmetics **(Sharma, 2003; Uphof, 1968)**.

49. *Chenopodium album* Linn. (Pl. 23A)

Syn. : *C. giganteum* Don, *C. nepalense* Hort., *C. viride* Linn., *C. laciniatium* Linn., *C. purpurascens* Ham.

Family : Chenopodiaceae.

Vern. : Bathu, Kunnah.

English, Sanskrit and Regional Names:

- **Eng.** : Bacon Weed, Dirty Dick, White Goose foot, Wild Spinach;
- **Sans.** : Agralohika, Chielika, Kankella, Mridapatri, Shakarata, Vastuki;

B. : Bathu sag, Chandan betu;
Bo. : Chakwit;
H. : Bathua, Khartua Sag;
P. : Bathu;
Tam. : Parupukire;
Tel. : Pappukuras.

Distribution : Throughout India, Tropical and Temperate Himalaya.

Description : Erect glabrous herbs having striped stems and oblong, rhombic, deltoid, obtuse, entire leaves. Flowers in cluster forming lux paniculate, often tinged with purple. Utricles orbicular, smooth shining.

Flowering and Fruiting : January-March.

Habitat Ecology : Cultivated fields. Wasteplaces.

Material Examined : EBH-HVP-701, 10.01.2001.

Parts Used : Stem. Leaves. Seeds.

Folk Uses : Plant extr. taken with sugar twice a day, for stones **(antilithic).** One tablespoon of seed powder with honey given to **expel worms**. Young shoots and leaves used as **vegetable** and considered good for **rheumatic pains*** and to relieve **constipation***. Plant also given as **fodder***.

Active Constituents : Constituents identified are ecdysteroids, β-ecdysone, polypodine B (roots), ascorbic acid, β-carotene, catechins, gallocatechin, caffeic acid, p-coumaric acid, ferulic acid, β-sitosterol, campesterol, xanthotoxin, stigmasterol, n-triacontanol, scopoletin, imperatorin, spinasterol and 7-stigmasterol (plant).

Biological Activity : Plant Et OH (50%) extr. CNS depressant.

Uses in Literature : Recorded in India as an aphrodisiac, anthelmintic, antiseptic, blood purifier, cardiac disorders, debility, diuretic, laxative, tonic and vegetable; and for anaemia, eye troubles, intestinal ulcers, piles, throat, skin diseases, spleen and bilious disorders, urinary complaints, worms, local drinks and reviving taste **(Ambasta, 1986; Asolkar *et al.*, 1992; Bondya *et al.*, 2006; Chhetri, 2005; Chopra *et al.*, 1956; Dey, 1973; Jain, 1991; Sen & Batra, 1997).**

50. *Cicer arietinum* Linn. (Pl. 23B)

Family : Fabaceae.

Vern. : Chollae.

English, Sanskrit and Regional Names:

Eng. : Bengal Gram, Chick Pea, The Common gram;
Sans. : Kalapai;
Bo. : Chenna;
G. : Harbara;

H. : But, Channa, Chola, Santali;
Kan. : Kadalai, Kadale;
Mar. : Harbara;
P. : Chana, Chola;
Tam. : Sannagala, Harimandhakam;
Tel. : Kadale, Karikadale, Kempu, Kudoly.

Distribution : Extensively cultivated throughout India.

Description : An annual herb with pinnate leaves. Flowers bluish. Pods small and green.

Flowering and Fruiting : March-April.

Habitat Ecology : Cultivated.

Material Examined : EBH-HVP-709, 12.03.2001.

Part Used : Seeds.

Folk Uses : Seeds consumed as '**dal**'. Powdered seeds ('basin') employed for various **culinary preparations**, and also for preparing **facial pack*** ('butna') in combination with sandal powder and sarson oil for groom and bride at the time of marriage. A handful of overnight soaked grams given in 'dona' to each woman attending ladies 'sangeet' (song and dance programme held for the marriage function) and considered **auspicious*.** Also grains **eaten** raw, roasted or boiled, and fed to horses; husk used as **cattle feed***.

Active Constituents : Principle constituents are oxalic, acetic, malic acid, isoaliquiritigenin, isoaliquiritigenin-4'-glucoside, 3,4,7,-trihydroxyflavone, daidzein, pratensein, p-coumaric acid, biochanin-7-glucoside, pyrimidine nucleoside-uracil-5-B-D fructofuranosyl 1-monophosphate and uracil-5-B-D ribofuranosyl, 2,-3-cyclic monophosphate and β–sitosterol (seeds), lecithin, phytin, saponin (grain), biochanin A, biochanin B, biochanin C (sprouting gram).

Uses in Literature : Seeds eaten as dal or after roasting and in form of basin, prophylactic feed to horses; and for acidity, constipation, dyspepsia, jaundice and snakebite **(Ambasta, 1986; Kirtikar & Basu 1935; Rastogi & Mehrotra 1990, 1991, 1993, 1995a, b).**

51. *Cinnamomum camphora* (L.) Nees and Ebesm (Pl. 23C)

Syn. : *Laurus camphora* L.

Family : Lauraceae.

Vern. : Kapoor, Kapur.

English, Sanskrit and Regional Names:

Eng. : Camphor, Camphor Cinnamomum, Official camphor;
Sans. : Karpura;
Mal. : Ghanarasam, Himavaluka;

Oriya : Chondrongyon, Romokorpura;
Tam. : Indu, Kirasemiran, Pugalvisurandiram;
Tel. : Chandramu, Ghanasaramu, Karpuramu.

Distribution : Cultivated in Himalaya up to 1,500 m.

Description : An aromatic evergreen tree possessing elliptic, long pointed 3 nerved leaves. Flowers tiny, greenish-yellow. Fruits globular with thickened club-shaped base.

Flowering and Fruiting : April-August.

Habitat Ecology : Planted.

Material Examined : EBH-HVP-834, 15.06.2004.

Parts Used : Wood. Fruit.

Folk Uses : 100g mixture of camphor and equal quantity of mustard oil applied to check **skin irritation*** and for suppuration of **boils.** Wood used for performing **'havan'*.**

Active Constituents : Eurolitsine, reticulate (roots), 1-b-elemene, 1-b-caryophyllene, humulene, selinene (leaf oil, twig), ylangene, b-santalene, calacorone, calamenene, d-cadinene (camphor oil) are the active ingredients.

Biological Activity : Plant sedative, anodyne, antispasmodic and diaphoretic.

Uses in Literature : So far reported as acro-necrotic poison, analeptic, anthelmintic, antileprotic, antiseptic, antispasmodic, carminative, natural camphor, psychoactive plant, sedative, stimulant, angina, pectoris and for bed sores, cancer, curing wounds, diarrhoea, inhoofs, fever, gout, hysteria, impotency, inflammation, malaria, nervous, headache, nymphomania, painful affections of uterus, perfumery, sleeplessness, smallpox, spermatorrhoea, whooping cough and wounds (vet.) (**Ambasta, 1986; Jain *et al.*, 1984; Mishra *et al.*, 1996b; Sharma, 2003; Sharma & Sood, 1997; Vihari, 1995; Watt, 1972).**

52. *Cissampelos pareira* Linn. (Pl. 23D)

Syn. : *Cissampelos pareira* L. var. *hirsuta* (Buch.-Ham. *ex* DC) Forman, *C. caapeba* L., *C. convolvulacea* Willd., *C. hirsuta* Buch.-Ham. *ex* DC, *C. orbiculata* Buch., *C. heradiflora* Wall., *C. obectata* Wall., *C. discolor* Buch., *C. diversa* Miers, *C. elata* Miers, *C. grallatoria* Miers, *C. eriantha* Miers, *C. subpeltata* Thwaites, *C. delicatula* Miers, *C. nepalensis* Rhodes, *Cocculus orbiculatus* DC, *C. villous* Wall., *Menispermum orbiculatum* L.

Family : Menispermaceae.

Vern. : Bhatindu, Pathindu.

English, Sanskrit and Regional Names:

Eng. : False Pareira Brava, Ice Vine, Periera Browa, Velvet leaf;
Sans. : Ambashtha, Ambashthika, Brihattika, Ruchaishya, Shishira, Vidhakarai;
A. : Saruthubukilata;

B.	:	Akanadi,Tejomalla;
Bo.	:	Pahadmul, Pahadvel;
H.	:	Akauadi, Dekhnirbissi, Pari;
Kan.	:	Padavali;
M.P.	:	Karepar;
Mal.	:	Kattuvalli, Pattuvalli;
P.	:	Bat bel, Batindu path, Haiyat, Pataki, Pilajur;
Tam.	:	Pomuskite, Punmushtie, Vatatirupi;
Tel.	:	Pata.

Distribution : Throughout India, Tropical and Subtropical parts of India up to 1,500 m.

Description : Climbing shrubs. Leaves orbicular to reniform. Flowers small, greenish-yellow. Male cymes long-peduncled, femaleones in racemes. Drupe ovoid, compressed, scarlet-red.

Flowering and Fruiting : May-November.

Habitat Ecology : Rocky slopes. Screes.

Material Examined : EBH-HVP-751, 10.07.2002.

Parts Used : Root. Leaves.

Folk Uses : 250 mg. powdered roots along with black pepper (*Piper nigrum* Linn.) given for seven days to cure **leucorrhoea***. 5-10 ml. decoction of leaves twice a day considered good for **constipation** and **fever**. Leaves along with wheat dough given to livestock for **dysentery***.

Active Constituents : Hayatine, 1-curine, iso-chondrodendrine, thymol, 11 quaternary, pareirine, hyatinin, hyatin and d-iso-chondrodendrine cycleamine, 1-bebeerine, hayatidine, hayatine-5 tertiary alkaloids and d-quercitol (leaves), hyatin, furnished hydrochloride, picerate, chloro-platirate, methiocide and methochloride (roots, vines), isoquinoline, pelosine, berberine, reserpine, cissam-petive, hyatin, hyatinin, quercitel and sterol (roots) are the active constituents.

Biological Activity : CNS depressant and muscle relaxant properties exhibited.

Uses in Literature : Known so far for abscess, antifertility, bodyache, boils, bronchitis, burns, carbuncle, catarrhal, child birth, cold, cough, cystitis, diarrhoea, dysentery, epilepsy, eye complaints, headache, heart burn, gastric complaints, indigestion, infection of bladder, inflammation, jaundice, laxative, local drinks, liver disorder, pimples, piles, scabies, spermatorrhoea, snake bites, stomach ulcer, urinary troubles, wounds; and as carminative, colic, diuretic, febrifuge, stomachic, sedative, tonic, symbolic **(Ambasta, 1986; Borthakur, 1993; Borthakur *et al.*, 2004; Brahmam & Saxena, 1990; Brahmam *et al.*, 1996; Chauhan, 1999; Datt, 1996; Dey, 1973; Drury, 1978; Jain, 1991; Katewa *et al.*, 2001; Manandhar, 1980, 1996; Masih, 2003; Maya *et al.*, 2003; Saxena *et al.*, 1981; Sharma, 2003; Sharma & Sood, 1997; Singh, 2003; Varma, 1997; Uphof, 1968; Verma & Chauhan, 2006; Watt, 1972).**

Major Ayurvedic Formulations : Agnimukh-churna, Brihatgangadhar-churna, Gangadhar-churna, Kutajastak-kwatha, Mahayograj-guggulu, Pathadi-kwatha, Pippalayasav, Pradarantaka-lauha, Pushayug-churna, Sarasvata-ghrita, Stanyasodhanakasaya-churna.

Trade : Roots sold in the drug market as 'Patha'.

Adulterant or Substitute : Root used as substitute for true Pareira Brava of commerce (*Chondodendron tomentosum* Ruiz. and Pav.).

53. *Citrus aurantium* L. (Pl. 24A)

Syn. : *C. aurantium* var. *bigardia* Hook. f.

Family : Rutaceae.

Vern. : Sangtara, Santra.

English, Sanskrit and Regional Names:

Eng. : Bigarde orange, China orange, Orange tree, Portugal orange, Seville Sour, Sweet orange;
Sans. : Chakradhivasi, Airavata, Kirmira, Nagar, Tragagandha;
B. : Kamalanembu, Narungi, Naranga, Varishtha;
Bo. : Cantara, Naranghi;
G. : Narangi;
H. : Amritphal, Narunge, Sangatra;
Kan. : Hearlay;
Mal. : Madhuranaranna;
Mar. : Nargungasala, Sakulimba;
Oriya : Nagoronga, Narongi;
P. : Naranj, Santara;
Tam. : Narangam, Puttansini;
Tel. : Battavinarinja, Narinja.

Distribution : Hot Valleys of Himalaya.

Description : Small trees. Leaves foliolate. Flowers pure white. Fruit globose, oblate, orange coloured. Rind loose, pulp sweet, yellow.

Flowering and Fruiting : October-March.

Habitat Ecology : Planted.

Material Examined : EBH-HVP-793, 12.10.2003.

Parts Used : Fruit. Rind.

Folk Uses : Fruit **edible** and excellent source of **vitamin C.** Rind powder constitutes one of important components of **face pack*** employed by the locals.

Active Constituents : These are 5-hydroxyauranetin, 5-o-demethylnobiletin, 6-7-dimethoxy-coumarin, aurantin, hesperidin (peel), hesperidin, naringin and poncirin, neo-hesperidin aglycone and limonene (fruits).

Biological Activity : α, β-pinenes, limonene, octanol, citronellal inhibit seed germination and seedling growth in *Amaranthus retroflexus*. Naingen and naringenin selectively inhibit *in vitro* growth of cancer cells in human cells (breast, colon, liver).

Uses in Literature : Reported earlier to be used as allxiphasmic, antispasmodic, blood purifier, carminative, contraceptive, disinfectant, flavouring agent, juice, psychoactive plant, refreshing drink, stimulating, stomachic, tonic; and for acne, allaying thirst in fever, bilious affections, cold, cough, edible purposes, eczema, gout, improving the appetite, indigestion, infusion, intestinal worms, jaundice, liquors, marmalades, perfumery, preparation of confections, psoriasis, reddish eye, skin cracks and vomiting **(Ambasta, 1986; Jain *et al.*, 1994; Khanna, 2000; Samati, 2004; Sharma *et al.* 2002; Sen & Batra, 1997; Sikarwar, 2002; Srivastava *et al.*, 2003; Watt, 1972).**

54. *Citrus grandis* (L.) Osbeck. (Pl. 24B)

Syn. : *C. maxima* (Burm.) Merr., *C. decumana* L., *Aurantium maximum* Burm.

Family : Rutaceae.

Vern. : Chakotra.

English, Sanskrit and Regional Names:

Eng. : Forbidden fruit, Paradies-Apple, Pommel-Moose Shadock, Pummelo;
Sans. : Madhukarkati;
B. : Batavinebu, Botornebu;
Bo. : Papanass;
G. : Obakotru;
H. : Batavinebu, Chakotra, Sadaphal;
Kan. : Chaoktre,Sakkota;
Mal. : Bambitinarakam;
Mar. : Panis;
Oriya : Batabilembu;
P. : Chakotra;
Tam. : Pambalimash;
Tel. : Pampanpanasa.

Distribution : Cultivated in India.

Description : Trees. Leaflets ovate-oblong. Flowers white. Fruits large-sized, pale yellow, globose. Rind thick, pulp varying in colour from crimson to pale pink.

Flowering and Fruiting : November-March.

Habitat Ecology : Fields. Near dwellings.

Material Examined : EBH-HVP-738, 18.11.2001.

Parts Used : Fruits. Leaves.

Folk Uses : Fruits **edible**. Leaves used for **scouring teeth***.

Active Constituents : Principle constituents are 5-geranyloxy-7-methoxy-coumarin, isoimperatorin, bergapten and 5-geranyloxypsoralen (bergamottin) in (fruits, leaves).

Uses in Literature : Used as nutritive, refrigerant, edible; and for cholera, coughs, constipation and inflamed nerves, diarrhoea (vet.), epilepsy, indigestion (vet.), jams and marmalades, nervous disorders, offering, scurvy, vomiting, waist and back pains (**Ambasta, 1986; Borthakur, 1996a, b; Gogoi & Borthakur, 1991; Kurian, 1995**).

55. *Citrus jambhiri* Lush. (Pl. 24C)

Family : Rutaceae.
Vern. : Jambirdi, Jhamirdi, Kimb.
English Name : Rough lemon.
Distribution : Hot Valleys of Himalaya.
Description : A small tree with stout spines. Leaf pale green. Flowers white, fragrant. Fruits yellow with thick rind.
Flowering and Fruiting : October-March.
Habitat Ecology : Cultivated.
Material Examined : EBH-HVP-763, 20.11.2002.
Parts Used : Fruit. Leaves.
Folk Uses : Fruit **edible** and also **pickled**. Its juice used as 'chukh' for imparting **acidic** taste to preparation. Leaves used for **scouring teeth**.
Active Constituents : A new flavone-tangeretin along with β-sitosterol, hesperidin (peel oil).
Uses in Literature : There is no mention of its use in literature.

56. *Citrus limon* (L.) Burm. f. (Pl. 24D)

Syn. : *C. medica* var. *leomonum* L. (Risso) Hook. f., *C. limonum* Risso., *C. medica* L. var. *limon* L.
Family : Rutaceae.
Vern. : Khatta.
English, Sanskrit and Regional Names:
Eng. : Lemon;
Sans. : Bara Nebu, Goranebu, Jambirah;
G. : Motalimbu;
H. : Bara Nibu, Jambhira, Paharinembu, Paharikagzi;
Kan. : Bijapura, Bijuri;
Mar. : Idalimbu, Thoralimbu;
Oriya : Periya Yelumichai;
Tel. : Bijapuram.
Distribution : Hot Valleys of Himalaya.
Description : An evergreen tree with scented-white flowers. Fruits ovoid, yellow, mammilate, very acidic. Rind thin, pulp abundant.
Flowering and Fruiting : November-February.
Habitat Ecology : Cultivated.

Material Examined : EBH-HVP-765, 22.11.2002.

Parts Used : Leaves. Fruit.

Folk Uses : Fruits **edible, pickled** and also used for **magico-religious*** purposes. Leaves used for **scouring teeth*** and **flavouring** desi ghee*.

Active Constituents : These are apigenin, luteolin, chrysoseriol, quercetin, isorhamnetin, new flavonols limcitrin and limocitrol-sinapic acid, p-coumaric acid, scopoletin, umbelliferone, hesperidin, citrusin A, B, C and D, coniferin, syringin and dehydro-deconeferyl alcohol-4-β-glucoside (fruit), eriocitrin (peel).

Biological Activity : Coniferin and syringin (1.0 mg/100 g i.v.) found to raise blood pressure of SHR-sp rats whereas dehydrodiconiferyl alcohol-4-β glucoside (i.v.) lowers it in spontaneously hypertensive rats.

Uses in Literature : Used earlier as an anthelmitic (vet.), carminative, disinfectant, juice (vet.), citric acid and for colds, corns, culinary purposes, dysentery, lowering body heat, peat, pickles, preparation of beverages and flavouring liqueurs, indigestion, lemon oil (obtained as by product), scabies, skin diseases, stomachache, throat diseases, vitality to body and warts **(Ambasta, 1986; Dagar & Dagar, 1999; Kadel & Jain, 2006; Kurian, 1995; Mishra *et al.*, 1996a, b; Retnam & Martin, 2006; Shome *et al.*, 1996; Tirkey *et al.*, 2003; Vishwanathan & Singh, 1996).**

57. *Citrus medica* L. (Pl. 25A)

Syn. : *C. medica* var. *medica*.

Family : Rutaceae.

Vern. : Nimbu.

English, Sanskrit and Regional Names:

Eng. : Adam's Apple, Cedrat, Citron, Melon Lime;
Sans. : Amlakshera, Bijaphalaka, Rochanaphala;
B. : Baranimbu, Nimbu;
Bo. : Bijapura, Limbu;
H. : Baranimbu, Nimbu;
Kan. : Madala, Mahaphala, Rusaka;
Mal. : Gilam, Matalanarakam;
Mar. : Mahalunga;
P. : Bajauri;
Tam. : Kogilachan, Kommattimadalai;
Tel. : Lungamu.

Distribution : Valley along the foot hills of Himalaya.

Description : An evergreen small tree. Leaves elliptic with round apex. Flowers scented, white tinged pink outside. Fruits obovoid, yellow with a leathery rind.

Flowering and Fruiting : April-September.

Habitat Ecology : Cultivated.

Material Examined : EBH-HVP-718, 12.05.2001.

Parts Used : Fruit. Leaves.

Folk Uses : Lemon squash ('sakanjabi') prepared by adding salt, sugar, black pepper (*Piper nigrum* Linn.) and lemon juice to water considered as a **refreshing drink** and believed to protect against heat strokes. Fruits **pickled** and its juice thrice daily checks **vomiting, indigestion** and **diarrhoea**. Leaves used for **scouring teeth***.

Active Constituents : New triterpenoid-citrusin along with palmitic and succinic acids, citropten, β-sitosterol and daulosterol have been isolated.

Biological Activity : Anthelmintic activity found to be positive.

Uses in Literature : So far reported as an anthelmintic, antidote to acronacrotic and animal poison, febrifuge, flavouring agent, refreshing drink, stomachic, vermifuge (vet.); and for abdominal pains, acnes, agricultural implements, biliousness, bleeding gums, cardiac diseases, checking constipation, diarrhoea, dysentery, essential oil, fatigue, foul breath, indigestion, improving general weakness and trembling of limbs, kidney disease, lemon syrup, marmalades, scorpion bites, urinary disorders and vomiting **(Ambasta, 1986; Borthakur & Sarma, 1996; Bhatt *et al.*, 2001; Chowdhery, 1996; Jain, 1991; Kurian, 1995; Lalramnghinglova, 1996; Mohanty & Padhy, 1996; Mohanty *et al.*, 1998; Negi & Gaur, 1991; Samati, 2004; Sen *et al.*, 2002; Sharma & Singh, 2001; Singh *et al.*, 1997; Tiwari & Tiwari, 1996; Watt, 1972).**

Major Ayurvedic Formulations : Gandhak-vati, Hingol-shodhan, Nimbuka-phala-rasa, Sankh-vati.

58. *Coccinia grandis* (L.) Voigt. (Pl. 25B)

Syn. : *C. cordifolia* Cogn., *C. indica* W. and A., *Cephalandra indica* Naud., *Brynia grandis* L.

Family : Cucurbitaceae.

Vern. : Kanduri.

English, Sanskrit and Regional Names:

Eng. : Ivy Gourd;
Sans. : Bimba, Bimbi;
B. : Tela kucha;
Bo. : Bhimb;
G. : Ghobe, Gluru;
H. : Kanduri, Kunduri;
Kan. : Tondekai;

P.	:	Kanduri;
Tam.	:	Kavaikai, Kovaikai;
Tel.	:	Donda, kaya.

Distribution : Throughout India.

Description : Large climbers having a woody base. Leaves penta-angled to entire, glabrous, denticulate, apex obtuse. Tendrils simple. Flowers solitary. Corolla white. Fruits oblong, narrow at both ends and blood red when ripe. Seeds oblong-ovoid, yellowish, compressed.

Flowering and Fruiting : June-September.

Habitat Ecology : Shrubberies.

Material Examined : EBH-HVP-815, 10.07.2004.

Part Used : Fruit.

Folk Uses : Fruit cooked as **vegetable** and considered good for **diabetes**.

Active Constituents : Isolated constituents are heperol and a bitter glycoside containing cucurbitacin B, taraxerone, taraxerol and 24 (R)-24-ethylcholest-5-en-3B-ol glucoside (fruits).

Uses in Literature : Known so far for diabetes, diarrhoea, dysentery, jaundice, mouth ulcer, rabies, skin eruption, tumours, wounds, and edible as vegetable, pickle, fodder, galactagogue and for gonorrhoea, skin diseases **(Ambasta, 1986; Katewa & Sharma, 1998; Kirtikar & Basu, 1935; Patil & Merat, 2003; Retnam & Martin, 2006).**

59. *Colebrookia oppositifolia* Smith. (Pl. 25C)

Syn. : *C. ternifolia* Roxb., *Elsholtzia oppositifolia* Poir.

Family : Lamiaceae.

Vern. : Dashane.

Regional Names:

Bo.	:	Bhahmini, Dasai, Dasari;
H.	:	Binda, Bindu, Pansra;
Kum.	:	Dulshat;
Oriya	:	Bosika, Darigopi;
P.	:	Bamera, Basuti, Dashane, Dhisbekkar, Sampru, Suali.

Distribution : Subtropical Himalaya.

Description : Hoary shrubs with oblong-elliptic, acuminate, crenate opposite leaves having soft tomentose. Flowers white in dense clusters.

Flowering and Fruiting : April-November.

Habitat Ecology : Common on dry slopes. Wastelands. Roadsides.

Material Examined : EBH-HVP-771, 12.04.2002.

Parts Used : Whole Plant. Wood.

Folk Uses : Plant juice mixed with mustard oil used as drops to relieve **earache***. Wood used as **fuel***.

Active Constituents : Active constituents are flavones-5-hydroxy-4', 6, 7, 8-tetramethoxy flavone and hydroxy-4', 6, 7-tetramethoxy flavone (aerial parts).

Uses in Literature : Used so far for asthma, bruises, burns, catarrh (vet.), cough, cuts, digestion, eye complaints, hysteria, epilepsy, sores, toothache, wounds, veterinary diseases; and as anthelmintic **(Ambasta, 1986; Jain, 1991; Jha & Varma, 1996; Kulkarni & Kumbhojkar, 1996; Rama Rao & Henry, 1996; Shrestha, 1985; Siwakoti & Verma, 1996; Varma, 1997; Watt, 1972).**

60. *Colocasia esculenta* (Linn.) Schott. (Pl. 25D)

Syn. : *C. antiquorum* Schott.

Family : Araceae.

Vern. : Kachalu.

English, Sanskrit and Regional Names:

Eng. : Cocoyam, Dasheen, Eddo, Taro;
Sans. : Kachwi, Kachu;
B. : Kachu;
H. : Arvi, Ghuiya, Kachalu;
Kan. : Kachhi, Shamagadde;
Mal. : Shembu;
Mar. : Alu;
Oriya : Sam, Saru;
Tam. : Seppan, Kizhangu;
Tel. : Chamadumpa, Chemagadda.

Distribution : Cultivated up to 2,500 m.

Description : Tall tuberous herbs. Leaves heart-shaped, peltate; petiole sheathing at the base. Flowers in spadix. Spathe yellow. Berries ovoid.

Flowering and Fruiting : July-October.

Habitat Ecology : Cultivated.

Material Examined : EBH-HVP-756, 16.09.2002.

Parts Used : Leaves. Rhizome.

Folk Uses : All parts **edible.** Leaves used for **'patroa'*** (a regional delicacy prepared by applying paste of powdered Bengal grams—'basin' along with salt and condiments on each leaf which is then rolled and steamed). Rhizome consumed as **vegetable**.

Active Constituents : Phytoalexin-9, 12, 13-trihydroxy-10 (E-octadecenoic acid, dihydrosterol, aliphatic compound, nonacosane, cyaniding-3-gluciside, β-sitosterol and stigmasterol (tuber) have been isolated.

Uses in Literature : Reported earlier as anodyne, astringent, congestion, demulcent, edible, lactagogue, laxative, rubefacient, stimulant, styptic, vegetable; and for abscess, bees sting, boils, chronic dysentery, increasing milk yield, gravies, haemorrhoids, insect bite, kidney stone, rheumatism, soups (**Ambasta, 1986; Augustine & Sivadasan, 2004; Girach *et al.*, 1998; Gogoi & Borthakur, 2001; Khanna, 2002; Kumar & Goel, 1998; Samati, 2004**).

61. *Cordia dichotoma* Forst. (Pl. 26A)

Syn. : *C. obliqua* Willd., *C. latifolia* Roxb., *C. myxa* Thwaites, *C. lowriana* Brandis, *C. indica* Lamk.

Family : Cordiaceae.

Vern. : Lasora, Lasura.

English, Sanskrit and Regional Names:

Eng. : Sebastan Plum;
Sans. : Bahuvarah, Kshudrashleshmataka, Madhubhu-tadruma;
B. : Bahubara, Chottabohnaari;
Bo. : Bargund;
G. : Rayagundo,Vadagundo;
H. : Chhotalasura, Lasura;
Kan. : Chikka Challe;
Mal. : Madaviriyasam, Vidi;
Mar. : Chokri, Goden;
Oriya : Domolo, Kophokoturi;
P. : Laswara;
Tam. : Naruvilli, Viriyam;
Tel. : Chinnanakkera, Nekkara.

Distribution : Throughout India.

Description : A medium-sized tree. Leaves coriaceous, broad ovate, toothed at the apex. Flowers whitish (greenish). Berries green, yellow when ripe.

Flowering and Fruiting : April-July.

Habitat Ecology : Drier areas. Wastelands. Open slopes. Moist forest.

Material Examined : EBH-HVP-803, 15.05.2004.

Parts Used : Inflorescence. Fruit. Fruit Gum.

Folk Uses : Inflorescence and fruits cooked as **vegetable** and considered good for **dysentery.** Fruits **pickled;** and its mucilaginous fluid used as **gum*** and also for curing **mouth ulcers**.

Active Constituents : β-sitosterol (roots), protein and oil (fruit kernel), mixture of palmitic, oleic, stearic, behenic, linoleic acids and other fatty acids (kernel), allantoin and b-sitosterol (stem bark), taxifolin-3-rhamnoside (seeds) are the active principles.

Uses in Literature : Used in India as anthelmintic, antidysentric, astringent, blood purifier, colic, diuretic, fibre yielding, purgative, vegetable; and for agricultural implements, boats, bone fracture (vet.), burning of throat, chest diseases, cholera, cooling, constipation, diarrhoea, dropsy, dysentery, edible purposes (fruits), expelling worms (vet.), fever, flatulence, gonorrhoea, headache, increasing sperm count, lung disease, pickling, ulcer, urination, urinary diseases urticaria and well curb **(Ambasta, 1972; Dagar & Dagar, 1996; Girach *et al.*, 1997; Hosagoudar & Henry, 1996a; Jain, 1991; Katewa & Sharma, 1998; Kumar & Pullaiah, 2003; Lalramnghinglova, 1996; Mokat & Deokule, 2004; Negi & Gaur, 1991; Parmar & Kaushal, 1982; Patole & Jain, 2002; Pawar & Patil, 2004; Rama Rao & Henry, 1996; Rao, 1996; Saini, 1996b; Singh, 1996; Tirkey, 2004; Watt, 1972).**

62. *Coriandrum sativum* Linn. (Pl. 26B)

Family : Apiaceae.

Vern. : Dhania.

English, Sanskrit and Regional Names:

- **Eng.** : Coriander;
- **Sans.** : Dhanyaka, Dhanyakam;
- **H.** : Dhaniya;
- **Mal.** : Kothambala;
- **Tam.** : Kotamalli;
- **Tel.** : Danyalu, Kotimiri.

Distribution : Throughout India.

Description : Erect herbs. Leaves ternetly lobed, ovate, base cuneate, apex obtuse. Flowers white, in umbels. Fruits sub-globose.

Flowering and Fruiting : October-December.

Habitat Ecology : Cultivated.

Material Examined : EBH-HVP-892, 12.11.2006.

Parts Used : Leaves. Seeds.

Folk Uses : Decoction of seeds (5 ml.) given thrice a day for **flatulence***. Leaves used for **flavouring*** and its paste used for **suppuration of boils***.

Active Constituents : Active constituents are umbelliferone, scopoletin, rutin, coumarins, quercetin, kaempferol, β-sitosterol, D-mannitol and flavonoid glycosides (seeds).

Biological Activity : Antimicrobial activities found positive.

Uses in Literature : Used so far as antipyretic, anthelmintic, aphrodisiac, aromatic, astringent, antimicrobial, carminative, diuretic, laxative, stimulant, stomachic, tonic; and for bronchitis, burning sensation, diarrhoea, dysentery, inflammation, jaundice, pain in eyes, piles, sore throat, stomatis, toothache, thirst **(Kirtikar & Basu, 1935; Sharma, 2003; Retnam & Martin, 2006).**

63. *Cucumis sativus* L. (Pl. 26C)

Syn. : *C. muricatus* Wall., *C. hardwickii* Royle.
Family : Cucurbitaceae.
Vern. : Kakri, Kheera.
English, Sanskrit and Regional Names:
Eng. : Cucumber, Musk Melon, Sweet Melon;
Sans. : Bahpuspha, Kandalu, Pitapulspa, Sudhavasa, Trapusha, Tundilaphala;
B. : Khira, Sasa;
Bo. : Kakri, Kankari;
G. : Khira;
H. : Khira;
Mar. : Kakdi, Khira, Tavase;
Oriya : Kavnai;
P. : Khira, Khiyar;
Tam. : Piping-kay;
Tel. : Dozakaya.
Distribution : Cultivated throughout India.
Description : A climbing herb. Tendrils simple. Leaves petaloid, palmately lobed, hispid on both sides. Flowers yellow. Fruits oblong, cylindric, covered with harsh hair of soft spines. Seeds many, oblong, compressed.
Flowering and Fruiting : May-September.
Habitat Ecology : Grown locally.
Material Examined : EBH-HVP-729, 12.09.2001.
Parts Used : Fruit. Seeds.
Folk Uses : Fruits **edible** and used in salads; its slice used in **facial cosmetics** to remove dark circle around eyes. Fruit juice prescribed thrice daily for 30 days to cure **kidney stones***. Dried seeds **eaten** and its extract considered a **refreshing drink***. Powdered decorticated seeds added to various **culinary** preparation and considered **nutritive**.
Active Constituents : These are palmitic, stearic and oleic acids, their glycerides, sterols, squalene (fruits).
Uses in Literature : Used as diuretic, fruit, salad, refrigerant, tonic; and for confectionary, dysentery, facial cosmetics, fever, gastric (vet.), pneumonia, sunstroke, sweets, throat affection, urinary infections and whooping cough **(Ambasta, 1986; Jain, 1991; Kulkarni & Kumbhojkar, 1993, 1996; Negi & Gaur, 1991; Samati, 2004; Satapathy & Brahmam, 1996; Vihari, 1995; Watt, 1972).**

64. *Cucurbita hispida* Thunb. (Pl. 26D)

Syn. : *Benincasa cerifera* Savi., *B. hispida* (Thunb.) Cogn.
Family : Cucurbitaceae.
Vern. : Dhuda Petha, Petha.

English, Sanskrit and Regional Names:

Eng. : Ash pumpkin, Tallow gourd, White Gourd, Melon;
Sans. : Gramya, Karkati, Kushmanda, Nagapuspha, Phala;
B. : Chalkumra;
Bo. : Kohala;
H. : Golkaddu, Petha;
Kan. : Budagumbala Kayi;
Mal. : Kumpalanna;
Mar. : Kohala;
P. : Chalkumra, Golkaddu;
Tam. : Kalytanppushinikkay;
Tel. : Burdagumuda, Pendigummadikaya.

Distribution : Cultivated.

Description : Climbers with rigid petiole having pungent hairs and large yellow flowers. Fruits flesh-white.

Flowering and Fruiting : June-October.

Habitat Ecology : Grown locally.

Material Examined : EBH-HVP-816, 12.08.2004.

Part Used : Fruit.

Folk Uses : Fruits used for **confectionary*** preparations, **religious worshipping** at the time of performing house warming ceremony and also employed for **symbolic*** animal sacrifices. Its vegetable considered good for **expelling worms**. Also used for **dying*** black hair into grey.

Active Constituents : β-sitosterol, lupeol, n-tricontanol, mannitol, arginine, aspartic acid, glutamic acid, asparagine, glutamine, praline, hydroxyl proline, isoleucine, cysteine, L-leucine, glucose and rhamnose have been characterised.

Uses in Literature : Reported earlier as anthelmintic, aphrodisiac, diabetic, diuretic, general tonic, laxative, vegetable (flowers); and for asthma, dysentery, epilepsy, intestinal worms, piles, preserving corpse, stomach problems and tuberculosis **(Ambasta, 1986; Bhardwaj & Gakhar, 2003; Ghosh *et al.*, 1996; Jha *et al.*, 1996; Lalramnghinglova, 1996; Mahato & Mahato, 1996; Rama Rao & Henry, 1996; Retnam & Martin, 2006; Sharma & Borthakur, 2003; Siwakoti & Verma, 1996).**

65. *Cucurbita pepo* L. (Pl. 27A)

Syn. : *C. pepo* Roxb.

Family : Cucurbitaceae.

Vern. : Kaddu.

English, Sanskrit and Regional Names:

Eng. : Pumpkin, Pompion, Squash red gourd vegetable Marrow;
Sans. : Karkaru, Kurkara;
B. : Kadimah, Kumra;
Bo. : Kaula;

G. : Dudia;
H. : Kadimah, Kaddu, Kadimah, Lal Kumra, Safed, Kaddu, Sitaphal;
Kan. : Kumbalakai;
Mal. : Mathan;
Mar. : Kohala;
Oriya : Panikakhara;
Tam. : Parangikayi, Suraikayi;
Tel. : Budadegummedi, Gummadi.

Distribution : Cultivated.

Description : A climber having stems and leaves with harsh prickly armour, a broad triangular-pointed outline. Flowers large, yellow. Fruit large-sized; peduncle woody, strongly grooved.

Flowering and Fruiting : June-September.

Habitat Ecology : Cultivated.

Material Examined : EBH-HVP-731, 16.06.2001.

Parts Used : Fruits. Seeds. Rind.

Folk Uses : Fruits cooked as **vegetable**. Fruits also **offered*** to appease Lord Shiva on the day of Shivratri. Roasted seeds **eaten**. Dried rind used as **utensil** to keep water for birds.

Active Constituents : Cholesterol, ergosterol, campesterol, β-sitosterol (seeds) are the active ingredients.

Biological Activity : Seeds exhibit anthelmintic activity *in vitro* and *in vivo* against *Hymenolepis nana* and *Dicrocoelium denddriticum*.

Uses in Literature : Known so far as anthelmintic, diuretic, galactagogue, poison (vet.), psychoactive plant, taeniacide, tonic, vegetable; and for boils, burns, diarrhoea, edible purposes, inflammation, scorpion sting, utensils, worshipping, **(Ambasta, 1986; Jain, 1991; Jain *et al.*, 1994; Jha *et al.*, 1996; Samati, 2004; Watt, 1972).**

66. *Curcuma domestica* (Medik.) Valeton. (Pl. 27B)

Syn. : *Curcuma longa* L.

Family : Zingiberaceae.

Vern. : Haldi.

English, Sanskrit and Regional Names:

Eng. : Turmeric;
Sans. : Aneshta, Aurukesafur, Bahula, Haridra, Gauri, Kurkum, Nisa, Rajani, Ralri, Sarsud;
B. : Halder, Halja;
H. : Haladu, Haldi, Halud, Hardi;
Kan. : Halede;
Mal. : Arishina;
Mar. : Halada;
P. : Haridru;
Tam. : Pasupu;
Tel. : Mannal, Marinal.

Distribution : Throughout India.

Description : A tall herb possessing long stiputate tubers which are orange coloured inside. Leaves large-sized, oblong-lanceolate, tapering to the base. Flowers in dense compound spikes, concealed by sheathing petiole.

Flowering and Fruiting : September-October.

Habitat Ecology : Cultivated.

Material Examined : EBH-HVP-735, 18.09.2001.

Parts Used : Rhizome. Leaves.

Folk Uses : Rhizome paste in combination with mustard oil considered good for **strengthening gums***. One teaspoon of powdered rhizome recommended with milk twice daily to cure **internal injuries***. Sweet 'laddos' with 'haldi' considered good for **rheumatism.** Paste prepared by taking equal proportion of turmeric, sandal wood and **'**neem' (*Melia azedarach* Linn.) leaves valued for **skin affections** and as **facial pack***. Turmeric considered **auspicious*** by local folks like other Hindus and used in all **religious ceremonies**. Leaves employed for wrapping '**patraos**'* before these are steamed to have its characteristic flavouring.

Active Constituents : Three antioxidants, curcumin-4 hydroxycinnamoy (feruloyl) methane and bis (4-hydroxycinnamoy) methane, α and β turmerones, zingiberene, β pinenes, camphene, limonene, terpinene, caryophyllene, curcumene, linalool, turmenone, artermenone, borneol, isoborneol, camphor, eugenol, curdione, curzerenone, curcumin, monodemethoxy-curcumin and didemethoxy curcumin, turmerone (rhizome) have been characterised.

Biological Activity : Curcumin on tropical application markedly inhibited TPA and arachidonic acid induced epidermal inflammation (ear oedema in mice). Turmerin exhibited efficient antioxidant and DNA protectant activities whereas curcumin-4-hydroxycinnamoy (feruloyl) methane and bis-(4-hydroxycinnamoy) methane showed anticoagulant activity.

Uses in Literature : Used earlier as antiarthritic, antibacterial, antiparasitic, antiperiodic, antiprotozoal, antiseptic, antispasmodic, carminative, cooling agent, condiments, CNS active, mastitis, spasmolytic, stomachic, tonic and for bad taste of mouth, bone fracture, chest pain, cough, cold, conjunctivitis, cuts, diarrhoea, deworming, gastric disorder, insect sting, joint diseases, liver disorder, rejuvenating milk veins, skin infection, sprain and bruises, swelling, urinary problems, whooping cough, wounds, vomiting **(Ambasta, 1986; Bhatt *et al.*, 2001; Borthakur *et al.*, 2004; Dwivedi *et al.*, 2005; Gogoi & Borthakur, 2001; Kadel & Jain, 2006; Khanna, 2002; Mohanty *et al.*, 1998; Mokat & Deokule, 2004; Retnam & Martin, 2006; Samati, 2004; Sen & Batra, 1997; Sen *et al.*, 2002; Sharma, 2003; Sharma & Singh, 2001; Sharma *et al.* 2002).**

67. *Cuscuta reflexa* Roxb. (Pl. 27C)

Syn. : *C. grandiflora* Wall., *C. verrucosa* Sweet., *C. hookeri* Sweet., *C. macrantha* G. Don., *C. elatior* Chois.

Family : Cuscutaceae.

Vern. : Akashbel, Amarvela, Halwan.

English, Sanskrit and Regional Names:

Eng. : Doddar;
Sans. : Akshspavana, Amaravallika, Akashabhavana, Akashavalli, Nilatar;
Ass. : Amarlati;
B. : Algusi, Haldi-algusi-lutta;
G. : Amarbel, Akaswel;
H. : Akashabela, Amarabela, Jaunsar, Sarabali;
Mar. : Nirmulia, Kashavela;
Oriya : Kolanirmuli;
P. : Amil, Nilathari, Niradhar, Zarbutti;
Tam. : Kodiya gundal;
Tel. : Lanjasavaramu, Savarap, Sitamapurgonallu;
U.P. : Kasbanwar, Banwar;
Urdu : Akashabel, Imalbel.

Distribution : Throughout India.

Description : Leafless parasitic twinners with fleshy yellowish stems. Flowers sweetly-scented, white, crowded in small clusters. Fruits fleshy, globose, 4-seeded, black.

Flowering and Fruiting : September-October.

Habitat Ecology : Parasitic climber on *Duranta*, *Citrus*.

Material Examined : EBH-HVP-760, 12.09.2002.

Parts Used : Aerial Plant Parts.

Folk Uses : 5-10 ml. decoction of the aerial plant (twice daily for 3 days) checks **dysentery.** A bath in warm decoction of stem considered good to relieve **swelling** and **rheumatic pains.** Poultice of the plant paste applied to cure **red spots*** due to blood clotting. Massage of the mustard oil boiled with powdered climber good for **internal injuries*** and **rheumatic pains.**

Active Constituents : Cerotic, linolenic, linoleic, oleic, stearic and palmitic acid, phytosterol (seeds); abscisic acid (leaves), dulcitol, luteolin, and its glycosides, quercetin, cuscutin (stem), amino acid, cusculatin, cuscutin (plant) have been isolated.

Biological Activity : Antiviral (aerial parts), relaxant and spasmolytic (aq. extr. of the plant) activities +ve. Spasmolytic and acetylcholine-like action in atropinised rabbit auricles and frog's heart also exhibited.

Uses in Literature : Earlier known to be used as an anodyne, anthelmintic, carminative, diaphoretic, diuretic, fodder, haematuria, haemostatic, purgative, stomachic, spermatorrhoea, biliousness; and for bodyache, bone fracture, burns, cuts, cold, cough, diarrhoea, diptheria, displaced knee cap, eye disorders, falling of hair, fever, headache, increase blood, indigestion, itching, jaundice, killing lice (vet.), liver complaints, nervine weakness, odema, rheumatism, sprain, skin diseases, swelling of legs, testicles, ulcers, urticaria, veterinary diseases and wounds (**Ambasta, 1986; Chauhan, 1999; Datt, 1996; Dixit & Pal, 1996; Girach, 1992; Girach & Aminuddin, 1995; Girach *et al.*, 1998; Hembrom, 1991; Jain, 1991; Jain *et al.*, 1991; Katewa & Arora, 1997; Kumar & Chauhan, 2005; Kumar & Jain, 1998; Mahato & Mahato, 1996; Manandhar, 1985, 1987a; Maya *et al.*, 2003; Sahoo & Mudgal, 1995; Saini, 1996a, b; Satapathy & Brahmam, 1996; Sen & Behera, 2003; Sharma, 2003; Singh, 1996; Singh & Prakash, 1994; Siwakoti & Verma, 1996 ; Srivastava *et al.*, 2003; Tirkey *et al.*, 2003b; Tiwari & Tiwari, 1996; Watt, 1972).**

Major Ayurvedic Formulations : Itriafal Ustakhuddus (Unani drug) beside being used in traditional drugs.

68. *Cynodon dactylon* (Linn.) Pers. (Pl. 27D)

Syn. : *C. stellatus* Willd., *Panicum dactylon* Linn., *Paspalum dactylon* DC., *Digitaria dactylon* Scop.

Family : Poaceae.

Vern. : Dhrub, Doob, Doorva.

English, Sanskrit and Regional Names:

Eng. : Bermuda Grass, Conch Grass, Dhub Grass, Doob Grass, Dubh, Dubla;
Sans. : Durva, Haritali;
B. : Dhobi-ghas, Durba, Dubh, Dubla;
H. : Dhub, Hariali;
Kan. : Kudi garikai, Garikaihallu;
Mar. : Arugam-pilla, Hariali;
P. : Barawa, Dhub, Khabbal, Tallu;
Tam. : Gherricha, Haryali;
Tel. : Gerichagaddi, Harvali.

Distribution : Throughout India.

Description : A prostrate grass; rooting at the nodes. Leaves linear-lanceolate, soft and smooth. Flowers in digitate spikes.

Flowering and Fruiting : August-November.

Habitat Ecology : Edges of fields. Open slopes.

Material Examined : EBH-HVP-879, 12.18.2005.

Parts Used : Leaves. Whole Plant.

Folk Uses : Plant considered **sacred*** and employed in all religious ceremonies, and also as good **fodder.** Its paste applied for suppuration of **boils***. Bare footed walk on its field early in the morning **improves eyesight***. 3-4 drops of its leaf juice check **nose bleeding***.

Active Constituents : Principle constituents are 2-(3-methoxy-4-hydroxy phenol propionic acid, 3-methoxy-4-hydroxy-benzoic acid, 4-

hydroxy-benzoic acid, 2-(4-hydroxy phenyl), phytol, β sitosterol, stigmasterol acetate (leaves), alkaloids, steroid, glycoside, flavone, terpenes, organic acid, alcohol, wax, sugars. β carotene, vit. C, palmitic acid, triterpenoids, alkaloids, ergonovine, p-coumaric acid and glucose.

Biological Activity : Plant exhibits antiviral properties.

Uses in Literature : Important as pasture and lawn grass, soil binder, anasarca, astringent, cooling, diuretic, expectorant, stomatitis, vulnerary; and for acrid, bad taste in mouth, biliousness, bruises, burning sensation, cuts, dropsy, dysentery, fatigue, fever, genito-urinary disease, hysteria, increasing lactation in animals, leucorrhoea, leprosy, odema, ophthalmia, piles, prickly heat, scabies, skin diseases, sweet thirst, vomiting, wounds, ulcers, and urticaria **(Ambasta, 1986; Gogoi & Borthakur, 2001; Kadel & Jain, 2006; Retnam & Martin, 2006; Sen & Behera, 2003; Sharma, 2003; Sharma *et al.* 2002; Upadhyay & Singh, 2005).**

Major Ayurvedic Formulations : Dooravadi-kwatha, Doorvadi-ghrita, Doorvadi-taila, Chandrakala-rasa.

Adulterant or Substitutes : Rhizome used as an adulterant or substitute to Couch grass (*Agropyron repens* Beauv.).

69. *Dalbergia sissoo* Roxb. (Pl. 28A)

Syn. : *D. pendula* Tenore.

Family : Fabaceae

Vern. : Shisham, Tahali.

English, Sanskrit and Regional Names:

Eng. : South Indian Red wood, The Sissoo;
Sans. : Aguru, Dhira, Kapila, Mandalapatri, Sasam, Sasim, Shisham, Sissu, Sissai, Sisam;
Ass. : Sissu;
B. : Sissu;
G. : Yette, Nukku-Kattai, Sissom, Tanach;
H. : Shisham, Shisu, Sisam, Sissai;
Oriya : Sisu;
Kan. : Sisapa, Singsupa;
Mal. : Iruvil;
P. : Shewa, Shin, Sisham, Tali, Tahli;
Tam. : Gette, Itti, Karra, Nukku, Sissu, Tali;
Tel. : Biridi, Cishmbage, Ettasissu, Sinsupa.

Distribution : Throughout India ascending to 1,650 m.

Description : A tree with numerous branches and alternate pinnate leaves having zig-zag leaf rachis. Flowers yellowish-white and in axillary panicles. Pods narrow at base into long stalk and strap-shaped.

Flowering and Fruiting : May-July.
Habitat Ecology : Open grassy slopes.
Material Examined : EBH-HVP-783, 21.06.2003.
Parts Used : Leaves. Wood.
Folk Uses : Wood employed for making **furniture, agricultural implements**, **building works**, **tool handles**, doors and windows. Leaves used as **fodder.** Dried twigs used as **fuel**.
Active Constituents : Biochanin A-7-glucosido (sissortin), isoflavone-o-glycoside (leaves), a new allylphenol of latifolin type-dalbergiphenol along with dalbergenone, OMe-dalbergin, hydroxyl and OMe-dalbergenone, methyl dalbergin (heart wood), biochanin A and tectorigenin (flowers) have been characterised.
Biological Activity : Spasmolytic (plant) activity confirmed.
Uses in Literature : So far described as an alternative, antiemetic, fuel wood, fodder, ornamental (ear pendant), sacred; and for agricultural implements, building works, boils, cabinet wood, carved work, charcoal, charpai legs, cuts, diarrhoea (vet.), dysentery, gonorrhoea, headache, high class furniture, leucorrhoea, musical instruments, railway sleepers, skin diseases, spermatorrhoea (**Ambasta, 1986; Borthakur & Sarma, 1996; Jain, 1991; Katewa *et al.*, 2001; Khanna & Kumar 2000; Kumar & Chauhan, 2005; Mandal & Basu, 1996; Siwakoti & Verma, 1996**).

70. *Datura innoxia* Mill. (Pl. 28B)

Syn. : *D. metel* auct., non Linn.
Family : Solanaceae.
Vern. : Datura, Dhatura.
Sanskrit and Regional Names:
Sans. : Dhustura;
Kan. : Ummatta;
Tam. : Vellumattai;
Tel. : Datturamu.
Distribution : N.W. Himalaya.
Description : An undershrub. Leaves broadly-ovate, pale green. Flowers creamish. Capsule armed with spines.
Flowering and Fruiting : June-August.
Habitat Ecology : Wastelands. Drier areas.
Material Examined : EBH-HVP-842, 13.08.2005.
Parts Used : Leaves. Twigs. Flowers. Seeds.
Folk Uses : Smoke of burnt dried leaves and twigs inhaled to cure **asthma**, **cough** and **cold***. Fruit juice applied on scalp to check **dandruff***. Poultice of its seeds mixed with mustard oil

applied externally for **rheumatic pains**. Flowers offered to appease 'Lord Shiva'; and its smoke considered a good remedy to ward off the effect of **magic** and **evil*** eyes to livestock.

Active Constituents : Important constituents are hyoscine (scopolamine) and hyoscymine (in all parts of plant); and scopolamine (seeds).

Biological Activity : Seeds antimicrobial; whole plant spasmolytic.

Uses in Literature : Known in India for asthma, boils, cough, dog bite, diarrhoea, high fever, insect bites, mumps, rheumatism, scabies, skin diseases, sprains, swelling, toothache, waist pain, wounds and offering to Lord Shiva; and as an anodyne, antihydrophobic, antiseptic, antispasmodic, country liquor, septic, sedative and cerebral depressant (**Ambasta, 1986; Agrawal, 1981; Bhatt *et al.*, 2001, 2002; Dwivedi *et al.*, 2005; Hajra, 1987; Jain *et al.*, 1991; Kadel & Jain, 2006; Katewa *et al.*, 2001; Kshirsagar & Singh, 2000; Kshirsagar *et al.*, 2003; Kumar & Pullaiah, 1998; Maliya & Singh, 2003; Mohanty *et al.*, 1998; Pandey & Verma, 2002; Saxena & Vyas, 1981; Shah & Joshi, 1971; Sharma, 2003; Shekhawat & Anand, 1984; Singh *et al.*, 1987; Venkataratnam & Venkata Raju, 2004**).

71. *Dendrocalamus hamiltonii* Nees and Arn. *ex* Munro (Pl. 28C)

Syn. : *Bambusa monogyna* Griff., *B. maxima* Ham., *B. falconeri* Munro.

Family : Bambusaceae.

Vern. : Baingh, Maggar.

Regional Names:

H. : Kokua; Pecha;

Tam. : Kagzi-Bans.

Distribution : N.W. Himalaya: 1,500-2,400 m.

Description : Sturdy shrubs. Leaves 3-nerved, long pointed with lower surface covered with white tomentose. Flowers yellow, unisexual. Drupes yellow, aggregated in small heads.

Flowering and Fruiting : May-August.

Habitat Ecology : Forests. Shrubberies (common).

Material Examined : EBH-HVP-784, 10.07.2003.

Parts Used : Leaves. Wood.

Folk Uses : Wood used as **fuel***, and leaves considered as good **cattle feed***. 10-20 ml. decoction of leaves once daily for 15-20 days considered good against **hernia***.

Uses in Literature : Used in India for cordage (bark), burning sensation, basket, edible purposes, expel placenta (vet.), hunting, fencing, fishing lines, sore throat, stomach pain, thread, traditional dwelling, ulcers and killing maggots in wounds of cattle and goats, and spermatorrhoea (**Ambasta, 1986; Chhetri, 2005; Jadhav, 2006; Kapur, 1996; Kumar *et al.*, 2004; Patel *et al.*, 2003; Patole & Jain, 2002; Punjani, 1998; Rama Rao & Henry, 1996; Samati, 2004; Sen & Batra, 1997; Singh, 2003; Watt, 1972**).

72. *Dioscorea bulbifera* Linn. (Pl. 28D)

Syn. : *D. sativa* auct. non Thunb., *D. pulchella* Roxb., *D. pulchella* Hoken., *D. tamnifolia* Salisb., *Helmia bulbifera* Kunth., *D. decemangularis* Herb., *D. thunga,* Herb., *Helmia bulbifera* Kunth.

Family : Dioscoreaceae.

Vern. : Tardi.

English, Sanskrit and Regional Names:

Eng. : Black bearing Yam;
Sans. : Badar Kachna, Vishvaksenapriya;
Ass. : Kathalu;
B. : Banalu;
G. : Shalvinavelya;
H. : Bhirvolikanda;
Oriya : Sisu;
Kan. : Onthalaigasu;
Mal. : Kattukachil;
P. : Zamikand;
Tel. : Pannukelangu.

Distribution : Subtropical Himalaya.

Description : Climbers with opposite, ovate-cordate leaves. Seeds 3 winged, included.

Flowering and Fruiting : June-August.

Habitat Ecology : Scrub Jungles. Hedges. Moist rocky slopes.

Material Examined : EBH-HVP-822, 17.08.2004.

Part Used : Tuber.

Folk Uses : Tubers consumed as **vegetable**.

Active Constituents : Active properties are due to diosgenin, diosbulbins A, diosbulbin B and diosbulbin C (Tuber).

Biological Activity : Anti-inflammatory and antifertility activities found +ve.

Uses in Literature : Known for its edible tubers, diabetes, intestinal worms, skin diseases and as contraceptives, psychoactive (**Ambasta, 1986; Gangwar & Ramkrishnan, 1990; Jain, 1991; Lal *et al.*, 1996; Sharma, 2003, Singh, 1996; Singh & Prakash, 1994; Singh & Zaheer, 1996; Sinha, 1996; Siwakoti & Verma, 1996; Watt, 1972**).

Major Ayurvedic Formulations : 'Chyawanaprasha', Garbhadharini-vati, Mahamash-tel, Narsingh-tel.

73. *Dioscorea deltoidea* Wall. *ex* Kunth. (Pl. 29A)

Family	:	Dioscoreaceae.
Vern.	:	Singli-Mingli.
Regional Hindi Name	:	Kirta.
Distribution	:	Temperate Himalaya.
Description	:	A climber with sub-opposite leaves. Stem glabrous. Male spikes often branched and seeds winged all around.
Flowering and Fruiting	:	August-September.
Habitat Ecology	:	Scrub Jungles. Hedges. Moist rocky slopes.
Material Examined	:	EBH-HVP-875, 15.08.2005.
Part Used	:	Tuber.
Folk Uses	:	Tuber paste applied externally for **rheumatism** and **joint diseases**.
Active Constituents	:	Active ingredients are due to cortisone, diosgenin, sapogenin (tuber).
Biological Activity	:	Anti-inflammatory and antifertility activities found +ve.
Uses in Literature	:	Known for rheumatics, gastric complaints, ophthalmic disorders, edible tubers killing lice, rheumatic diseases **(Ambasta, 1986; Chauhan, 1999; Jain, 1991; Kapur *et al.*, 1996; Sharma, 2003, Watt, 1972).**
Remarks	:	Diosgenin is used in the synthesis of modern drugs like cortico-steroids, which enter into preparation of sex hormones, contraceptive pills, and the anabolic agents.
Trade	:	Singli-mingli.

74. *Dodonaea viscosa* Jacq. (Pl. 29B)

Syn.	:	*D. angustifolia* L., *D. dioca* Roxb., *D. burmanniana* DC., *D. pallida* Miq., *D. micropoda* DC., *D. wightiana* Blume., *D. pentandra* Griff., *D. heterophylla* Hortul, *D. spathulata* Sm., *D. arabica* Hochst. and Steud., *Ptelea viscosa* L.
Family	:	Sapindaceae.
Vern.	:	Mehandru.

English, Sanskrit and Regional Names:

Eng.	:	Jamaca Switch Sorrel;
Sans.	:	Rasna;
B.	:	Bandari, Zakhlumi;
Bo.	:	Jakhmi, Lutchmi;
H.	:	Alliar, San-atta, Sonalta, Walaytinahndi;
Kan.	:	Bandara;
Mal.	:	Unnataruvi, Virali;
Mar.	:	Lutchmi;
Oriya	:	Mohra;
P.	:	Banmehndu, Mendhar, Sonatta;

Tam. : Velari, Virali;
Tel. : Bandaru, Pullenu.

Distribution : Afghanistan-H.P. up to 1,500 m.
Description : Shrubs. Leaves alternate, viscid with yellow resinous dots. Flowers greenish-yellow. Capsule yellowish-brown, compressed.
Flowering and Fruiting : February-April.
Habitat Ecology : Open slopes. Roadsides; often gregarious.
Material Examined : EBH-HVP-775, 12.02.2003.
Parts Used : Leaves. Wood.
Folk Uses : Poultice of boiled leaves applied to heal **fractured bones** and **sprains***. Leaves also used as **cattle feed.** Wood used as **fuel.**
Active Constituents : These are β-sitosterol, stigmasterol and isorhamnetin, a leucocyanidin (bark), two prosapogenins-RI barrigenol-21, 22 diangelate and jegosapogenol-21-(2, 3, dihydroxy-2-methyl-butyroyl) and 22 angelates (stem bark).
Biological Activity : Plant extr. exhibits cardio-inhibitory and spasmolytic activity on smooth muscle and intestine. It also has sedative effect on virgin Guinea pig uterus and hypotensive effect unaffected by atropine.
Uses in Literature : Used as aromatic, astringent, carminative, febrifuge, fire wood, fodder, manure; and for backache, bone fracture, broken legs of livestock, burns, cold, cuts, dandruff, dental diseases, epilepsy, intestinal worms, fracture, intestinal worms, muscular pains, paralysis, sprains, waist pain, worms (vet.) and wounds **(Ambasta, 1986; Geetha *et al.*, 1996b; Goud & Pullaiah, 1996a; Jain 1991; Kumar & Pullaiah, 1998; Rama Rao & Henry, 1996; Retnam & Martin, 2006; Sharma, 2003; Sharma & Sood, 1997; Vishwanathan, 1997; Watt, 1972).**
Adulterant or Substitute : Fruits used as substitute for True Hops (*Humulus lupulus* Linn.).

75. *Duranta repens* L. (Pl. 29C)

Syn. : *D. plumieri* Jacq.
Family : Verbenaceae.
Vern. : Neel Kanthi (Neel Kanta).
English, Sanskrit and Regional Names:
Eng. : Creeping Skin Flower, Golden Dew Drop, Pigeon Berry;
Sans. : Vriti Kanta;
H. : Daranta.

Distribution : W. Himalaya: 200-1,600 m.

Description : A spiny shrub. Leaves opposite, broadly elliptic, entire. Spines stout. Flowers bright, lavender-blue with pale centers. Fruits golden yellow, globular.

Flowering and Fruiting : April-August.

Habitat Ecology : Common as a hedge plant.

Material Examined : EBH-HVP-719, 13.04.2001.

Parts Used : Aerial parts. Leaves.

Folk Uses : Plant grown as hedges for **fencing***. Gargles of decoction of its leaves cure **toothache***.

Active Constituents : β-sitosterol, α-amyrin, scuttellarien, 4' methoxy-scutellarein, scutellarein 7-O-rhamnoside, pectolinarigenin and 7-O-rutinoside are the active constituents.

Uses in Literature : Recorded earlier as an insecticidal and for turnery **(Ambasta, 1986)**.

76. *Elettaria cardamomum* Maton (Pl. 29D)

Syn. : *Amomum cardamomum* L., *Alpinia cardamomum* (L.) Roxb.

Family : Zingiberaceae.

Vern. : Elaichi.

English, Sanskrit and Regional Names:

Eng. : Lesser cardamom, Cardamom;
Sans. : Bahoola, Chundruvala, Ela, Elo;
B. : Ilachi;
G. : Elachi;
H. : Chottielachi.
Mal. : Malabari elachi;
Mar. : Veldode, Velloda;
P. : Illachi;
Tam. : Ellakay, Aila-cheddi, Ela-kaya;
Tel. : Ellakay, Elaki chettu, Ela-kaya.

Distribution : Throughout India.

Description : Perennial herbs. Leaves distichous. Flowers in panicle, white or pale green in colour. Fruits trilocular, pale green to yellow in colour.

Flowering and Fruiting : April-August.

Habitat Ecology : Cultivated.

Material Examined : EBH-HVP-747, 15.05.2002.

Part Used : Seeds.

Folk Uses : Seeds used as a **spice** and for **flavouring** tea and coffee. It is considered good for **flatulence, indigestion, bronchitis** and **upper respiratory tract infections***.

Active Constituents : Active constituents are α, β-pinene, sabinene, myrcene, limonene, cineol, cymene, linalool, borneol, nerylacetate, geraniol, nerol, neolidol.

Uses in Literature : Known to be used as aromatic, aphrodisiac, carminative, cooling, diuretic, emmenagogue, purgative, stimulant; and for asthma, bronchitis, indigestion, liver disorders, piles, scabies, toothache, vomiting, flavouring, confectionary, beverages and liquors (**Ambasta. 1986; Sharma, 2003; Retnam & Martin, 2006**).

Major Ayurvedic Formulations : Chyawanaprasha, Elachi Bati.

77. *Emblica officinalis* Gaertn. (Pl. 30A)

Syn. : *Phyllanthus emblica* Linn., *Cicca emblica* Kurz., *Dichelastina nodicaulis* Hance.

Family : Euphorbiaceae.

Vern. : Ambla, Amla.

English, Sanskrit and Regional Names:

Eng. : Emblic Tree, Myrobalan Tree, Indian Goose Berry;

Sans. : Adiphala, Akara, Amalaka, Amalaki, Amlika, Amaraphala, Amritaphala, Bahuphali, Dhatri, Dhatrika, Jatiphala, Kayastha, Pancharasa, Rochani, Shirphala, Shiva, Triphala,Vrishya, Vrittaphala;

A. : Amlaki, Sohmyrlain;

B. : Amla, Amlaki;

H. : Amalaki, Amla, Amlika, Anola, Anuli, Aungra, Bhoza, Daula;

Kan. : Amalaka, Nelli, Nilika, Kayi;

Mal. : Amalakam;

Mar. : Anvala, Anoli, Avala, Arola, Bhuiawali;

Oriya : Amlaki, Gondhona, Ohalu;

Tam. : Amalagam, Nelli, Sirottam;

Tel. : Amalakamu, Amalaki, Usiriki;

Urdu : Anwala.

Distribution : Throughout India; often cultivated 1,400 m.

Description : A deciduous tree with crooked trunk. Leaves subsessile, light green, glabrous, obtuse. Flowers greenish-yellow, in axillary fascicles on the leaf bearing branchlets. Fruit succulent, globose, pale yellow, acidic, 6-lobed.

Flowering and Fruiting : April-September.

Habitat Ecology : Common along borders of agricultural fields. Dry soil.

Material Examined : EBH-HVP-778, 10.09.2003.

Parts Used : Fruit. Branches. Plant. Wood.

Folk Uses : One teaspoon of fruit powder taken with honey (twice a day) for seven days to cure **cough** and to increase **appetite.** Fresh fruits considered **refrigerant***, source of **vitamin C** and good for checking **constipation** and to enhance **memory***. Pickle and jam of its fruits prescribed in case of **jaundice***. Paste of the mixture of its powdered dry fruits and leaves of 'henna' (*Lawsonia alba* Linn.) applied as **dye** to blacken hairs and for their growth. Plant considered **sacred** and worshipped on **'Panch Bhikma'** (specific religious day of Hindus). Small pieces of its wood along with sandal wood are offered by one and all in last rites to the **funeral pyre*** as a token of their respect to the departed soul.

Active Constituents : Active constituents are due to tannins, polyphenolic compounds 1, 3, 6-trigalloyl glucose, terchebin, corialgin, ellagic acid, phyllantidine and phyllantine, terchebin (fruits, leaves), vit. C, I.A.A., four other auxinsal, a3, a4, a5, two growth inhibitors R1 and R2.

Biological Activity : Leaf extr. antibacterial. 50% Et OH extr. of fruits anticancerous, antiviral, carminative and stomachic. Water soluble fraction of methanol extr. of leaves exhibits anti-inflammatory activity in carrageenin and dextran induced rat hond pan odema.

Uses in Literature : Recorded to be anticancerous, aphrodisiac, appetiser (vet.), astringent, detergent, diuretic, edible (eaten raw or cooked), expectorant, hair dye, hair shampoo, important gredient of 'Chyawanaprasha' and 'triphala', inferior quality furniture, laxative, magico-therapeutic to treat toothache, spermatorrhoea, rich source of vit. C, symbolic abode of God (sacred), tonic; and for agricultural implements, anthrax, blood purifier, cold, cough, diarrhoea, dysentery (vet.), gonorrhoea, head-ache, inflammation, rinder pest (vet.), tanning, garments, ulcer, vaginal complaints, vomiting, washing head, well rings, wound maggots **(Ambasta, 1986; Aminuddin & Girach, 1991, 1993; Bhardwaj & Gakhar, 2003; Banerjee & Pal, 1996; Borthakur, 1996a; Borthakur & Gogoi, 1994; Chauhan, 1999; Ghosh *et al.*, 1996; Girach, 1992; Gogoi & Borthakur, 2001; Hosagoudar & Henry, 1996a; Jain, 1968, 1991; Jain *et al.*, 1994; Khanna *et al.*, 1996b; Kothari & Moorthy, 1996; Kulkarni & Kumbhjokar, 1996; Kumar & Pullaiah, 2003; Mandal & Basu, 1996; Mehrotra *et al.*, 1995; Mohanty *et al.*, 1998; Negi *et al.*, 1993; Pandey & Verma, 2003; Patole & Jain, 2002; Pawar & Patil, 2004; Retnam & Martin, 2006; Reddy *et al.*, 1998; Sahoo & Mudgal, 1995; Sharma, 2003; Sharma & Singh, 2001; Sharma & Sood, 1997; Singh, 2003; Tirkey *et al.*, 2003; Upadhyay & Singh, 2005; Vihari, 1995; Vishwanathan & Singh, 1996; Watt, 1972).**

Major Ayurvedic Formulations : Agnitundi-vati, Ambalyadi-churna, Amlaki-churna, Amlakyadi-churna, Amlaki-rasayan, Arogya-vardhini, Brahma-rasayan, Brahmi-amla-taila, Brahmi-rasayana; fresh fruit pulp is the main ingredient of 'Chyawanaprasha', Sanjivani-vati, Triphala-churna, Triphala-ghrita, Triphala-Karanjaadi yog, Kabz-care.

Trade : Amla, Emblic Myrobalan.

78. *Eriobotrya japonica* Lindl. (Pl. 30B)

Syn. : *Mesilpus japonica* Banks.
Family : Rosaceae.
Vern. : Laukhat, Lukat.
English and Regional Names:
Eng. : Chinese Medlar, Japan Medlar, Loquat;
H. : Logat;
Kan. : Lokkte;
Tam. : Ilakota;
Urdu : Lakhota.
Distribution : Throughout India; often cultivated.
Description : An evergreen, small woolly cultivated tree. Leaves obovate, coriaceous, strongly ribbed and toothed. Flowers white, fragrant in terminal panicles. Fruits pyriform, succulent orange yellow.
Flowering and Fruiting : July-September.
Habitat Ecology : Along borders of agricultural fields. Roadsides.
Material Examined : EBH-HVP-862, 10.07.2005.
Parts Used : Fruit. Leaves.
Folk Uses : Fruits **edible** and considered as good **appetiser***. Leaves used as **fodder***.
Active Constituents : These are loquatoside, glucose, fructose and sucrose (fruit pulp), sorbitol-6-phosphate dehydrogenase, ascorbic acid oxidase and vit. B (leaves).
Biological Activity : Ursolic acid exhibits potent anticarcinogenic activity against sarcoma 180 cells and significantly suppresses myconcogene 0 expression. Leaf extr. exhibits marked antifungal activity against *Ustilago tritici*.
Uses in Literature : So far described as an aphrodisiac, antiscorbutic, diuretic, expectorant, ruler and drawing material, sedative; and for skin eruptions, allaying vomiting, jams, jellies, thirst, toothache, piles and desert (**Ambasta, 1986; Chopra *et al.*, 1956; Watt, 1972**).

79. *Eruca sativa* Lam. (Pl. 30C)

Syn. : *Brassica eruca* Linn., *B. erucoides* Rox.
Family : Brassicaceae.
Vern. : Taramira.
English, Sanskrit and Regional Names:
Eng. : Rocket;
Sans. : Tambeh;
B. : Duan, Sahwan, Tira, Tura, Taramira;
H. : Suffed shorshi, Shwet-surska, Taramir;
P. : Mandao.

Distribution : N.W. Himalaya.
Description : Erect annuals. Petals veined. Seeds globose.
Flowering and Fruiting : February-April.
Habitat Ecology : Wastelands. Roadsides. Cultivated.
Material Examined : EBH-HVP-844, 12.02.2005
Parts Used : Seed. Whole Plant.
Folk Uses : Seeds with oil cake given to cattle to increase **lactation***. ½ teaspoon of seed powder given for 5 days, twice a day, to check muscular **pains***. Plant used as **fodder***.
Active Constituents : Therapeutic properties are due to isorhamnetin and its 3-glucosides (leaves); (4-Me-thio)-Bu-glucosinolate (glucoerucin) (seeds, whole plant).
Biological Activity : Seeds antibacterial and also agglutinate human red cells without specificity to any particular group.
Uses in Literature : Used so far as aphrodisiac, antiscorbutic, diuretic, stimulant, stomachic; and for epilepsy, inflammation, itching, leucorrhoea, piles, skin eruptions, toothache, vomiting and ulcers (**Chopra *et al.*, 1956; Kirtikar & Basu, 1935**).

80. *Eucalyptus umbellata* Domin. (Pl. 30D)

Syn. : *E. globulus* L.; *E. tereticornis* Sm.
Family : Myrtaceae.
Vern. : Safeda.
English, Sanskrit and Regional Names:
Eng. : Blue Gum, Eucalyptus;
Sans. : Tailaparnah;
H. : Safeda, Yikeliptas;
Tam. : Karupuramaram.
Distribution : Introduced in India up to 1,200 m.
Description : A tree having bark peeling in long thin layers. Leaves alternate. Flowers white and in umbels.
Flowering and Fruiting : March-April.
Habitat Ecology : Wastelands. Slopes. Commonly planted along Roadsides.
Material Examined : EBH-HVP-854, 12.03.2005.
Parts Used : Leaves. Wood. Twigs.
Folk Uses : Poultice of boiled leaves checks **headache.** Smoke of burning leaves good for **repelling mosquitoes***. Wood used for **construction, poles** and **furniture***. Dried twigs used as **fuel***.
Active Constituents : Isolated constituents are cis- and trans-pinacarveol, a pinene, an unsaturated a-ketone, oleanoic, gallic, ferulic and maslinic acids (leaves), gallic acid, 7-B-D-glucosyloxy-5-hydroxy-2 methylchromone, β-sitosterol and (+) catechin (wood).
Biological Activity : Oil astringent, thermogenic, antiseptic, stimulant, anodyne,

cardiotonic and diuretic.

Uses in Literature : Used so far for asthma, burns, cardiotonic, diuretic, dysentery, dyspepsia, flatulence, intermittent fever, headache, pneumonia, pyorrhoea, sleepers, engineering construction, ship building and flooring; and as antiseptic, deodorant **(Ambasta, 1986; Khanna, 2002; Kumar & Jain, 1998; Sharma, 2003).**

Major Ayurvedic Formulations : Panchgun-taila, Nilgiri-taila.

81. *Eugenia cerasoides* Roxb. (Pl. 31A)

Syn. : *Syzygium cerasoideum* (Roxb.) Raizada; *S. cerasoides* var. *obovatum* (Kurz.) Santapau and Jar.; *E. operculata* Duthie.

Family : Myrtaceae.

Vern. : Kathman.

Regional Names:

- **Ass.** : Godhajam, Mokra-Jam;
- **H.** : Dugdugia, Piaman;
- **Oriya** : Pitajam, Monisiagamo.

Distribution : Sub-Himalayan tract.

Description : Medium-sized evergreen trees. Leaves elliptic. Flower greenish-white, sessile, in cymes. Berries globose or ovoid.

Flowering and Fruiting : April-July.

Habitat Ecology : Border of agricultural fields.

Material Examined : EBH-HVP-864, 13.04.2005.

Parts Used : Leaves. Twigs.

Folk Uses : Leaves used as **fodder***. Twigs and leaves employed as **'pattri'*** (bedding for animals).

Uses in Literature : Useful for bronchitis, dysentery, joint pains, building works, boards; and as acrid, astringent **(Ambasta, 1986).**

82. *Eugenia jambolana* Lam. (Pl. 31B)

Syn. : *Syzygium cumini* (L.) Skeels, *S. jambolanum* DC, *Calyptranthes jambolana* Willd., *C. capitellata* Ham., *Myrtus cumini* Linn., *E. frondosa* Wall.

Family : Myrtaceae.

Vern. : Jaamun, Jaman.

English, Sanskrit and Regional Names:

- **Eng.** : Black Plum, Java Plum;
- **Sans.** : Jambula, Jambu, Mahaskandha;
- **Ass.** : Jamu;
- **B.** : Kala-jam;
- **Bo.** : Jambul;

	G.	: Jambura, Jamli, Ravanam;
	H.	: Jaman, Jam, Jamun, Paiman, Phalinda;
	Mal.	: Naga, Nava, Perinnaral;
	Oriya	: Jamkuli, Jamo;
	Tam.	: Naval, Narvel, Naga;
	Tel.	: Nairuri, Naredu, Nasudo.

Distribution : Throughout India.

Description : A large evergreen glabrous tree with elliptic-oblong, acuminate, coriaceous leaves which are smooth and shining above. Flowers greenish-white, fragrant, sessile. Fruits oblong, black with pink juicy pulp.

Flowering and Fruiting : May-July.

Habitat Ecology : Roadsides. Wastelands. Borders of agricultural fields.

Material Examined : EBH-HVP-874, 12.06.2005.

Parts Used : Fruit. Seed. Leaves. Wood.

Folk Uses : Fruits **edible.** One teaspoon of powdered fruit and seed taken twice a day with water to control **diabetes**. One teaspoon decoction of its fruit with onion (*Allium cepa* Linn.) considered **antiemetic** (twice a day). Wood used for **construction works*** and making **agricultural implements**. Leaves used as **fodder**.

Active Constituents : Active principles are due to jambolin, ellagic acid, tannin (19%), gallic acid, chlorophyll, fatty oil, starch, resin, sugar, traces of oil (seed), anthocyanin, delphinidin-3-gentiobioside, malvidin-3-laminaribioside-petunidin-3-gentiobioside, glucose, fructose, gallic acid, malic acid, eugenia tri-terpenoids A and B, oleanic acid (fruit), isoquercitrin, quercetin, kaempferol, myricetin (flowers), myristic, linoleic, stearic, palmitic, vernolic, lauric, sterculic, malvalic acids (seed oil), betulinic acid, friedelin, friedelinol, kaempferol, quercetin and sitosterol (stem-bark).

Biological Activity : Seeds hypoglycaemic, whereas aerial parts hypotensive.

Uses in Literature : Known till todate to cure asthma, anaemia, blood dysentery, blisters in mouth, cancer, diarrhoea, diabetes, digestive disorders, dysentery, liver disorders, leucorrhoea, pains, spermatorrhoea, ulcers, vomiting, agricultural implements, thatching and as an astringent, fodder, carminative, symbolic and tonic (**Ambasta, 1986; Chopra *et al.*, 1956; Borthakur *et al.*, 2004; Gogoi & Borthakur, 2001; Jain, 1991; Kadel & Jain, 2006; Katewa *et al.*, 2001; Khanna & Kumar 2000; Kumar & Chauhan, 2006; Mohanty *et al.*, 1998; Ram *et al.*, 2002; Retnam & Martin, 2006; Sharma, 2003; Sharma & Singh, 2001; Upadhyay & Singh, 2005**).

Major Ayurvedic Formulations : Jambarista, Jambuadi-taila, Panchapallara-yoga, Pancha-pallava-yoga, Pusyanuga-churna, Usirasava.

Trade : Jaamun-guthli, Jambul-stone.

83. *Euphorbia royleana* Boiss. (Pl. 31C)

Syn. : *E. pentagona* Royle.

Family : Euphorbiaceae.

Vern. : Chparchoo, Chui, Chu.

Regional Names:

H. : Shakar-pitan, Sohund, Surai;
Kum. : Sohund, Suli;
P. : Chula, Chu, Suli, Thar, Thor, Tordanda.

Distribution : Himalaya, up to 2,000 m.

Description : Erect, cactus-like fleshy shrubs having angled branches. Leaves alternate, sessile, spathulate with thorny, persistent stipules. Involucres yellow. Capsule pale grey, pedunculate.

Flowering and Fruiting : June-July.

Habitat Ecology : Wastelands. Drier areas. Rocky slopes.

Material Examined : EBH-HVP-884, 16.07.2005.

Parts Used : Whole Plant. Latex. Pith.

Folk Uses : Latex applied for suppuration of **boils**. Also a few of its drop applied on the proximal end of **infected tail*** to check infection **('nanain')** in livestock. Part of crushed stems considered **antiseptic***. Pith slices mixed with wheat dough fed to cattle for stimulating **ovulation***; and also to avoid **abortion*** after conception. Plant considered good for **fencing*** purposes.

Active Constituents : Ellagic acid (flowers), taraxerol (bark), euphrol, cycloeucalenol, taraxerol, sitosterol, coumarin, aromatic carboxylic acid (latex); 2, 3-di-O Me ellgic and succinic acid (plant); taraxeryl-OAc, glut-5-en-3β-yl-OAc, β-sitosterol, friedelan-3-α-ol, lupeol, β-amyrin and epitaraxerol (stem) have been isolated.

Biological Activity : 50% EtOH extr. of stem exhibits antiviral and spasmogenic properties.

Uses in Literature : Used as abortifacient, anthelmintic, cathartic (latex), fish poison, fuel wood, vegetable; and for boils, burns, congestion, cough, cuts, diarrhoea, dysentery, ear complaints, increasing milk yield, leprosy, leucorrhoea, skin diseases, sprains, warts and wounds **(Ambasta, 1986; Binoj & Balakrishnan, 1996; Gogoi & Borthakur, 2001; Jadhav, 2006; Jain, 1984, 1991; Jain *et al.*, 1991; Kadel & Jain, 2006; Kalpesh *et al.*, 2004; Kapur & Nanda, 1996; Manandhar, 1987; Patil, 2004; Shah & Joshi, 1971; Sharma, 2003; Singh, 1996; Upadhyay & Singh, 2005; Watt, 1972).**

84. *Ficus benghalensis* Linn. (Pl. 31D)

Syn. : *F. elastica* Roxb., *F. indica* L., *Urostigma benghalense* Gasp. *Urostigma elasticum* Miq., *Visiana elastica* Gasp.

Family : Moraceae.

Vern. : Bad, Barh, Bare.

English, Sanskrit and Regional Names:

Eng. : Banyan Tree.

Sans. : Avaroha, Bahupada, Bhandira, Bhringi, Chira, Dhruva, Jatala, Karmaja, Mana, Nila, Nandi, Rohina, Sungi, Vata, Vanaspati, Yamapriya;

A. : Bot;

B. : Bar, Bot, But;

G. : Vad, Vadlo, Vor;

H. : Bar, Bargad, Bargat, Ber, Bor;

Mal. : Ala, Nyagrodham, Peral, Vatam;

Mar. : Vada, Wad, War;

P. : Bargad, Bera, Bohar, Bohir, Bor;

Tam. : Al, Ala, Eguvasam, Kadavam, Kavasugam, Pali, Parul, Vadam;

Tel. : Bhandiramu, Mari, Matti, Pedaimatti, Rohi, Vati, Vitapi.

Distribution : Throughout India.

Description : A large spreading tree with numerous aerial prop roots. Leaves leathery, ovate, obtuse, entire. Figs pubescent, sessile and axillary; red on ripening.

Flowering and Fruiting : February-June.

Habitat Ecology : Usually planted along with *Ficus religiosa* as a shade tree.

Material Examined : EBH-HVP-851, 17.02.2005.

Parts Used : Leaves. Wood.

Folk Uses : Oil smeared heated leaves applied externally for suppuration of underarm **boils** ('kachhali'). Dried latex chewed as a substitute for **chewing gum*** by children. Tree considered **sacred** and worshipped; its wood used for performing **'havans'**; the garland of its leaves hung on the main entrance of house at the time of **house warming*** ceremony.

Active Constituents : Known to contain leuco-anthocyanin (stem bark), quercetin-3-galactoside, rutin, b-sitosterol, taraxasterol, lupeol; β-amyrin along with psoralen and bergapten (leaves), and glycoside (ripe fig).

Biological Activity : Bark exhibits hypoglycaemic activity in both normal and alloxan diabetic rats.

Uses in Literature : Reported earlier as an antidote, aphrodisiac, astringent, edible, fodder, tonic, tooth brush, vegetable (leaf bud), well curbs; and for atrophy, bangy poles, bleeding and swelling of gums, boils, cart shafts, cart yokes, diabetes, diarrhoea, dysentery, fever, furniture, gastric (vet.), gonorrhoea, kibe, leprosy, menorrhagia, nausea, piles, plates, scabies, sexual diseases, sores in mouth, snake bites, spermatorrhoea, toothache, urinary complaints, vomiting, wounds, wood in tent poles and religious ceremonies **(Ambasta, 1986; Bondya**

& Sharma, 2003; Chauhan & Kumar, 2005; Datt, 1996; Girach, 1992; Hembrom & Goel, 2005; Huidrom, 1996; Jain, 1991; Jain & Sharma, 1996; Khanna & Kumar 2000; Kulkarni & Kumbhojkar, 1996; Maliya & Singh, 2003; Manandhar, 1996; Masih, 2003; Mohanty & Padhy, 1996; Mohanty *et al.*, 1998; Mukherjee & Namhata, 1990; Pandey & Verma, 2002; Rao & Pullaiah, 2001; Sahoo & Mudgal, 1995; Saini, 1996a; Sen & Behera, 2003; Singh, 2003; Singh & Pandey, 1996; Sinha, 1996; Tirkey *et al.*, 2003; Upadhye *et al.*, 1994; Vihari, 1995; Watt, 1972).

Major Ayurvedic Formulations : Nyagrodhadi-kvetha-churna, Nyagrodhadi-churna, Vata-jatadi-taila.

Adulterant or Substitute : Fruit powder used as an adulterant for 'Kamala' (*Mallotus phillippensis* (Lam.) Muell-Arg.).

85. *Ficus carica* Linn. (Pl. 32A)

Syn. : *F. palmata* Forssk., *F. caricoides* Roxb., *F. pseudo-sycamorus* Decne., *F. virgata* Roxb.

Family : Moraceae.

Vern. : Dura, Anjir.

English, Sanskrit and Regional Names:

Eng. : Common Fig;
Sans. : Anjira, Ten;
B. : Fagari, Fagu, Faguri;
Bo. : Anjir;
G. : Anjura, Anjuri;
H. : Anjir;
Kan. : Anjura, Tie-thie;
Mal. : Simayatti;
P. : Anjira;
Tam. : Simaiyatti, Tenatti;
Tel. : Anjuru, Manjimedi.

Distribution : Cultivated in many parts of India, Punjab, W. Himalaya.

Description : Small deciduous trees. Leaves cordate, deeply lobed, lobes obtuse, rough on upper side. Figs in pairs on long leafless scaly shoot, globose, reddish brown when ripe.

Flowering and Fruiting : March-August.

Habitat Ecology : Wastelands. Edges of fields. Drier areas.

Material Examined : EBH-HVP-857, 15.07.2005.

Parts Used : Leaves. Fruit.

Folk Uses : Fruits **edible** and leaves used as **fodder.**

Active Constituents : Calotro phenyl acetate, lupeil acetate, oleanolic acid, β-sitosterol and bergapten (leaves) are the active constituents.

Uses in Literature : Reported for curing boils, constipation, delivery, diabetes, dysentery, indigestion, leucorrhoea, liver disorders, lung and bladder disease, fodder, building purposes; and as nutritive, edible, demulcent, emollient, laxative, wine, brandy, and anthelmintic (**Ambasta, 1986; Khanna & Kumar 2000; Kirtikar & Basu, 1935; Maliya & Singh, 2003; Negi *et al.*, 1985; Shah & Joshi, 1971; Sikarwar, 2002**).

86. *Ficus glomerata* Roxb. (Pl. 32B)

Syn. : *F. chittagonga* Miq., *F. racemosa* Wall., *F. mollis* Miq., *F. goolereca* Roxb., *Covellia glomerata* Miq., *C. mollis* Miq.

Family : Moraceae.

Vern. : Umara.

English, Sanskrit and Regional Names:

- **Eng.** : Cluster Fig;
- **Sans.** : Jantuphala, Phalasambendha, Suprastish: thida;
- **Ass.** : Jangedumuru, Yagyadimru;
- **B.** : Jagyadumar, Loa, Lowa;
- **Bo.** : Umbar;
- **G.** : Umbro;
- **H.** : Gular, Pa Rao, Umar, Umarai;
- **Kan.** : Athi;
- **Kash.** : Rumbal;
- **Mal.** : Rumadi, Umbaram;
- **Oriya** : Dimri, Odumboro;
- **P.** : Batbar, Kakallam, Kathagular;
- **Tam.** : Adavu, Anjiram, Malaiyinmmunivam, Tumbaram;
- **Tel.** : Atti, Medi.

Distribution : Throughout India.

Description : An evergreen tree. Leaves ovate-oblong, tapering to a bluish point at apex with entire margin, glabrous above and beneath. Reddish receptacle possesses depressed umbilicus, edible but usually full of worms.

Flowering and Fruiting : April-July.

Habitat Ecology : Mostly found growing in moist localities and bordering streams.

Material Examined : EBH-HVP-871, 22.07.2005.

Parts Used : Leaves. Stem.

Folk Uses : Plant considered **sacred** and worshipped. Wood used as shuttering for **house construction,** and also used for performing **'havan'*** during religious ceremonies. Paste of leaves applied for healing **open wounds**; also used as **fodder**.

Active Constituents : Lupeol, friedelin, β-sitosterol, stigmasterol, cerylbehe-nate, lupeol, its acetate, α-amyrin acetate (trunk bark), a tetracyclic triterpene, glucanol acetate (leaves) have been identified.

Biological Activity : A glycoside rich fraction from leaves reported to have hypotensive, vasodilator and cardiac depressive effect. Also found to be antiprotozoal, hypoglycaemic and anticancerous.

Uses in Literature : Useful as an astringent, febrifuge, fodder, edible (fruits), powerful tonic, sacred, stomachic, tooth brush, vegetable, wood for shopping wells; and for asthma, blood dysentery (vet.), boils, bronchitis, curing piles, diarrhoea, disease of kidney and spleen, dislocation of joints, door, dry cough, fusebox, gonorrhoea, gum, healthy foetus (vet.), leucorrhoea, matchbox, menorrhagia, mouth sores, mumps, intestinal worms, shutters, rinderpest (vet.), smallpox, urinary and venereal complaints, well frames and wounds **(Ambasta, 1986; Hosagoudar & Henry, 1996a; Jain, 1991; Jain *et al.*, 1991; Jha *et al.*, 1996; Khanna & Kumar 2000; Khanna *et al.*, 1996b; Kumar, 1996; Kumar & Chauhan, 2005; Kumar & Jain, 1998; Kumar & Pullaiah, 1998; Kurian, 1995; Mishra *et al.*, 1996a; Mohanty *et al.*, 1998; Mokat & Deokule, 2004; Painuli & Maheswari, 1991; Rama Rao & Henry, 1996; Saini, 1996a; Sharma, 1996; Singh & Pandey, 1996; Sinha, 1996; Watt, 1972).**

Major Ayurvedic Formulations : Mutrasamgrahaniya—kasaya-churna, Nyagrodhi-kwatha-churna, Udumbara-kavaleha, Udumbaramata, Udumba-rasara.

87. *Ficus religiosa* Linn. (Pl. 32C)

Syn. : *F. affinior* Griff., *Urostigma religiosum* Gasp., *U. affine* Miq.
Family : Moraceae.
Vern. : Pipal.
English, Sanskrit and Regional Names:

Eng. : Pepal Tree, Peepul Tree, Pipul Tree;
Sans. : Asvatthah, Chaityavriksha, Pavitraka, Vriksharaja, Yojnika;
A. : Ahot;
B. : Ashathwo, Hesar, Hesak, Pipar;
Bo. : Arli, Jari, Pimpala;
G. : Arasa, Aswartham;
H. : Ashathwa, Aswat, Asud, Asvattha, Pipal;
Kan. : Arali, Nyanugleaudi, Nyoungbaude, Nyoungbaudi, Nyaungbawdi;
Mal. : Arayal, Chlodalam;
Mar. : Ashvatha, Pipul, Pimpala;
Oriya : Bor, Burr;
Tam. : Achuvattam, Podoroganam, Rai, Raiga, Ragi, Ravi, Vangardi;
Tel. : Arali, Arle, Asvalta, Basri, Bogaha, Haspath, Rangi.

Distribution : Throughout India. Widely cultivated in Himalaya up to 1,400 m.

Description : A large glabrous tree with grey bark. Leaves entire, long petioled, tip narrowed into linear-lanceolate tail, shining above. Figs axillary, in globular pair, dark purple when ripe.

Flowering and Fruiting : March-October.

Habitat Ecology : Usually planted along with *Ficus benghalensis*.
Material Examined : EBH-HVP-853, 12.03.2005.
Parts Used : Stem. Leaves.
Folk Uses : Tree considered **sacred** and worshipped. Wood used for performing **'havan'** ceremonies. Garland of leaves hung at the main door of house during **'grih-pravesh'*** (house warming ceremony) and as **fodder***.
Active Constituents : Tannins, β-sitosterol and D-glucose, laurosterol (bark) have been isolated.
Biological Activity : Stem bark exhibits antiprotozoal activity against *E. histolytica*, anthelmintic activity against *Ascardia galli* and antiviral against Ranikhet disease virus.
Uses in Literature : So far reported for its use as anthelmintic, abortifacient, absorbent in inflammatory swelling, antidote to snake bite, astringent, blood purifier, colouring material, gum, fixative, contraceptive, edible (fruits and tender leaf buds eaten in times of scarcity), fodder, spoons, laxative, psychoactive plant, sacred tree (worshipped-cutting tree more sinful than killing a saint), tonic, tree of fertility, vermifuge (vet.); and for after pains, asthma, boils, conception, constipation (vet.), dysentery, expelling foetal membrane, falling of hairs, foot and mouth diseases (vet.), gonorrhoea, hiccups, jaundice, leucorrhoea, menstrual disorders, painting (Mithila), tanning, snake bite, sores, throat infection, toothache, urinary complaints, whooping cough and wounds **(Ambasta, 1986; Bhatt *et al.*, 2001; Borthakur, 1996a; Borthakur *et al.*, 2004; Gogoi & Borthakur, 1991; Gogoi *et al.*, 2003; Hosagoudar & Henry, 1996a; Jadhav, 2006; Jain, 1991; Jain *et al.*, 1991; Jha & Basak, 1994; Kalpesh *et al.*, 2004; Khanna, 2002; Khanna & Kumar 2000; Khanna *et al.*, 1996a, b; Kumar & Chauhan, 2006; Kumar & Jain, 1998; Masih, 2003; Mishra *et al.*, 1996a; Pawar & Patil, 2004; Rama Rao & Henry, 1996; Satapathy & Brahmam, 1996; Sikarwar & Kaushik, 1992; Singh & Rao, 2003; Srivastava *et al.*, 2003; Virjee *et al.*, 1984; Watt, 1972).**
Major Ayurvedic Formulations : Nyagrodhadi-churna, Nyagrodhadi-kwatha-churna.
Adulterant or Substitute : Stem bark and roots used as substitute for Soma (*Ephedra gerardiana*).

88. *Ficus roxburghii* Wall. *ex* Miq. (Pl. 32D)

Syn. : *F. auriculata* Lour., *F. macrophylla* Griff., *F. regia* Miq., *F. scleroptera* Griff., *Covellia macrophylla* Miq.
Family : Moraceae.
Vern. : Rumal, Trayambalu, Triamble.

Regional Names:

B.	:	Demur, Doomoor, Sapai;
H.	:	Tirmal, Trimmal;
P.	:	Ber, Burh, Dadari, Timbal, Tirnal, Tirmal, Trimbal.

Distribution : Sub-Himalayan tract. Outer Himalaya ascending up to 1,600 m.

Description : Trees having brown bark and 5-7 leaves nerved with cordate base. Figs stalked, ridged.

Flowering and Fruiting : April-August.

Habitat Ecology : Commonly found growing in moist localities and bordering streams.

Material Examined : EBH-HVP-873, 12.06.2005.

Parts Used : Leaves. Fruits.

Folk Uses : Leaves used as **fodder**. Fruits **edible**.

Active Constituents : Tetracyclic triterpene, glucanol acetate (leaves), β-sitosterol, friedelin and epifriedelanol (plants), lupeol, friedelin, β-sitosterol and stigmasterol, cerylbehenate, lupeol, its acetae, amyrin acetate (trunk bark) have been isolated.

Biological Activity : Known as galactagogue (vet.) and for leucoderma and skin diseases.

Uses in Literature : Known in India as galactagogue (vet.), edible, vegetable, fodder; and for breast ulcers, delivery (vet.), leucoderma, meal plates for feasts in village, ring worm, ropes (bark), skin diseases **(Jain, 1991; Parmar & Kaushal, 1982; Rama Rao & Henry, 1996; Watt, 1972).**

89. *Flacourtia ramontchi* L. Herit (Pl. 33A)

Syn. : *Flacourtia indica* (Burm.f.) Merr., *F. latifolia* Cooke., *F. occidentalis* Blatter, *F. sapida* Roxb., *F. sepiara* Roxb., *Gmelina indica* Burm f.

Family : Flacourtiaceae.

Vern. : Kangu.

English, Sanskrit and Regional Names:

Eng.	:	Governor's Plum, Madagascar Plum;
Sans.	:	Swadukantaka;
B.	:	Binja, Katail, Tambat;
Bo.	:	Kakad, Swadu;
H.	:	Bhanber, Bincha, Iambat, Kakein, Katai, Katar;
Kan.	:	Hattarimullu, Humunki;
Kol.	:	Merlee;
M.P.	:	Kankar;
Mar.	:	Kaikun, Kaker, Paker;
Oriya	:	Bonisha, Potnabonisho;
P.	:	Bhutankas, Bavachi, Kakao;
Tam.	:	Mullikkarai, Sottaikala;
Tel.	:	Nakkaneredu;
U.P.	:	Katai.

Distribution : Dry hills throughout India up to 1,000 m.

Description : Small deciduous trees armed with axillary thorns. Leaves broadly ovate-crenate, apex acute, pubescent. Flowers greenish-yellow. Fruits globose, red or dark brown, endocarp hard.

Flowering and Fruiting : March-June.

Habitat Ecology : Wasteslopes. Rocky slopes.

Material Examined : EBH-HVP-862, 10.06.2005

Parts Used : Bark. Stem. Fruits. Wood.

Folk Uses : Gargles of bark decoction useful for **mouth sores***. Fruits **edible**. Wood used as **fuel,** and for making **agricultural implements**. Leaves used as **fodder**.

Active Constituents : Important constituents are tannin, flacourtin, phenolic glucoside ester (bark), 4-o-(β-D-glucosyl), (1-4) -β-2-3-di-o-methyl-D-xyloside (heart wood).

Biological Activity : Aerial parts diuretic and spasmogenic. Hypotensive activity and effect on CNS found +ve.

Uses in Literature : Known so far as an antidote to snake bite, appetiser, astringent, diuretic, bite for mad dog, edible; and for burning sensation in chest, carbuncle, cholera, diarrhoea, digestion, dysentery, eczema, facilitating child birth, fever, jaundice, liver disorder, skin diseases, sores, stab wounds, veterinary foot and mouth diseases, cattle fodder, agricultural implements, excess body heat, edible purposes, gout, indigestion, rheumatic pain after delivery in woman, tanning, tonsillitis, turnery, wounds; and also to reshape abdomen after delivery **(Ambasta, 1986; Aminuddin & Girach, 1996; Brahmam & Saxena, 1990; Jain, 1991 ; Jain *et al.*, 1991; Katewa & Arora, 1997; Maheshwari *et al.*, 1986; Mohanty *et al.*, 1998; Nazarudeen *et al.*, 1996; Parmar & Kaushal, 1982; Patole & Jain, 2002; Rama Rao & Henry, 1996; Ramachandran & Nair, 1981; Rao & Pullaiah, 2001; Saini, 1996a; Shah & Singh, 1990; Sharma & Singh, 2001; Sikarwar & Kaushik, 1992; Singh, 1996).**

90. *Foeniculum vulgare* Mill. (Pl. 33B)

Syn. : *Foeniculum panmorium* DC., *F. officinale* Allion., *Anethum foeniculum* Linn., *A. panmorium* Roxb., *F. capillaceum* Gillb., *Ozodia foeniculum* W. and A.

Family : Apiaceae.

Vern. : Meethi-Saunf.

English, Sanskrit and Regional Names:

Eng. : Fennel;

Sans. : Madhurika;

B.	:	Bari-shopha, Panmohuri;
Bo.	:	Bari-Shopa, Variari, Variyali, Wariaree;
G.	:	Badishep.
H.	:	Badhesoppu, Bari-saunf, Mauri, Pan-muhori;
Kan.	:	Aspa, Badi-Sopu, Badyan;
Mar.	:	Badisapu, Badishep;
Tam.	:	Pedda-jila-kura, Sohikire;
Tel.	:	Smadhurika, Pedda-jila-kurra.

Distribution : Throughout India up to 2,000 m.

Description : Tall glabrous herbs. Leaves tripinnate. Umbel having 15 or more rays. Vittae conspicuous.

Flowering and Fruiting : August-September.

Habitat Ecology : Cultivated. Escape in waste places.

Material Examined : EBH-HVP-852, 15.08.2005.

Part Used : Seeds.

Folk Uses : Seeds used as **condiments** and its infusion used as **refrigerant***.

Active Constituents : Principle constituents are antehole, fenchone, anisalehyde and feniculine (plant), columbianetin, 7 hydroxy and 6, 7-dihydroxy coumarins, marmesin, osthenol, psoralen, scoparone, seselin, xanthotoxin, α and β-amyrins, oleanolic acid, Δ5 avenasterol, campesterol, Δ7 stigmasterol, stigmastadienol, linoleic, oleic, palmitic and petroselinic acid (fruit, seeds).

Biological Activity : Antimicrobial and haematinic activities found to be confirmed.

Uses in Literature : Considered useful in India as an aromatic, stimulant, carminative, colic, emollient, febrifuge, galactagogue, haemorrhoids, ophthalmic, stomachic; and for burning sensation in body, chest diseases, cough, diarrhoea, dysentery, dyspepsia, emollient, fever, flatulence, flavouring, headache, kidney problems, menstruation and spleen complaints, retention of urine, thirst, toothache and wounds **(Ambasta, 1986; Bhalla *et al.*, 1982; Kumar & Chauhan, 2005; Retnam & Martin, 2006; Sharma, 2003; Uniyal, 1968).**

Adulterant or Substitute : Oil from fruit used as an adulterant for Anise oil (*Pimpinella anisum* Linn.).

91. *Fragaria chiloensis* (L.) Duchesne. (Pl. 33C)

Syn. : *F. vesca* auct., non Linn.

Family : Rosaceae.

Vern. : Strawberry.

English Names : Alpine strawberry, Perpetual strawberry.

Distribution : Throughout India.

Description : A creeping herb with trifoliate long stalked leaves. Flowers white. Fruits red.

Flowering and Fruiting : May-June.
Habitat Ecology : Cultivated.
Material Examined : EBH-HVP-848, 24.05.2005.
Part Used : Fruit.
Folk Uses : Fruit **edible**.
Active Constituents : Fruit contains isofucosthol, sitosterol, compesterol, cholesterol and stigmasterol.
Uses in Literature : Used so far as edible; and for desert and ice-creams (**Ambasta, 1986; Rastogi & Mehrotra, 1990, 1995a, b; Watt, 1972**).

92. *Fumaria indica* (Haussk.) Pugsley. (Pl. 33D)

Syn. : *F. parviflora* Lam., *F. vaillantic* Loisel.
Family : Fumariaceae.
Vern. : Papra, Pitpapda, Pitpapra.
English, Sanskrit and Regional Names:
- **Eng.** : Fine-leaved-Fumitory;
- **Sans.** : Araka, Charaka, Kalapanga, Krishnashakha, Nakra, Pansha, Panshu;
- **B.** : Bansulpha;
- **Bo.** : Pitpara;
- **Guj.** : Khasudlio, Pitpapda;
- **H.** : Shatra, Pitpapada, Pittapapara, Pitpapra;
- **Mar.** : Pitpapda;
- **Tam.** : Chata-rashi, Tura, Tusa;
- **Tel.** : Bagaltulmulk, Chatarashi.

Distribution : Throughout India.
Description : An annual diffuse herb having much-divided glaucous leaves with linear segments. Flowers sessile, pale pink, irregular and in lux raceme. Fruits globose, indehiscent.
Flowering and Fruiting : February-March.
Habitat Ecology : Moist slopes. Weed in cultivated fields.
Material Examined : EBH-HVP-843, 04.03.2005.
Part Used : Whole Plant.
Folk Uses : 50 ml. decoction of plant taken once every morning for 2-5 days to cure **fever**. Plant also used as **fodder***.
Active Constituents : Protopine, nonacosanol, sitosterol, n-alkanes and sterols (leaves, stem); protopine and 1-tetrahydrocopitsine, d-8-OMe-dihydroxy sanguinarine, fumarilinc and oxy sanguinarine (seeds), sanguinarine, protopine, cryptopine, fumaridine, fumaramine, perfumidine, perfumine, d and dl-bicuculline, dl-tetrahydrocopyisine, fumariline, fumarilicine, narceimine, dehydrocheilanthi-foline, copticine and narlumidine have been isolated.
Biological Activity : Herb extr. antiarrhythmic in rabbit.
Uses in Literature : Used earlier for bodyache, diarrhoea, dyspepsia, fever, flu,

indigestion, liver complaints, mouth ulcers and skin diseases; and as an anthelmintic, aperient, astringent, blood purifier, diaphoretic, diuretic, laxative, cool drink and vegetable **(Ambasta, 1986; Chopra *et al.*, 1956; Dhiman, 1976; Drury, 1978; Hembrom & Goel, 2005; Sharma, 2003; Uphof, 1968; Watt, 1972).**

Major Ayurvedic Formulations : Arvindasava, Chandrakala-rasa, Chandanasava, Maha-sudarshan-kwatha, Livomyn drops, Parpatadi-kwatha, Parpatadyarista, Sadanga-paniya.

Trade : Plant sold as 'Pitpara' in drug markets of India.

93. *Gossypium arboreum* Linn. (Pl. 34A)

Syn. : *G. indicum* Tod., *G. neglectum* Tod., *G. sanguineum* Hassk., *G. intermedium* Tod., *G. obustifolium* Roxb., *G. nanking* Meyen.

Family : Malvaceae.

Vern. : Kapah.

English, Sanskrit and Regional Names:

Eng. : Tree Cottan;
Sans. : Karpasamu;
Bo. : Deokapas;
H. : Deokapas, Nurma;
Kan. : Hathi;
Mal. : Chemparutti;
Mar. : Devakapusa;
Oriya : Karposa, Kopa, Kapas;
P. : Kapas;
Tam. : Sembaruti;
Tel. : Patti, Karpasmu.

Distribution : Indian plains up to 1,800 m.

Description : A shrub possessing palmately 5-7 lobed leaves. Flowers purple or yellow with a crimson centre. Capsule ovoid, acute.

Flowering and Fruiting : July-September.

Habitat Ecology : Cultivated; planted near houses and garden.

Material Examined : EBH-HVP-867, 15.07.2005.

Parts Used : Seeds. Seed coat.

Folk Uses : Cotton fibres used for **stuffing*** pillows, quilts and making **sacred thread** for religious purposes, **wick*** for the oil lamps, and also **clothes** for wearing. Seeds **fed*** with oil cake to livestock.

Active Constituents : These are 1-gossypol (seeds), steroids, organic acid, pigment and vit. B.

Biological Activity : Gossypium-8-o-rhamnoside along with quercetin, 3-7-o-glucoside of quercetin exhibit antibacterial activity against *Pseudomonas maltophila* and *Enterbacter cloacae*. Seed extr. spasmolytic and hypotensive.

Uses in Literature : Known so far as emollient, soothing agent and symbolic for making sacred thread; and for asthma, arthritis, conjuctivitis of infants, easy delivery, earache, fever, fibre, gonorrhoea, insanity, labour complaints, leucorrhoea, regulating fertility, regulating menses, skin eruptions, stomach pains, stranguary and patellar dislocations (**Bhatt *et al.*, 2001; Jadhav, 2006; Khanna *et al.*, 1996b; Pawar & Patil, 2004; Sharma, 2003; Sharma & Borthakur, 2003; Sikarwar, 2002; Tirkey, 2004; Watt, 1972**).

94. *Grevillea robusta* A. Cunn. *ex* R. Br. (Pl. 34B)

Family : Proteaceae.
Vern. : Silver oak.
English and Regional Names:
Eng. : Silky oak, Silver Oak;
Tam. : Savukkumaram.
Distribution : Planted in gardens.
Description : An evergreen tree. Leaves spinnate. Flowers orange in cluster. Fruits brown, obliquely-oblong, compressed.
Flowering and Fruiting : March-July.
Habitat Ecology : Often planted along Roadsides.
Material Examined : EBH-HVP-847, 12.04.2005.
Parts Used : Plant. Wood.
Folk Uses : Planted for **ornamental** purposes and its wood considered good for **ornamental work**.
Active Constituents : These are grevillo, sitosterol, epilyeniresinol, its 2-o-glucoside, rhamnocitrin-3-o-rutinosid and kaemferol-3-o-rutinoside.
Uses in Literature : So far known as edible, green manure, ornamental, and for furniture, gum, panelling, paraquet, plywood, tannin, toys and veneering (**Ambasta, 1986; Sinha, 1996**).

95. *Grewia oppositifolia* Roxb. (Pl. 34C)

Syn. : *G. optiva* Drumm.
Family : Tiliaceae.
Vern. : Beul.
Regional Names:
H. : Biul, Biung, Bewal, Bahul, Bemal, Bhimal, Bhengul;
Kan. : Thidsal;
Kum. : Bimal, Dhamnu;
P. : Bahul, Biul, Behal, Dhaman, Pharwa.
Distribution : N.W. Himalaya from Indus to Nepal ascending to 2,000 m.

Description : A moderate-sized tree. Leaves distichous; stipules longer than the petioles. Flowers large, yellowish and in leaf-opposed umbellate cymes. Drupes smooth, olive black, succulent when ripe.

Flowering and Fruiting : May-June.

Habitat Ecology : Edges of agricultural fields. Roadsides. Wasteplaces.

Material Examined : EBH-HVP-856, 15.06.2005.

Parts Used : Bark. Leaves. Stem.

Folk Uses : Crushed bark kept in buccal cavity and mouth for **inducing chewing*** of **cud** in livestock. Leaves lopped for **fodder** for increasing **fat content** of milk. Stem retting yields **fibre** for making ropes and 'bhallathu' (fuelwood) for making **fire**. Wood used for **cot frames*** and **tool handles**.

Active Constituents : Properties attributed to crude proteins (10.1%), fat (6.8%), crude fibre (14.1%), N-free extr. (54.8%), total carbohydrates (68.9%), ash (14.1%), calcium (4.18%), phosphorus (0.25%) (dry mature leaves).

Uses in Literature : Used in India as detergent, dye, fuel, cordage, fodder and torch in religious ceremonies; and for constipation, eruptions, facilitating child birth, agricultural implements, axe handles, clothing, increasing lactation, oarshaft, ropes **(Ambasta, 1986; Hosagoudar & Henry, 1996a; Jain, 1991; Kapur, 1996; Lal & Dube, 1984; Lata, 1996; Rajwar, 1983; Rana *et al.*, 1996; Shah & Joshi, 1971; Sharma, 2003; Singh, 1996; Watt, 1972)**.

96. *Helianthus annuus* L. (Pl. 34D)

Family : Asteraceae.

Vern. : Sooraj mukhi.

English, Sanskrit and Regional Names:

Eng. : Sunflower;
Sans. : Adityabhakta;
H. : Surajmukhi.

Distribution : Throughout India.

Description : An annual herb with rough hairy stems and large ovate leaves. Flowers yellow. Achenes cylindrical and compressed.

Flowering and Fruiting : April-July.

Habitat Ecology : Cultivated.

Material Examined : EBH-HVP-869, 15.05.2005.

Parts Used : Flower. Seed.

Folk Uses : Paste of its flower prepared in oil considered good for external application in **skin diseases***. Seed oil applied externally to **bruises**.

Active Constituents : Active principles are due to kaurenic, trachylobanic, enttrachylobanic acids, grandiflorolic acid, ψ-taraxene derivatives-heliantriol C and heliantriol F (flowers), palmitic acid (10.0), stearic (5.7) and linoleic acid (40.0) (seed oil).

Uses in Literature : Used so far as moisturizer, diuretic, cooking oil, edible; and for bruises, pulmonary and laryngeal afflictions, whooping cough **(Kirtikar & Basu, 1935; Watt, 1972).**

97. *Hibiscus rosa-sinensis* L. (Pl. 35A)

Family : Malvaceae.

Vern. : China Rose, Shoe Flower.

English, Sanskrit and Regional Names:

Eng. : Chinese Hibiscus, Shoe Flower;
Sans. : Hariballabha, Odhrapushpa;
Ass. : Jobra;
B. : Jiwa, Oru;
Bo. : Jasavanda;
G. : Jasuva;
H. : Gulhar, Jasum, Jasut;
Kan. : Dasavala;
Mal. : Ayamparutti, Jumpa;
Mar. : Dasindachapula, Jasavanda;
Oriya : Mondaro, Ongribollika;
P. : Jasum;
Tam. : Irattaichegappuchembrat.

Distribution : Cultivated throughout India.

Description : Shrubs. Leaves coarsely toothed at apex and entire at base. Corolla red. Capsule many seeded.

Flowering and Fruiting : February-October.

Habitat Ecology : Cultivated in gardens.

Material Examined : EBH-HVP-846, 12.03.2005.

Parts Used : Flower. Plant.

Folk Uses : Cultivated as an **ornamental***. Paste of its flower applied on **boils**.

Active Constituents : Active properties are due to quercetin-3-diglucoside 3, 7-diglucoside, cyanidin-3, 5-diglucoside and cyanidin-3-sophoroside-5-glucoside (deep yellow flowers), and kaempferol-3-xylosylglucoside (ivory white flowers).

Biological Activity : Hypotensive and antispasmodic activities found (+)ve.

Uses in Literature : Known to be used for abdominal pains, boils, bronchitis, cough, diabetes, leucorrhoea, menstrual irregularity, menorrhagia, piles, whooping cough; and as emollient, refrigerant **(Gogoi & Borthakur, 2001; Katewa & Sharma, 1998; Maliya, 2004; Pawar & Patil, 2004; Retnam & Martin, 2006; Sen & Behera, 2003; Sharma, 2003; Sharma *et al.*, 2001).**

Major Ayurvedic Formulations : Japa-taila.

98. *Holarrhena antidysenterica* (Buch.-Ham.) Wall. (Pl. 35B)

Syn. : *H. codaga* G. Don., *H. pubescens* Wall. *ex* Don., *Echites antidysentrica* Roxb., *E. pubescens* Buch.-Ham., *Wrightia antidysentrica* Grah.

Family : Apocynaceae.

Vern. : Inderjaun, Keur.

English, Sanskrit and Regional Names:

Eng. : Antidysentric Holarrhena, Conessi Bark, Kurchi;
Sans. : Girimallika, Indra, Indrayavaphala, Kalinga, Kutaja, Mallikapushpa;
Ass. : Dhutkhuri, Dudcory;
B. : Kurchi, Titaindaraju;
Bo. : Dolokura, Dowla, Kura, Pandhrakura;
G. : Dhowda, Hath, Indrajwannujad;
H. : Dhudi, Karchi, Karra, Kora, Kureya;
Kash. : Keor, Kewar;
Mal. : Kaippakkotakappala, Venpala;
Mar. : Bedaki, Pandharakuda;
Oriya : Kherwa, Krya, Potrokrya;
P. : Kewar, Koeva, Kogur;
Tam. : Erukkalaipalai, Kudasam, Vattagam;
Tel. : Amkudu, Kakakodise, Peddankuda.

Distribution : Throughout India.

Description : A small deciduous tree. Leaves opposite, subcoriaceous. Flowers creamy-white, sweet-scented. Bark pale brown. Follicles white dotted. Seeds glabrous, tipped with brown hair.

Flowering and Fruiting : June-February.

Habitat Ecology : Field borders. Open slopes. Drier areas.

Material Examined : EBH-HVP-856, 12.06.2005.

Parts Used : Bark. Leaves. Fruit. Seeds.

Folk Uses : Poultice of bark with 'taramira' (*Eruca sativa* Lam.) applied for suppuration of **boils**. Also crushed bark mixed with wheat dough fed to buffaloes for treating **prolapsed uterus***. Leaves used for performing **'shraadha' rituals*** (religious ceremony for remembering departed souls). Watery extr. of seeds taken to relieve **body heat***. Fruits in 'halwa' (sweet porridge) prescribed to check **irregular menstrual cycle**.

Active Constituents : Active properties attributed to arginine, aspartic acid, concuressine, 3-epihetero-conessine, 11-α-20-dihydroxy-18, 20-epoxypregna, 1-4-dien-3-one kurchamine, kurchiminine, kurcholessine, protein, holarrifine, holadienine, lupeol, β-sitosterol, tannin (plant), kurchiphylline, kuchiphyllamine, kurchaline, holadysine and holadysamine (leaves), conessine, kurchicine, kurchine, conamine, conkurchine, holarrhenine, holarrhinine, holarrhin, lettocine (bark), linoceric, linolenic, oleic, palmitic and stearic acid, holarricine and 9-D-hydroxy-cis-12-octadecenoic acid (seeds).

Biological Activity : Hypotensive (bark), antiprotozoal (fruit), anticancerous, hypoglycaemic and antifungal (seed oil) activities +ve.

Uses in Literature : So far used as an anthelmintic, antidote to snake bite, appetiser, CVS active, edible, expectorant, haematic, lactagogue, ornamental; and for anaemia, asthma, bronchial fever, cold, colic, constipation, cough, dental caries, diarrhoea, digestive complaints, dysentery, eczema, epilepsy, fever, gastric disorder, gastroenteritis, gout, jaundice, leprosy, leucoderma, malaria, menstrual complaints, nose bleeding, piles, skin diseases, stomachache, spoons and veterinary problems, reviving taste, curdling milk and making small articles **(Ambasta, 1986; Chhetri, 2005; Gogoi & Borthakur, 2001; Jain, 1991; Pandey & Rout, 2006; Retnam & Martin, 2006; Sharma, 2003; Sharma *et al.*, 2002; Venkataratnam & Venkata Raju, 2004; Verma & Chauhan, 2006).**

Major Ayurvedic Formulations : Kutajarista, Kutaghan-vati, Kutawleha, Krimikuthar-rasa, Karpur-rasa, Punarnava-mandur, Sarvajawarhava-laghuganga-dhar-churna, Sudarshan-churna, Yograja-guggl u.

Trade : Bark and seeds traded in Indian markets as 'Koora Chhal' and 'Karwa Inderjau'.

99. *Holoptelea integrifolia* (Roxb.) Planch. (Pl. 35C)

Syn. : *Ulmus integrifolia* Roxb.

Family : Ulmaceae.

Vern. : Pardesi.

English, Sanskrit and Regional Names:

Eng. : Indian Elm;
Sans. : Chirabilva;
G. : Kanjho, Waola;
H. : Banchilla, Begana, Chilbil, Chilmil, Dhamina, Kanju, Papri;
Kan. : Kaladri, Nilavahi, Rosbija, Thavasai;
M.P. : Karanji, Chirhol, Karanjalam;
Mal. : Aval;
Mar. : Papara; Vavli;
Oriya : Dauranja, Turuda;
P. : Arjan, Rajain, Khulan;
Tam. : Aya, Ayil, Kanchi, Vellaya;
Tel. : Nemali, Pedanevili, Tha-pasi.

Distribution : Outer ranges of Himalaya ascending to 700 m; often planted in N. India.

Description : Large deciduous trees. Leaves distichous, elliptic, apex acute, flattened above. Fruits winged.

Flowering and Fruiting : March-May.

Habitat Ecology : Wastelands. Drier areas.

Material Examined : EBH-HVP-905, 12.05.2005.
Parts Used : Leaves. Wood. Seed.
Folk Uses : Leaves used as **fodder*** and **bedding*** for animals (**'pattri'**). Wood employed for **construction purposes.** Seeds **edible*.**
Active Constituents : Chief ingredients are hexacosanol, octacosanol, b-sitosterol, b-amyrin (leaves), friedelin-3-b-ol, 2 amino napthaqunone (0.001), friedelin (0.002), epifriedetinol (0.0005), β sitosterol (0.004) and its β-D-glucoside (0.01%), triterpenoid fatty acid esters-holoptelin A and B (stem bark).
Biological Activity : Oxytocic activity +ve.
Uses in Literature : Known for abortion, asthma, brush, back handles of dusting brooms, boils, building purposes, cheap furniture, cabinet work, carving, carts, combs, fractured bone, fertility, jaundice, intermittent fever, ploughs, mathematical instruments, meningitis, menstrual disorders, piles, plywood, match box, rheumatism, scabies, swelling, wounds and yokes **(Ambasta, 1986; Henry & Rao, 1996; Kirtikar & Basu, 1935; Kumar & Chauhan, 2005, 2006; Kumar & Goel, 1998; Kumar & Jain, 1998; Rastogi & Mehrotra, 1991, 1993; Watt, 1972).**
Trade : Kanju, Indian Elm.

100. *Impatiens balsamina* Linn. (Pl. 35D)

Family : Balsaminaceae.
Vern. : Teur.
English, Sanskrit and Regional Names:

Eng. : The Garden Balsam;
Sans. : Dushpatri-jati;
B. : Dupati;
Bo. : Terada;
Guj. : Gulmendi; Pantambol;
H. : Gulmendi;
Mal. : Mecchingoon;
Mar. : Terada;
Oriya : Mujethi, Haragaura;
P. : Bantil, Halu, Tatura, Tilphar;
Tam. : Kasittumbai.

Distribution : Throughout India.
Description : Annual pubescent herbs. Leaves alternate. Flowers axillary, spur long. Seeds reticulate.
Flowering and Fruiting : July-September.
Habitat Ecology : Wasteplaces. Field borders.
Material Examined : EBH-HVP-858, 12.09.2005.
Part Used : Leaves.
Folk Uses : Leaves used as a substitute for **henna**.
Active Constituents : These are 3-mono-glucosides of cynidin, delphindin,

kaempferol, pelargonidin and quercetin (stem), caffeic, p-coumaric, ferulic, gentisic, p-syringic and vanillic acid, 4-2 (2-carboxyphenyl)-4-oxobutyric acid, 2-hydroxy-1, 4-napthaquinone and napthoglycosidase (plant).

Biological Activity : 50% EtOH extr. of flowers exhibit marked antibiotic activity against *Sclerotinia fructicola* and *Colletotrichum lindemuthianum*.

Uses in Literature : Reported in India as an edible, emetic, cathartic, cooling, diuretic, demulcent, laxative, substitute for henna and tonic; and for burns, lumbago, and intercostal neuralgia **(Ambasta, 1986; Chopra *et al.*, 1956; Kirtikar & Basu, 1935).**

101. *Ipomoea cairica* (L.) Sweet (Pl. 36A)

Syn. : *I. palmata* Forssk., *I. senegalensis* Lamk., *I. stipulaceae* Jacq., *I. pendula* Br *I. pulchella* Roth, *I. tuberculata* Roem. and Sch., *Convulvulus cairicus* L., *C. tuberculatus* Lamk., *C. bellus* Spreng., *C. digitalis* Roxb., *C. heptaphyllus* Rottl., *C. lupulifoilus* Griff.

Family : Convolvulaceae.

Vern. : Railway creeper.

English Name : Railway creeper.

Distribution : Deccan peninsula. Nepal. H.P.

Description : A twinner. Leaves palmately lobed. Flowers whitish pink.

Flowering and Fruiting : June-November.

Habitat Ecology : Roadsides. Forest edges. Wasteplaces. Drier areas.

Material Examined : EBH-HVP-868, 24.06.2005.

Parts Used : Leaves. Plant.

Folk Uses : Leaves used as **fodder***. Plant used for **fencing*** agricultural field borders.

Active Constituents : Leaves possess ergosine, ergocornine and ergocistrine.

Uses in Literature : Known todate as purgative and for cough and pain **(Ambasta, 1986; Jain, 1991).**

102. *Ipomoea carnea* Jacq. (Pl. 36B)

Syn. : *I. crassicaulis* (Benth.) Robinson, *I. fistulosa* Mart. *ex* Choisy, *Balatas crassicaulis* Benth.

Family : Convolvulaceae.

Vern. : Basuti.

Distribution : W. Himalaya up to 1,400 m.

Description : A shrub with hollow stem. Leaves long-stalked, ovate triangular. Flowers light pink. Seeds brown.

Flowering and Fruiting : May-July.
Habitat Ecology : Roadsides. Forest edges. Wasteplaces. Drier areas.
Material Examined : EBH-HVP-859, 30.06.2005.
Part Used : Whole Plant.
Folk Uses : Plant used for **fencing*** agricultural fields and as hedge plant to check **soil erosion***.
Active Constituents : Rich in anthracene glucoside, β-sitosterol, gum, ipomose, jalapin, triacontane and saponin (leaves).
Biological Activity : Ethanol extr. of plant CNS depressant.
Uses in Literature : Known to be used as poison (vet.), purgative; and for arthritis, boils, cuts, sprain, swelling, ulcer and wounds **(Ambasta, 1986; Jain, 1991; Khanna, 2002; Kumar & Chauhan, 2005; Kumar & Jain, 1998; Singh & Prakash, 1994).**

103. *Jasminum multiflorum* Roth. (Pl. 36C)

Syn. : *J. hirsutum* Willd., *J. pubescens* (Retz.) Willd., *Morgorium pubescens* Lamk., *Nyctanthes mutiflora* Burm., *N. pubescens* Retz.
Family : Oleaceae.
Vern. : Chamaeli, Malti.
English, Sanskrit and Regional Names:
Eng. : Downy Jasmine;
Sans. : Kunda;
B. : Kunda;
H. : Kundphul;
Kan. : Kaak-avenako, Katta-vanakka;
Mal. : Katutsijiregam-mulla;
Mar. : Mogra;
Tam. : Malligai;
Tel. : Gujari, Kundamu, Malle.
Distribution : India, Common.
Description : A scandent diffuse shrub. Leaves opposite, hairy. Flowers white, in dense cyme. Fruits berry.
Flowering and Fruiting : February-April.
Habitat Ecology : Roadsides. Open slopes.
Material Examined : EBH-HVP-849, 21.03.2005.
Parts Used : Leaves. Flowers.
Folk Uses : Leaves and flowers used for **offering*** to deities on religious ceremonies. Tender twigs along with leaves of 'garnae' (*Carissa spinarum* Hook. f.), fruits of 'chilli' (*Capsicum annuum* Linn.) and 'kali mirach' (*Piper nigrum* Linn) and wheat dough given to livestock as **febrifuge***.
Active Constituents : Properties attributed to 10-hydroxyligustroside, 10-hydroxy-oleoside-11-methyl ester mutifloroside and multrioside

(plant), α-amyrin, betulin, friedelin, lupeol, betulinic, oleanolic and ursolic acid and β-sitosterol (leaves).

Biological Activity : 50% ethanol extr. of aerial plant diuretic.

Uses in Literature : So far used as an antidote to cobra venom, emetic; and for boils, burns, skin diseases, ulcer and wounds **(Ambasta, 1986; Chopra *et al.*, 1956; Maheshwari *et al.*, 1981; Sharma, 2003; Retnam & Martin, 2006).**

104. *Jasminum sambac* (L.) Ait. (Pl. 36D)

Family : Oleaceae.

Vern : Motia.

English, Sanskrit and Regional Names:

Eng. : Arabian jasmine, Tuscan jasmine;
Sans. : Mallika;
H. : Banmallika;
Mal. : Cheru pitchakam;
Tam. : Adukkumalli, Gundumalli;
Tel. : Boddumalle, Gundumalli.

Distribution : Subtropical regions of India.

Description : A scandent shrub with broadly ovate-elliptic leaves. Flowers white, fragrant.

Flowering and Fruiting : February-April.

Habitat Ecology : Roadsides. Open slopes.

Material Examined : EBH-HVP-863, 12.04.2005.

Part Used : Flowers.

Folk Uses : Flowers used for **offering*** to deities on religious ceremonies. Flower head applied over forehead to treat **headache**.

Uses in Literature : Known so far for headache, fever, skin infection, syphilis, wounds (**Retnam & Martin, 2006).**

105. *Jatropha curcas* Linn. (Pl. 37A)

Syn. : *J. moluccana* Herb. Russ.

Family : Euphorbiaceae.

Vern. : Rattanjot, Jamnotta.

English, Sanskrit and Regional Names:

Eng. : Purging Nut, Phusic Nut;
Sans. : Dravanti, Kananaerand, Nepala;
B. : Maraharalu, Mogali eranda, Rana-yerandi;
G. : Erundi, Ehandrajot, Jangli-arundi, Jangli-Yarandi, Ratanjota;
H. : Bagbherenda, Ban-bheranda, Eranda-gachh, Pahari errand, Safed-arnad;
Kan. : Kaak-avenako, Katta-vanakka;

Mal. : Kaak-avenako, Kisu-gi, Thinbau-kyeksu, Thinbau-kyetsu;
Mar. : Mogalieranda;
Oriya : Kaddam;
P. : Jamalgota, Japhrota, Kurukarlu, Mogalieranda, Maraharalu, Rattanjot;
Tam. : Kaat-amunck;
Tel. : Bettadaharlu, Maranarulle, Maraharalu.

Distribution : Cultivated and naturalized in most parts of India.

Description : Small deciduous trees with alternate 3 or 5-lobed leaves. Flowers in terminal cymes, yellowish-green. Fruits ovoid, black. Seeds ovoid-oblong, dull brownish black.

Flowering and Fruiting : February-May.

Habitat Ecology : Frequent in wasteplaces. Roadsides. Field borders.

Material Examined : EBH-HVP-866, 10.05.2005.

Parts Used : Fruit. Twigs. Seeds. Latex.

Folk Uses : Twigs used as **tooth brush**. ½ teaspoon of roasted seeds with water prescribed as **purgative**. Paste of powdered seeds with ghee considered a good **emollient*** for chapped skin and feet. Latex good for **mouth sores***, **ulceration***, **decaying teeth*** and healing **injuries**.

Active Constituents : Isolated ingredients are β-sitosterol, β-amyrin, taraxerol (stem bark), Ca, Mg, Na, K traces of P, arachilic, linoleic, myristic, oleic, palmitic and stearic acid, arabinose, galactose, glucose, rhamanose, xylose, galacturonic acid and β-sitosterol (seeds).

Biological Activity : 50% EtOH extr. of aerial parts CNS depressant and diuretic.

Uses in Literature : Reported as an abortifacient, anasarca, lactagogue, purgative, rubefacient, fish poison, dye, tooth brush, symbolic to ward off evil spirits; and for body swelling, bloat, burns, cancer, carbuncle, cramps, diarrhoea, dropsy, dye, dysentery, earache, eczema, fencing, headache, herpes, inflammation, piles, pleurisy, purgative, pneumonia, rheumatism, ringworm, sciatica, scabies, skin troubles, sores, syphilis, tooth trouble, whitlow and wounds **(Ambasta, 1986; Chopra *et al.*, 1956; Dey, 1973; Jain, 1991; Kadel & Jain, 2006; Mohanty *et al.*, 1998; Mokat & Deokule, 2004; Punjani, 1998; Retnam & Martin, 2006; Sharma, 2003).**

Adulterant or Substitute : Seed oil used as an adulterant for ground nut, olive oil and linseed. Oil also substituted for diesel.

106. *Juglans regia* L. (Pl. 37B)

Syn. : *J. regia* L. var. *kumaonica* Cas. DC, *J. arguta* Wall.

Family : Juglandaceae.

Vern. : Akhrot.

English, Sanskrit and Regional Names:

Eng. : Common Walnut Tree, European Walnut, Himalayan Walnut, Walnut Tree;

Sans. : Akhoda, Akschota;
Ass. : Kabsing;
B. : Akhrut;
G. : Akhrot;
H. : Akhor, Akhrot, Khor;
Kash. : Akhor, Krot;
Mar. : Akroda;
P. : Akhori, Charmagz, Kabotang;
Tam. : Akrottu;
Tel. : Akrotu.

Distribution : Kashmir-S. E. Tibet: 1,500-3,000 m.

Description : Trees with imparipinnate, aromatic leaves. Male and female flowers on the same tree. Drupe ovoid with green, thick, fleshy rind enclosing a woody, wrinkled two-valved nut.

Flowering and Fruiting : April-October.

Habitat Ecology : Forests. Widely cultivated in W. Himalaya.

Material Examined : EBH-HVP-876, 10.08.2005.

Parts Used : Fruit. Leaves. Bark. Wood.

Folk Uses : Fruits **edible**. Powdered burnt fruit added in the preparation of one of the local delicacies-'**mahani**'*, served especially to guests attending the marriage in the family. Wood used for **ornamental works**. Leaves used for **scouring teeth** and bark for **reddening lips** as substitute for lip-stick.

Active Constituents : Principle constituents are 5-hydroxy-1, 4-napthoquinone, berberine (0.08%) (stem bark), oxalic acid, resin (fruit).

Biological Activity : Juglons inhibit growth of liver ascites carcinoma and 5-180 solid tumour in mice at 0.1 mg/ml *in vitro*.

Uses in Literature : Recorded earlier as an alexipharmic, alternative in scrofula, anthelmintic, aphrodisiac, astringent, cattle feed, detergent, edible (fruits), insect repellent, laxative, nerve tonic, psychoactive plant, tooth brush; and for bed wetting, carving, cough, dysentery, reddening the lips, confectionary (pastries), decolouring charcoal and active carbon, diarrhoea, dispelling flies, drawing instruments, dyeing, eczema, food and mouth diseases (vet.), frames, furniture, gunstocks, hoof (vet.), ice creams, kidney stones, lacquer work, menorrhagia, musical instruments, culinary purposes, illuminant, oil paints, ploughs, printing, rheumatism, skin disease, sores of toes, syphilis and tanning **(Ambasta, 1996; Chhetri, 2005; Kirtikar & Basu, 1935; Lalramnghinglova, 1996; Negi & Pant, 1990; Negi *et al.*, 1993; Rana *et al.*, 1996; Saini, 1996b; Sharma, 2003; Singh, 1991; Verma & Chauhan, 2006; Vishwanathan & Singh 1996; Watt, 1972).**

Major Ayurvedic Formulations : Amrtipras-ghrit, Ashoka-taila, Dandasa (bark), Dantasodhak, Yarin.

Trade : Dandasa, Akhrot, Akhor, Akhor, Akrut, Krot.

107. *Lagenaria siceraria* (Molina) Standl. (Pl. 37C)

Syn. : *Cucurbita siceraria* Molina; *Lagenaria leucantha* (Duchesne) Rusby; *Lagenaria vulgaris* Ser.

Family : Cucurbitaceae.

English, Sanskrit and Regional Names:

- **Eng.** : Bottle gourd, Calbash gourd;
- **Sans.** : Alabu;
- **Ass.** : Lau-bogalau;
- **B.** : Lau;
- **G.** : Dudhi, Tumada;
- **H.** : Lauki, Tum
- **Kan.** : Halagumbala, Sorekayi;
- **Mar.** : Bhopala, Dudh;
- **P.** : Ghiya;
- **Tam.** : Shorakkai;
- **Tel.** : Sorrakaya.

Distribution : Commonly cultivated throughout India.

Description : A softly pubescent herb with angled stem and 3-5 lobed, hairy leaves. Flowers large, white, solitary.

Flowering and Fruiting : July-September.

Habitat Ecology : Cultivated.

Material Examined : EBH-HVP-878, 14.08.2005.

Part Used : Fruits.

Folk Uses : Fruits cooked as **vegetable** and also employed for making '**pakoras**'*. Fruit juice twice a day considered good for **heart problems***.

Active Constituents : Active properties are due to rhamnose, fructose, glucose, galactose, sucrose, raffinose and stachyose (seed).

Uses in Literature : Used so far as anthelmintic, vegetable, water bottles, bowls, diuretic; and for dropsy **(Ambasta, 1986)**.

108. *Lagerstroemia indica* L. (Pl. 37D)

Syn. : *L. elgans* Wall., *Velaga globosa* Gaertn.

Family : Lythraceae.

Vern. : Har-Shingar.

English and Regional Names:

- **Eng.** : China Privet, Crape, Myrllle;
- **B.** : Phurush, Telinga Ghina;
- **Bo.** : Chinai-Mendhi, Dhayti;
- **P.** : Saoni, Dhaura;
- **Tam.** : Pavalakurinji, Sinappu.

Distribution : Throughout India.

Description : Trees with glabrous leaves and bright pink flowers. Capsule nearly globose with calyx teeth erect on them.

Flowering and Fruiting : June-October.

Habitat Ecology : Grown in Indian Gardens.
Material Examined : EBH-HVP-888, 25.10.2005.
Parts Used : Flowers. Plant.
Folk Uses : Grown as an **ornamental**. Flowers **offered*** especially to appease Lord Shiva. 15-20 ml. of its decoction given thrice daily to cure **back pain***.
Active Constituents : Delphinidin-3-arabinoride, petnnidin-3-arabinoside, malridin-3-arabinoside, gallic acid, methyl gallate and ellagic acid (flowers) have been characterised.
Uses in Literature : Known to be used as an astringent, febrifuge, hydragogue, narcotic, ornamental, purgative, stimulants; and for apathe of mouth, dysentery (**Ambasta, 1986; Bhardwaj & Gakhar, 2003; Sharma, 2003; Watt, 1972**).

109. *Lannea coromandelica* (Houtt.) Merrill. (Pl. 38A)

Syn. : *L. grandis* (Dennst) Engl., *Odina wodier* Roxb.
Family : Anacardiaceae.
Vern. : Kehmble.
English, Sanskrit and Regional Names:

Eng. : Egyptian Privet, Henna;
Sans. : Hinna, Yoranna;
Ass. : Jia;
B. : Bhadi, Jiyal, Jeol;
G. : Mavedi;
H. : Jhingan, Kaimil, Mohin;
Kan. : Ajasringi, Kuratige, Gojal;
Mal. : Odiya, Marum;
Mar. : Moi, Shimti;
Oriya : Indramai, Moi;
Tam. : Wodier, Kalasan;
Tel. : Appriyada.

Distribution : Throughout hotter parts of India.
Description : A medium-sized deciduous tree. Leaves odd-pinnate. Flowers greenish-yellow. Drupes oblong, red.
Flowering and Fruiting : July-September.
Habitat Ecology : Edges of fields.
Material Examined : EBH-HVP-841, 10.07.2005.
Part Used : Bark.
Folk Uses : Powdered bark applied to **cuts** and **wounds** as antiseptic.
Active Constituents : Bark contains Dl-epicatechin and leucocyanidin.
Uses in Literature : Known to be used as astringent, dye, wood for building, packing cases, furniture, wheel spokes, flocculating agent for clarification of cane juice, astringent, cattle yolks, oil processors, plough and spear shaft, combs; and for carving and turnery, railway sleepers, yielding paper pulp, dyeing and

textile printing; and for child birth, cuts, gum trouble, toothache, wounds **(Ambasta, 1986; Augustine & Sivadasan, 2004; Jain & Patole, 2001; Retnam & Martin, 2006; Sharma, 2003).**

Trade : Jhingan, Wodier.

110. *Lantana camara* Linn. (Pl. 38B)

Syn. : *L. aculeata* Linn., *L. myxa* L., *L. camera* L. var. *aculeata* Moldenku.

Family : Verbenaceae.

Vern. : Phulnu.

English and Regional Names:

Eng. : Lantana, Wild Sage;
G. : Ghanidalia;
H. : Desi-Lantana, Raimunia;
Kan. : Hesike, Kakke, Nata-hu-gida;
M.P. : Raimuniya;
Mal. : Arippu;
Mar. : Chadurang;
Oriya : Naga airi;
P. : Desilantana;
Tam. : Arippu;
Tel. : Pulikampa.

Distribution : Very Common weed throughout India.

Description : A straggling aromatic shrub having prickly branches. Leaves simple, opposite, ovate and acute. Flowers yellowish or orange red and in dense umbel. Drupes small, black, fleshy, shining.

Flowering and Fruiting : June-October.

Habitat Ecology : Wasteplaces. Rocky slopes.

Material Examined : EBH-HVP-837, 11.09.2004.

Parts Used : Whole Plant. Wood.

Folk Uses : Planted as hedge plant along **farm boundaries**. Wood used as **fuel***. Plant decoction applied on fresh **cuts** to stop bleeding.

Active Constituents : Isolated principle are arachidic, linoleic, myristic and palmitic acids, 1-triacontanol, 1, 8 cineole, citral, eugenol, furfural, geraniol, linalool, α-phellandrene, phellandrone, α-amyrin, β-sitosterol, lantadene (flowers, leaves, stem), lantadene A, lantadene B, lantanolic acid, lantic acid, leucine, ursonic acid, valine, tyrosine and glycine (leaves).

Biological Activity : Acid extr. of shoot exhibits antibacterial activity. Antitumour activity of verbascoside *in vitro* might partly be due to inhibition of protein.

Uses in Literature : Used as an antiseptic, carminative, diaphoretic, fish bait, insect repellent, stimulant, tooth brush and vulnerary; and for curing itching, jaundice, malaria, mumps, rheumatic pains, ulcers of mouth, veterinary problems, basketry, bleeding gums, burns, cold, cough, cut, decaying teeth, diabetes, dyspepsia (vet.), fencing, fever, fits, green manure, gum care, malaria, sprains, swelling, tetanus, tumours, ulcers, vulnerary, whooping cough and wounds (**Ambasta, 1986; Aminuddin & Girach, 1991; Augustine & Sivadasan, 2004; Bhattacharya, 1996; Dwarakan & Ansari, 1996; Geetha *et al.*, 1996a; Girach & Aminuddin, 1995; Goud & Pullaiah, 1996a; Henry *et al.*, 1996; Hosagoudar & Henry, 1996b; Jain, 1991; Kshirsagar *et al.*, 2003; Kumar & Goel, 1998; Kurian, 1995; Lal *et al.*, 1996; Lalramnghinglova, 1996; Masih, 2003; Mishra *et al.*, 1996a, b; Pandey & Verma, 2002; Punjani, 1998; Retnam & Martin, 2006; Saini, 1996b; Upadhyay & Singh, 2005; Watt, 1972**).

Adulterant or Substitute : Fruits used as an adulterant to black pepper (*Piper nigrum*).

111. *Lawsonia alba* Lam. (Pl. 38C)

Syn. : *L. inermis* Rox., *L. spinosa* Linn., *L. purpurea* Lamk.
Family : Lythraceae.
Vern. : Mehendi.
English, Sanskrit and Regional Names:

Eng. : Egyptian Privet, Henna;
Sans. : Hinna, Yoranna;
B. : Manghati, Muljuyati;
G. : Aivanam, Marithondi, Marutonri;
H. : Mehedi, Mendi, Shudi;
Kan. : Daunlacca, Pontaletshi, Mayilanchi, Marutonni;
Mal. : Dan, Danbin;
Mar. : Midi, Mindi;
Oriya : Mihndi;
P. : Mehendi;
Tam. : Goranta, Iveni, Pachcha, Pedda;
Tel. : Gorantlu, Gorante.

Distribution : Cultivated throughout India; Wild in W. regions.
Description : A shrub with sessile, elliptic-lanceolate leaves. Flowers creamish-white, fragrant, in corymbose. Seeds many, angular, brown.
Flowering and Fruiting : June-September.
Habitat Ecology : Edges of fields.
Material Examined : EBH-HVP-850, 10.09.2005.
Parts Used : Leaves.
Folk Uses : Paste of leaves alone (50 g powdered 'mehendi') or in combination with 25 g Amla (*Emblica officinalis* Gaertn.) and 25-50 ml. of milk applied on hair twice a week as a **dye** and for **cuts** between toe. Also henna paste applied on hands and feet

of young bride and other married womenfolk at the time of marriage in the family and on other ceremonial occasions, and also to check **burning sensation**. Gargles of decoction of its leaves considered good for **gum diseases***.

Active Constituents : Henna contains coumarins, naptha-quinones, flavonoids, sterol and tannin.

Biological Activity : Antibacterial, anti-inflammatory and possesses antihyaluronidase activities confirmed.

Uses in Literature : Known for cuts between toes, colouring hair, diarrhoea with blood, boils, contraceptive, gynaecological diseases, headache, jaundice, leucoderma, burning feet, cooling effect, foot and mouth diseases, fungal infection, gonorrhoea, skin inflammation, spleen enlargement, syphilis, piles, fissures, pimples, loose motions, headache, conjunctivitis, debility; and as hair dye and tonic, prophylactic, spermatorrhoea and tool handles (**Ambasta, 1986; Bhatt *et al.*, 2001; Bennet, 1985; Chopra *et al.*, 1956, 1983; Gogoi & Borthakur, 2001; Kirtikar & Basu 1984; Kumar & Chauhan, 2005; Maya *et al.*, 2003; Mohanty *et al.*, 1998; Pawar & Patil, 2004; Rao & Pullaiah, 2001; Rastogi & Mehrotra, 1993; Retnam & Martin, 2006; Sen & Behera, 2003; Sharma, 2003; Singh & Pandey, 1998; Vihari, 1995; Watt, 1972**).

112. *Leucaena leucocephala* (Lam.) de Wit. (Pl. 38D)

Syn. : *Mimosa leucocephala* Lam., *L. glauca* DC., *Acacia frondosa* Willd., *A. caringa* Ham., *A. biceps* DC., *A. leucocephala* DC.

Family : Fabaceae.

Vern. : Alseenia.

English and Regional Names:

- **Eng.** : White popinae, Lead Tree;
- **G.** : Lasobaval, Pardesi Baval, Vilaytibaval;
- **Mal.** : Takaranniram;
- **Oriya** : Rajokasundiri;
- **Tam.** : Tagarai-virai.

Distribution : Throughout India.

Description : Trees. Leaves bipinnate. Flowers white and in short peduncled heads. Pods linear flat, shining, dehiscent.

Flowering and Fruiting : September-November.

Habitat Ecology : Common on dry slopes. Wasteplaces, Roadsides. Along boundaries of fields.

Material Examined : EBH-HVP-890, 12.10.2005.

Part Used : Leaves.

Folk Uses : Leaves employed as **fodder** for milching cattle.

Active Constituents : Active principles are due to kaempferol-3-arabinoside, kaempferol-3-xyloside, quercetin-3-arabinoside, quercetin-3-

rhamnoside and quercetin-3-galactoside.

Uses in Literature : Described earlier for its edible tender leaves and pods, baskets, carotene, fancy bags, ornaments; and as a source of protein, stimulant, tonic, vegetable, fibre, fodder, poultry **(Ambasta, 1986; Gill & Nyawuame, 1994; Sinha, 1986; Singh** ***et al.*****, 1987; Watt, 1972).**

113. *Litchi chinensis* Sonner. (Pl. 39A)

Syn. : *Nephelium litchi* Camb.
Family : Sapindaceae.
Vern. : Litchi.
English and Regional Names:
Eng. : Litchi, Lychee;
Bo. : Lichi;
H. : Litchi;
Oriya : Lishi;
Tam. : Illichi.
Distribution : Cultivated.
Description : Evergreen small trees. Leaves alternate, exstipulate, shining above, glaucous beneath. Flowers greenish and in terminal panicles. Fruits ovoid, red when ripe. Pericarp dry, thin, brittle, tuberculate, containing one large ovoid brown seed enclosed in a fleshy succulent sweet edible aril.
Flowering and Fruiting : April-June.
Habitat Ecology : Cultivated in gardens.
Material Examined : EBH-HVP-832, 15.06.2004.
Part Used : Fruit.
Folk Uses : Fruits **edible** and considered good for **heart*** and **neuralgic disorders.**
Active Constituents : Fruit contains cyanidin-3 rutinoside, cyanidin-3 glucoside and malvidin-3 acetylglucoside.
Uses in Literature : Known in India as anodyne in neuralgic disorders and orchitis, edible (fruits); and for offering (**Ambasta, 1986; Gogoi & Borthakur, 1991**).

114. *Luffa aegyptica* Mill. (Pl. 39B)

Syn. : *Luffa pentandra* Roxb., *L. racemosa* Roxb., *L. clavata* Roxb., *L. petola* Seringe., *L. cattu-picinna* Seringe., *L. parvala* Wall., *L. gosa* Wall. *L. hederaces* Wall., *L. satpatia* Wall., *L. acutangula* W. and A., *L. cylindrica* (L.) Roem. *Bryonia cheirophylla* Wall., *Momordica cylindrica* L., *M. luffa* L.
Family : Cucurbitaceae.
Vern. : Ghiya Tori.

English, Sanskrit and Regional Names:

Eng. : Gourd Towel, Sponge Gourd, Vegetable, Sponge, Washrag Sponge;
Sans. : Aibhi, Brihatkoshataki, Dhamargava, Dirgha-koshataki, Luff, Mahaphala, Rajakoshataki, Sapitaka;
Ass. : Palo;
B. : Bhol, Bhatkerela, Bhatkarkel;
Bo. : Turia;
G. : Ghutti-bira, Neti-bira;
H. : Dun-dul, Dhundul, Ghiatarui, Nenua, Purula;
P. : Liasada, Turi;
Tam. : Pichukku, Pikku;
Tel. : Guttibira, Netibira, Nunebira, Thabwot, Thapotkha.

Distribution : Throughout India.

Description : Climbing herbs. Stem stout, twisted, glabrous. Tendrils trifid. Leaves orbicular-reniform, palmately lobed. Flowers yellow. Fruit cylindric, blunt at the ends, marked with longitudinal lines. Seeds grey, much compressed, narrowly winged.

Flowering and Fruiting : May-September.

Habitat Ecology : Cultivated. Wasteplaces.

Material Examined : EBH-HVP-895, 10.09.2006.

Part Used : Fruit.

Folk Uses : Fruit cooked as **vegetable** and also used in **culinary preparations**, 'pakoras', 'kofta'. Its juice considered good for **heart ailments***. Dried fruits yield a sponge used as a **scrubber** for bathing.

Active Constituents : Saponins-lucyosides A, E, F, I, J, K, L and M.

Uses in Literature : Found to be as anthelmintic (vet.), carminative, cathartic, diuretic, edible, emetic, fibre, galactagogue, purgative, scrubber, vegetable; and for abnormal menstruations, anaemia, cold and cough (vet.), diarrhoea, dog bites, dyspepsia (vet.), expulsion of worms, insect bites, killing white ants, liver diseases, piles, premature ejaculation, prickly heat, scabies, snake bite and tonsillitis (**Ambasta, 1986; Jain, 1991; Jha *et al.*, 1996; Katewa *et al.*, 2001; Kulkarni & Kumbhojkar, 1993; Kumar & Chauhan, 2006; Kurian, 1995; Maya *et al.*, 2003; Samati, 2004; Sarangi & Sahu, 2004; Satapathy & Brahmam, 1996; Singh & Maheshwari, 1992; Upadhye & Kulkarni, 2004; Watt, 1972**).

Adulterant or Substitute : Seed oil substituted for olive oil. The sponge is employed in place of cellulose sponges.

115. *Lycopersicum esculentum* Mill. (Pl. 39C)

Syn. : *L. lycopersicon* (Linn.) Brit and Brown, *L. lycopersicum* (Linn.) Farewell, *L. lycopersicum* (Linn.) Karsten, *Solanum lycopersicum* Linn., *S. humboldtii* Willd., *S. pseudolycopersicon* Jacq.

Family : Solanaceae.

Vern. : Tamatar.

English and Regional Names:

Eng. : Tomato, Love Apple;
Ass. : Belatibengana;
B. : Gur-begun, Tamati;
Bo. : Vilaytivengan;
G. : Vilaytivengun;
H. : Tamatar;
Kan. : Chapar-badne;
Mar. : Velvangi;
P. : Bhatte;
Tam. : Simie-takalie-palam.

Distribution : Cultivated or an escape over the greater parts of India.

Description : Pubescent herbs. Leaves pinnate. Flowers yellow and in cymes. Berries red when ripe. Seeds compressed.

Flowering and Fruiting : May-June.

Habitat Ecology : Cultivated.

Material Examined : EBH-HVP-881, 09.06.2005.

Part Used : Fruit.

Folk Uses : Fresh ripe fruits eaten raw with black pepper (*Piper nigrum* Linn.) and salt early in the morning as **wormicides***. Pieces of ripe fruits fried along with onion in ghee or oil and cooked as **vegetable**. Fruits considered **appetiser** and consumed raw in salads; also used in the preparation of **juices, ketchup, sauces, soups, 'chutney'** etc.

Active Constituents : Founds rich in tomatine, traces of solanin, glucose, fructose and some sucrose and ketoheptose, all essential amino acids except tryptophan and citric acid (ripe) and narcotine (unripe).

Biological Activity : Leaves antifungal.

Uses in Literature : Considered useful as antiseptic, aperient, blood purifier, cholagogue, digestive, edible, renal stimulant, salad, soup, juices and fodder; and for asthma, bronchitis, child birth, dyspepsia and apthous stomatitis. **(Ambasta, 1986; Augustine & Sivadasan, 2004; Chatterjee & Pakrashi, 1997; Maji & Sikdar, 1982; Shah, 1984).**

Adulterant or Substitute : Roots used as an adulterant for that of Periwinkle (*Catharanthus roseus* G. Don).

116. *Mallotus philippinensis* (Lam.) Muell.-Arg. (Pl. 39D)

Syn. : *Rottlera tintorii* Roxb., *R. aurantiaca* Hook and Arn., *R. affinis* Hassk., *R. montana* and *mollis* Wall., *Croton philippinensis* Lam., *C. punctatus* Retz., *C. coccineus* Vahl., *C. montanus* Willd., *C. cascarilloides* Rauesch. *ex* Steud., *C. distans* Wall.

Family : Euphorbiaceae.

Vern. : Kambhal.

English, Sanskrit and Regional Names:

Eng. : Monkey Face Tree, Kamala Tree;
Sans. : Bahupushpa, Kampilla, Kapila, Pikaksha, Rechanaka;
Ass. : Chinderpang, Gangai, Jaggaru, Machugan, Puddum;
B. : Bosonto-gundi, Kumala, Sundragundi, Tung;
Bo. : Kamala, Ruhin, Shendri;
G. : Kapli, Kapila, Kapilo, Kamella-mavu;
H. : Dhola Sindur (Birbhum), Kamala, Kambhal Kambila, Rohni, Rula, Rulu, Sinduri, Rori;
Kan. : Hullchelu, Kasari, Kurku, Rangamali;
Mal. : Chenkolli, Piponnakam, Poonagam;
Mar. : Kapilo, Shendri;
Oriya : Bosntogundi, Kumala, Rora, Sundragundi;
Tam. : Chendrasinduri, Kapilla, Kapli, Kunkuma, Vassuntagunda;
Tel. : Kunkuma, Kurku, Rangamale, Sarankasari.

Distribution : Throughout tropical India.

Description : A small branched tree. Leaves alternate, ovate-lanceolate, acuminate, entire toothed, orbicular red glands beneath. Flowers red, small. Capsule trilobed, covered with a bright red powder. Seeds subglobose, black.

Flowering and Fruiting : November-April.

Habitat Ecology : Dry slopes. Moist forests.

Material Examined : EBH-HVP-891, 12.11.2006.

Parts Used : Leaves. Fruit. Seeds. Bark.

Folk Uses : ½ teaspoon of fruit powder with 'ajwain' (*Trachyspermum ammi* Linn.) and black pepper (*Piper nigrum* Linn.) taken for 12-15 days to **purify blood**. One teaspoon of 'Kamila' powder (fruit) prescribed with cow milk or curd twice a day to **expel worms.** Red glandular powder coating on fruit surface used as **dye** for animals and taken along with jaggery for **stomach pains**, or with butter milk to check **constipation**. An amulet of pieces of its bark tied around neck, wrist and feet for treating **skin rashes** ('badalbis'). Leaves lopped for **fodder** and bedding for cattle ('**pattri**'*). Wood used as **fuel*** and for making **tool handles***.

Active Constituents : Properties attributed to rottlerin, isorottlerin, homorotteler and octacosanol, traces of wax, volatile oil, starch, tannins, sugar, oxalic and citric acid (plant), crotoxigenin and its rhamnoside coroglaucigenin, rottlerin, isorottlerin (seeds), betulin-3-acetate, lupeol, lupeol acetate, sitosterol, bergenin (heartwood), rottlerin, isorottlerin and red compound (leaves).

Biological Activity : Alc. extr. of fruits show antibacterial activity against *Micrococcus pyogenes* var. *aureus* whereas aq. extr. of red resinous powder (0.03 g) exhibits lithotropic effect in adult albino rats.

Uses in Literature : Known so far as alternative to sendu, antibacterial, anticancer, anthelmintic, antioxidant for ghee, aphrodisiac, blood purifier, contraceptive, cathartic, dye, fuel, fodder, purgative, manure, styptic, red dye from it (adorned by women as vermilion on forehead as a sign of being married), tonic for married woman, vermilion, rapid drying of varnish (seeds); and for eczema, jaundice, rheumatism, scabies, skin diseases, snake bite, sores, stomachache, tannins, thoracic pains, ulcers, and wounds (vet.) **(Ambasta, 1986; Banerjee & Pal, 1996; Chhetri, 1994; Hosagoudar & Henry, 1996b; Jain, 1968, 1991; Kapur & Singh, 1996; Kumar & Jain, 1998; Mukherjee & Namhata, 1990; Rama Rao & Henry, 1996; Rana *et al.*, 1996; Retnam & Martin, 2006; Sharma, 2003; Singh, 1996; Singh & Prakash, 1994; Tirkey, 2004; Tirkey *et al*, 2003a, b; Tiwari & Tiwari, 1996; Vihari, 1995; Watt, 1972).**

Major Ayurvedic Formulations : Dhanvantara-ghrita, herbenol cream, Kampilla-ka-churna, Kampilaka-taila, Kashisadi-ghrita, Krimighatini-vati, Krimi-kuthar-rasa, Misraka-sneha, Tinctura kamla.

Trade : Kamila powder, Kanikaa, Kainda.

Adulterant or Substitute : Red powder from fruits adulterated with Annatto Dye (*Bixa orellana* Linn.).

117. *Malvaviscus conzattii* Greenm. (Pl. 40A)

Family : Malvaceae.

Vern. : Gudhal.

Description : A shrub with serrated leaves, red orange flowers and staminal column.

Flowering and Fruiting : Major part of the year.

Habitat Ecology : Hedges; near houses.

Material Examined : EBH-HVP-899, 18.11.2006.

Part Used : Flowers.

Folk Uses : Flowers considered **sacred*** and offered to various deities. Also 5-10 ml. of its juice prescribed in **impotency**, once daily for a few months.

Active Constituents : Found to contain betaine, β-sitosterol, and its glucoside, allantoin, kaempferol and their 3 β-glucoside and 3 β-sophoroside.

Biological Activity : Betaine exhibits sperm immobilising action *in vitro*.

Uses in Literature : Used for premature ejaculation (**Khanna, 2002**).

118. *Mangifera indica* Linn. (Pl. 40B)

Syn. : *M. domestica* Gaertn.

Family : Anacardiaceae.

Vern. : Amb, Aam.

English, Sanskrit and Regional Names:

Eng. : Cuckoo's Joy, Cupid's Favourite, Mango Tree, Spring Tree;
Sans. : Adisairrabha, Amrah, Manmathalaya, Parapush-tama hotsava;
Ass. : Am, Gharium;
B. : Ambra, Uli;
Bo. : Amba, Amb;
G. : Am, Amb, Ambo, Amri, Anb;
H. : Am, Ambra, Amb;
Kan. : Mampalam, Manna, Mava, Mavu;
Mal. : Amram, Gomanne, Nattumavu;
Mar. : Ambo;
Oriya : Amb, Boula, Chuto;
P. : Amb, Amu;
Tam. : Adishelara yam, Elamavi, Guggu-mamidi, Mamadi, Mamid, Mamidi, Makandamu;
Tel. : Amba, Gujjumamidi, Makandamu, Mavina, Mavu.

Distribution : Tropical Himalaya, cultivated in tropics generally.

Description : Large evergreen trees. Leaves oblong, acute shining, entire, margin undulate. Flowers yellow, odorous arranged in many flowered pubescent panicles. Drupes large, fleshy, obliquely pyriform, yellow. Stone compressed, hard.

Flowering and Fruiting : March-June.

Habitat Ecology : Field borders. Roadsides (planted).

Material Examined : EBH-HVP-831, 12.06.2004.

Parts Used : Leaves. Fruit. Wood.

Folk Uses : Young and unripe fruits acidic and used for making various types of **pickle** and '**chutney**'*. Decoction of unripe fruits with a pinch of salt, sugar and spices prescribed to check **sun stroke***. **'Aamchur'** (powdered drupe except kernel) used in various culinary preparations for imparting acidic taste and also important component of one of local delicacies- **'mahani'*** served during feasts and marriage ceremonies. Ripe fruits **edible** and its pulp used for making **candies ('aampapar'*)**. A garland of leaves hung on the main door at the time of house warming ceremony and considered **sacred***. Wood used as **'samadas' (sacred wood)** for performing '**havan**'* and also for **furniture** and construction of **house**.

Active Constituents : Isolated constituents from different plant parts are homomangiferin and characterised as 2-β-D-glucopyranosyl-3-methoxy-I, 6-7-trihydroxyxanthone, leucine, tyrosine and valine (leaves), alanine, glycine, aminobutyric acid, kinic and shikimic acids (leaves, bark), triterpenoids (mangiferin, taraxerone, b-sitosterol), flavonoids, phenolic compounds and tannins.

Biological Activity : Gallic acid (1 mg./ml.) and quercetin (4 mg./ml.) showed marked anti-influenza virus activity. Antifungal and anti-diabetic activity also confirmed.

Uses in Literature : Considered as abortifacient, anthelmintic, antisyphilitic, appetiser, astringent, cremation wood, diuretic, edible, febrifuge, kibe, laxative, magico-therapeutic, tonic, rheumatism, sacred (Lord Buddha reposed under this tree), stomachic, symbolic (abode of god), vegetable; and for abdominal pain, agricultural implements, anasarca, anthrax (vet.), antifertility, arthritis, asthma, blood cancer, burns, cholera, cold, constipation (vet.), cough, cracking of foot, cut, diarrhoea, diabetes, dysuria, dysentery, excessive urination, fever, hiccups, house construction, hepatitis, indigestion, intestinal worms, jaundice, liver disorders, piles, pyorrhoea, musical instrument, sun stroke, swelling of throat (vet.), wounds, writing inks, and for decoration in marriage and festivals (**Ambasta, 1986; Aminuddin & Girach, 1991; Bhatt *et al.*, 2001; Borthakur & Gogoi, 1994; Borthakur *et al.*, 2004; Chowdhery, 1996; Ghosh *et al.*, 1996; Gill *et al.*, 1993, 1997; Gogoi *et al.*, 2003; Goud & Pullaiah, 1996a; Henry *et al.*, 1996; Hosagoudar & Henry, 1996a, b; Jain, 1991; Jain *et al.*, 1994; Jha *et al.*, 1996; Kadel & Jain, 2006; Kumar & Goel, 1998; Kapur *et al.*, 1996; Lalramnghinglova, 1996; Maliya & Singh, 2003; Manandhar, 1996; Masih, 2003; Mishra *et al.*, 1996b; Mohanty & Padhy, 1996; Mohanty *et al.*, 1998; Pandey & Rout, 2006; Pandey & Verma, 2002, 2003; Retnam & Martin, 2006; Samati, 2004; Sen & Batra, 1997; Sen & Behera, 2003; Shah & Singh, 1990; Sikarwar, 1996; Tirkey *et al.*, 2003b; Verma & Chauhan, 2006; Viswanathan *et al.*, 2001; Watt, 1972**).

Trade : Mango.

119. *Martynia annua* L. (Pl. 40C)

Syn. : *M. diandra* Glox.
Family : Martyniaceae.
Vern. : Kaun.
English and Regional Names:

- **Eng.** : Devil's Plant, Snake Head, Tiger Claw, Unicorn Plant;
- **B.** : Bagh-noki;
- **Bo.** : Vinchu;
- **H.** : Bichu, Hathajori;
- **Mar.** : Vinchu;
- **P.** : Bichu;
- **Tam.** : Thelkodukk-ukai, Nagavidakkai;
- **Tel.** : Garudamukku.

Distribution : Tropical weed. In Himalaya up to 1,200 m.
Description : Tall viscid herbs. Leaves cordate, sticky, sinuate toothed with

unpleasant smell. Flowers pink, blotched with purple. Fruits black with a pair of hooked spinous beak on maturity.

Flowering and Fruiting : June-October.
Habitat Ecology : Shrubberies. Wastelands.
Material Examined : EBH-HVP-828, 25.07.2004.
Parts Used : Fruit. Leaves.
Folk Uses : Fruits paste in water applied externally over **scorpion-stings**. Gargles of leaf juice prescribed twice daily for **sore throat*** till cure.
Active Constituents : Pelargonidin-3-5 diglucoside and cyanidin-3-galactoside have been isolated.
Uses in Literature : Used as antidote for blisters, boil, snake and scorpion stings; and for epilepsy, inflammation, tuberculosis, ulcers, wounds (**Ambasta, 1986; Goud & Pullaiah, 1996a; Kirtikar & Basu, 1935; Patil, 2004; Reddy *et al.*, 1998; Retnam & Martin, 2006**).

120. *Melia azedarach* Linn. (Pl. 40D)

Syn. : *Melia sempervirens* Sw., *M. bukayun* Royle.
Family : Meliaceae.
Vern. : Draek.
English, Sanskrit and Regional Names:

Eng. : Barbados Liliac, Bead Tree, Indian Lilac, Persian Lilac, Pride of India;
Sans. : Akshadru, Arishta, Brihannimba, Dreka, Gairia, Hinadruma, Karmuka, Mahanimba, Nimbaka, Parvatta-nimba-vrikshha, Vikshaha;
Ass. : Bakainu, Thamage;
B. : Gara Nim, Ghornim, Mahanim;
Bo. : Bakayam, Drek, Vilayatinim;
G. : Bakan, Bakanlimbodo, limbodo, Malai, Malai: veppam, vembu;
H. : Bakain, Bakarja, Drek, Deikna, Ghoranim, Mahanim, Mahanimba;
Kan. : Arevebu, Hutchu Bevu;
Mal. : Ka-ma-ka, Malaveppu, Ta-ma-ka;
Mar. : Bakananimb, Bakan limbodo, Limbara;
P. : Bakain, Chein, Dek, drek;
Tam. : Konda-vepa, Malaivembu, Makanim, Tarakavepa;
Tel. : Turakavepa,Vettiveppa;
Urdu : Bakayan.

Distribution : Common Throughout India.
Description : A tree having bipinnate, opposite leaves. Leaflets ovate, obtusely serrate, deciduous. Flowers fragrant, lilac in long axillary panicle. Drupes subglobose, wrinkled.
Flowering and Fruiting : March-May.
Habitat Ecology : Roadsides. Edges of cultivated fields.
Material Examined : EBH-HVP-898, 15.05.2006.
Parts Used : Leaves. Wood.
Folk Uses : Poultice of leaves considered good against **dandruff***. Leaves

used as **fodder.** Wood used for making **furniture** (doors, window) and **agricultural implements***.

Active Constituents : Principle constituents are kaemferol-3-l-rhamno-D-glucoside, rutin (leaves), kuline, tetracyclic triterpenoides kulactone, kulalactone and kulinone (tree, bark), azardine, aresin, tannin, meliotannic acid, benzoic acid, bakayanin sterol, a bitter principle margosine and a fixed oil containing sulphur (fruits).

Biological Activity : Stem bark anticancer, antiviral and spasmolytic. Azedarchin A and 12-o-acetylazedarchins A and B exhibit antifeedant activity on larvae of *Spodoptera exigua*.

Uses in Literature : Reported earlier as an antilithic, antiseptic, astringent, blood purifier, brain tonic, diuretic, emmenagogue, fodder, fuel, wood, laxative, psychoactive plant, stomachic, symbolic abode of god; and for cold, fever (vet.), furniture, gout, hair fall, hypertension, hysteria, malarial fever, making idols, spleen enlargement, sprain (vet.), spasm (vet.), skin disease, tumours, ulcer (vet.) wounds, musical instruments, boxes and ornamental plywood **(Ambasta, 1986; Borthakur & Sarma, 1996; Chhetri, 1994; Dey, 1973; Goud & Pullaiah, 1996b; Jain, 1991; Jain *et al.*, 1994; Jha & Basak, 1994; Jha *et al.*, 1991; Kapur & Singh, 1996; Kurian, 1995; Lalramnghinglova, 1996; Rana *et al.* 1996; Retnam & Martin, 2006; Sharma, 1996; Siwakoti & Verma, 1996; Srivastava *et al.*, 2003; Watt, 1972).**

Trade : Persian lilac.

Adulterant or Substitute : Leaves and bark substituted for true Neem (*Melia azedirachta* Linn.).

121. *Melia azadirachta* Linn. (Pl. 41A)

Syn. : *M. parviflora* Moon., *M. indica* Brandis., *Azadirachta indica* Adr.

Family : Meliaceae.

Vern. : Neem.

English, Sanskrit and Regional Names:

Eng. : Indian Lilac, Margosa Tree, Neem Tree;
Sans. : Arishta, Avkapadapa, Chhardana, Dzaad-darakhte;
H. : Hingu, Kaitarya, Malaka, Neta, Nimba, Nimbaka, Pitsara, Swita, Sumana;
B. : Nim, Nimgachh;
G. : Nim;
H. : Nim, Nimgachh;
Kan. : Ariya-veppa, Vippa;
Mal. : Bawtamaka, Kamaku, Tamdbin, Thamaka;
Mar. : Danujhada, Kohumba, Libado, Limba, Limbado;
Tam. : Arulundi, Kinji, Nimbamu, Taruka, Vepa, Vepu, Vembu, Yapa, Yeppa;

Tel. : Bevina-marakadbevina-mara, Heb-bavu, Nimbamu, Vemu, Vepa, Yapa, Yeppa.

Distribution : Cultivated throughout greater part of India.

Description : Trees. Leaves alternate, imparipinnate. Flowers white, in axillary panicles. Drupe endocarp woody.

Flowering and Fruiting : June-September.

Habitat Ecology : Edges of cultivated fields.

Material Examined : EBH-HVP-880, 15.09.2005.

Parts Used : Leaves. Twigs. Fruits.

Folk Uses : Twigs used as **dentrifice***. Bath of Neem leaves recommended for **skin ailments**. ¼ teaspoon of powdered leaves given once a day for 30 days to **purify blood**. Its poultice good for suppuration of **boils**. Ripe fruits also considered good against **skin ailments**.

Active Constituents : Found to contain nimocinol, 2'-3' dehydrosalano, meliacins, nimocinolide, isoazadirolide and nimbocinolide (leaves).

Biological Activity : Leaves anthelmintic.

Uses in Literature : Useful in India as antiseptic, anthelmintic, antidote, astringent, blood purifier, diuretic, insecticidal, insect repellent, mastitis, purgative, tonic; and for boils, bronchitis in children, chicken pox, chest pain, constipation, dermatitis, diarrhoea, eczema, epilepsy, foot and mouth diseases, fever, itching, insect bites, malaria, miscarriage, mosquito repellent, piles, pimples, pyorrhoea, skin disease, small pox, stimulant, sores, small pox, suppuration (vet.), syphilitic sore, wounds **(Ambasta, 1986; Bhatt *et al.*, 2001; Bondya *et al.*, 2006; Gogoi & Borthakur, 2001; Jadhav, 2006; Kadel & Jain, 2006; Kalpesh *et al.*, 2004; Khanna, 2002; Khanna & Kumar 2000; Kumar & Chauhan, 2006; Kumar & Goel, 1998; Kumar & Jain, 1998; Kumar *et al.*, 2004; Mohanty *et al.*, 1998; Pandey & Rout, 2006; Retnam & Martin, 2006; Sharma, 2003; Sharma *et al.*, 2002; Silori & Rana, 2000).**

Major Ayurvedic Formulations : Shankhapuspi-taila, Jvarasamhara-rasa, Mahamarichyadi-taila, Arsoghna-vati, Pathyadi-kwatha, Maha-manjistha-dyarista, Punarnavadi-kwatha, Raktasodhaka-vati.

122. *Mentha piperita* Linn. (Pl. 41B)

Family : Lamiaceae.

Vern. : Pipermint.

English and Regional Names:

Eng. : Brandy Mint, Peppermint;

H. : Gamathi Pudina, Paparaminta;

Tel. : Vilayeti, Pudneh.

Distribution : Throughout India. Commonly cultivated in native gardens all over plains of India.

Description : A perennial herb with pungent smell. Leaves smooth, lanceolate, acute, coarsely dentate. Flowers lilac, in loose cylindrical slender spikes.

Flowering and Fruiting : July-September.

Habitat Ecology : Cultivated.

Material Examined : EBH-HVP-900, 15.09.2006.

Parts Used : Leaves. Whole Plant.

Folk Uses : Decoction of leaves considered **stomachic** and checks **vomiting.** Poultice of whole plant cures **headache***. Leaves also used as a **flavouring** agent in **'chutney'*** preparation.

Active Constituents : These are methyl acetate 20%, menthol 9.1%, menthone and hydrocarbon 15% (leaves).

Uses in Literature : Used so far as carminative, cosmetic, digestive, stimulant; and for bronchitis, diarrhoea, dysentery, flatulence and vomiting (**Ambasta, 1986; Chopra *et al.*, 1956; Sharma, 2003; Jain, 1991; Uphof, 1968**).

123. *Mentha spicata* Linn. (Pl. 41C)

Syn. : *M. spicata* var. *viridis* Linn., *M. viridis* Linn.

Family : Lamiaceae.

Vern. : Pudina.

English and Regional Names:

- **Eng.** : Brown Mint, Garden Mint, Lamb Mint, Makaerel Mint, Spear Mint;
- **B.** : Pahari Pudina;
- **Bo.** : Pudina;
- **G.** : Pudino;
- **H.** : Pudina, Paharipudina;
- **Mar.** : Pudino;
- **Tel.** : Nagbb, Pudneh, Shahsufian.

Distribution : Throughout India. Commonly cultivated in native gardens all over plains of India.

Description : A perennial herb with pungent smell. Leaves smooth, lanceolate, acute, coarsely dentate. Flowers lilac, in loose cylindrical slender spikes.

Flowering and Fruiting : March-July.

Habitat Ecology : Cultivated.

Material Examined : EBH-HVP-904, 15.05.2006.

Parts Used : Leaves.

Folk Uses : Decoction of leaves considered **stomachic** and checks **vomiting.** One tablespoon of its powdered leaves, 'ajwain' (*Trachyspermum ammi* Linn.) and 'harar' (*Terminalia chebula* Retz.) powder given twice a day for a week to cure **gastric problems**. Leaves also used as a **flavouring** agent in **'chutney'** preparation.

Active Constituents : These are menthol, carvone (66.9%), limonene (10.84%), phellandrene, dihydrocarvone, pinen, phellandrene, 1-limoene, octyl alcohol, dihydrocarveol, diosmetin and diosmin (leaves).

Uses in Literature : Used so far as antispasmodic, carminative, cosmetic, digestive, flavouring agent, making chutney, perfume, stimulant; and for bronchitis, diarrhoea, indigestion, fever, hysteria and vomiting (**Ambasta, 1986; Kirtikar & Basu, 1935; Sharma, 2003**).

124. *Mimosa pudica* Linn. (Pl. 41D)

Family : Fabaceae.
Vern. : Lajalu.
English, Sanskrit and Regional Names:

Eng. : Touch-me-not;
Sans. : Lajjali;
Kan. : Laji;
H. : Lajwanti;
Mal. : Tintarmani;
Tam. : Thottal vadi;
Tel. : Attapati.

Distribution : Throughout India.
Description : Herbs with bipinnate leaves. Flower heads pink, long-stalked.
Flowering and Fruiting : July-September.
Habitat Ecology : As an escape.
Material Examined : EBH-HVP-794, 25.09.2005.
Parts Used : Plant. Leaves.
Folk Uses : Planted as an **ornamental***; children love to **play*** and enjoy by touching its pinnules. Leaf paste applied to heal fresh **cuts** and **wounds**.
Active Constituents : β-amyrin, β-sitosterol, friedelin (leaves).
Uses in Literature : Used so far in India as an astringent, carminative; and for cuts, diabetes, dysentery, joint pains, scorpion sting, wounds (**Retnam & Martin, 2006; Sharma, 2003**).

125. *Mirabilis jalapa* Linn. (Pl. 42A)

Syn. : *Mirabilis multiflora* (Torr.) Gray.
Family : Meliaceae.
Vern. : Gulal.
English, Sanskrit and Regional Names:

Eng. : Four O'Clock Flower, Marvel of Peru;
Sans. : Krishnakeli, Sandhyakali;
B. : Gulabas, Krishnakeli;
Bo. : Gubhaji;
H. : Gulabbas;

Mal. : Antimalari;
P. : Abasi, Gulabas;
Tam. : Antinarulu, Pattarachi;
Tel. : Batharachi, Chandrakanta;
Urdu : Guleabbas.

Distribution : Throughout India.

Description : An erect leafy herb. Leaves cordate. Flowers pink, long tubular. Fruits elliptical, leathery black.

Flowering and Fruiting : June-September.

Habitat Ecology : Moist localities. Wasteplaces. Roadsides.

Material Examined : EBH-HVP-901, 25.09.2005.

Parts Used : Flowers. Root.

Folk Uses : Flowers used as **offering** to deities. Root paste given orally to cattle for **cooling effect*** against sun stroke.

Active Constituents : Important constituents of flowers are miraxanthins I, II, III and IV, indicaxanthin and vulgaxanthin.

Biological Activity : Spasmolytic activity +ve.

Uses in Literature : Used so far in India as an aphrodisiac, purgative, tonic, ornamental and vegetative; and for boils, blisters, cold at child birth, earache, scorpion bites, piles, scabies and urticaria **(Ambasta, 1986; Dastur, 1970; Jain, 1991; Retnam & Martin, 2006; Sharma, 2003; Upadhye & Kulkarni, 2004).**

126. *Momordica charantia* L. (Pl. 42B)

Syn. : *M. humilis* Wall., *M. muricata* DC., *M. senegalensis* Lamk., *Cucumis africanus* Bot.

Family : Cucurbitaceae.

Vern. : Karela.

English, Sanskrit and Regional Names:

Eng. : Bitter Gourd;
Sans. : Brihadavalli, Karavellah, Karavellaka;
Ass. : Kakiral;
B. : Baramasiya, Potikakkar;
Bo. : Karla;
G. : Karela, Karola;
H. : Karela, Kareli, Karola;
Kan. : Hagal;
Mal. : Kaipavalli, Pavakacheti;
Mar. : Karale;
Oriya : Karena;
Ind. : Pavakkachedi;
Tel. : Mattakakata, Tellakakra.

Distribution : Throughout India.

Description : An annual climbing herb with simple tendrils. Stem long, branched and hairy. Leaves 5-7 lobed, orbicular, pubescent on both sides, cordate at base. Flowers yellow. Fruits bright green coloured, long, pendulous, fusiform, pointed. Seeds long, compressed, sculptured with corrugate margins.

Flowering and Fruiting : June-October.

Habitat Ecology : Cultivated.

Material Examined : EBH-HVP-902, 12.11.2005.

Part Used : Fruits.

Folk Uses : Fruit cooked as **vegetable.** 50 ml. of its juice taken daily to purify **blood** and as **anthelmintic.** 1 teaspoon of powdered fruit taken daily in the morning with water to cure **diabetes** and **skin affections**.

Active Constituents : Fruits contain acylglucosyl sterols, lipids, charantin, β-sitosterol, glucoside, ascorbigen and a new stigmost 5, 25-diene-3 β-O-glucoside. charantin (momordicin M = phytosterolin = β-sitosterol, moriodosides G, F1, F2, I, K and L.

Biological Activity : Positive hypoglycaemic and cytotoxic activity of fruits attributed to charantin-β-sitosterol-glucoside and stigmost 5-25-diene-3 β-O-glucoside.

Uses in Literature : Used so far as an abortifacient, anthelmintic, astringent, blood purifier, carminative, emetic, febrifuge, laxative, stomachic, tonic, vegetable; and for arthritis, bilious affection, bleeding piles, blood cancer, burning of the sole of the feet, colic, cooling, cutaneous disease, diabetes, disease of spleen and liver, dysentery, eczema, gout, gummosis, headache, hypertension, jaundice, leprosy, malaria, mouth ulcer, night blindness, odema, peptic ulcer (vet.), poisonous bite (vet.), poultry disease, rheumatic pains, scabies, snake bite, sore, toothache, ulcers, wound maggots **(Ambasta, 1986; Bajpayee & Dixit, 1996; Bondya & Sharma, 2003; Ghosh *et al.*, 1996; Gill *et al.*, 1993; Jain, 1991; Jain *et al.*, 1994; Kothari & Moorthy, 1996; Kshirsagar *et al.*, 2003; Kulkarni & Kumbhojkar, 1993, 1996; Kumar & Chauhan, 2006; Lalramnghinglova, 1996; Mahato & Mahato, 1996; Negi & Gaur, 1991; Pandey & Verma, 2002; Rama Rao & Henry, 1996; Retnam & Martin, 2006; Sharma, 2003; Sen & Batra, 1997; Sen & Behera, 2003; Sen *et al.*, 2002; Shome *et al.*, 1996; Singh & Pandey, 1996; Watt, 1972).**

Major Ayurvedic Formulations : Mahavisagarbha-taila, Karneme.

Trade : Karela.

127. *Momordica dioica* Roxb. *ex* Wall. (Pl. 42C)

Syn. : *M. cochinchinensis* Spreng, *M. mixa* Roxb.

Family : Cucurbitaceae.

Vern. : Kakron.

English, Sanskrit and Regional Names:

Eng. : Small bitter gourd;
Sans. : Karkota, Karkataka, Vahisi;
B. : Adavi Kakara, Go-Ka-Kara, Kakrol;
G. : Karapata;
H. : Adavi, Gulkakra, Gangerua, Jangli Karela, Kakara, Kukrol;
Kan. : Karlikai;
Mar. : Kakana, Kartoli;
P. : Kakura, Kirara;
Tel. : Adavi-Kakara, Agakara, Samongnway;
Tam. : Tholoopavari.

Distribution : Throughout India.

Description : Perennial climbing herbs. Leaves broadly ovate, cordate, tri-lobed, denticulate, glabrous. Flowers yellow. Fruits ovate, pointed, green and bright red when ripe, softly spinous. Seeds many, horizontal, irregular, ovate, compressed, black corrugate on the margins.

Flowering and Fruiting : July-October.

Habitat Ecology : Shrubberies.

Material Examined : EBH-HVP-903, 12.11.2005.

Part Used : Fruit.

Folk Uses : Fruits cooked as **vegetable.** 20 ml. of its juice prescribed once a day to control **diabetes*** and **stones*** in **kidney.**

Active Constituents : These are oleanolic acid glucoside and oleanolic acid, arabinoglucoside, chondrillasterol.

Biological Activity : Oleanolic acid glucoside and oleanolic acid, arabinoglucoside, chondrillasterol exhibited hypoglycaemic activity in streptozotocin-induced diabetic rats at a dose of 25 mg./kg.

Uses in Literature : So far known as vegetable, illuminant, paints, varnishes, aperient, ulcer, lumbago, fracture; and for bleeding piles, bowl affections, cough, chest complaints, stimulant, ulceration, urinary complaints and uterine discharge **(Ambasta, 1986, Kirtikar & Basu, 1935)**.

128. *Morus alba* Linn. (Pl. 42D)

Syn. : *M. serrata* Wall., *M. tatarica* L.

Family : Moraceae.

Vern. : Toot.

English, Sanskrit and Regional Names:

Eng. : Mulberg or White Mulbery;
Sans. : Tula;
B. : Tut;
Bo. : Chinni, Satur, Tut;
G. : Shetur;
H. : Chun, Chinni, Tut;

Kan. : Hipnerle, Uppunute;
Kash. : Tut;
Mar. : Ambat, Tut;
Oriya : Tuticoli, Tuto;
P. : Tut, Tutri;
Tam. : Kambali, Pattuppuchi;
Tel. : Pippalipanduchettu, Reshmechattu;
U.P. : Shahtoot.

Distribution : Cultivated extensively in drier areas of W. Himalaya up to 3,300 m.

Description : A medium-sized tree having brown rough bark and ovate, toothed cordate leaves. Flowers greenish. Fruits ovoid black when ripe.

Flowering and Fruiting : March-May.

Habitat Ecology : Common along agriculture fields.

Material Examined : EBH-HVP-802, 14.05.2004.

Parts Used : Leaves. Fruit.

Folk Uses : Leaves used for **rearing silkworm** and as **fodder** for cattle. Fruits **edible** and considered **refrigerant.** Bark used as **cordage*** to tie pile of grasses.

Active Constituents : Main ingredients are N-butanol, β-γ-hexenol, methyl-ethylacetaldehyde, hexaldehyde, isobutylaldehyde, methyl-ethyl-ketone, methyl-hexyl-ketone, butylamine and acetic, propionic isobutyric acid, astragalin (leaves), heptanoic (0.047), caprylic (0.21), pelargonic (0.047), capric (0.07), myristic (0.105), palmitic (11.85), palmitolic (0.188), methylheptadeconic (0.14), stearic (4.46), oleic (12.67), linoleic (68.3), nonodecanoic (0.641), linilenic (0.98%) acids (fruit), two phytoalexins-morasin C and moracin D (shoot), mulberrin, mulberrochromene and cyclomulberin (root, bark, stem).

Biological Activity : Mulberrifuran A exhibits antibacterial activity against *Staphylococcus aureus* and *Fusarium roseum* whereas Moran A exhibits hypoglycaemic effect in normal and alloxan induced hyperglycaemic mice.

Uses in Literature : Reported to be used as anthelmintic, diaphoretic, edible, fodder, haemorrhoids, purgative, refrigerant, vermifuge and host for silk worms; and for agricultural implements, boats, building, diarrhoea, dysentery, dyspepsia, goitre, inducing oestrus cycle, feeding silk worms, fever, furniture, melancholia, paper, sore throat (**Ambasta, 1986; Bhardwaj & Gakhar, 2003; Jain, 1991; Kshirsagar & Singh, 2000; Maliya, 2004; Negi & Gaur, 1991; Rajwar, 1983; Retnam & Martin, 2006; Saini, 1996a; Shah, 1984; Sharma, 2003; Sharma *et al.*, 2001; Vishwanathan *et al.*, 1981; Watt, 1972**).

Trade : Mulberry.

129. *Mucuna pruriens* (L.) DC (Pl. 43A)

Syn. : *M. purita* Hook., *M. utilis* Wall., *Caryopogon pruriens* Roxb., *Dolichos pruriens* L.

Family : Fabaceae.

Vern. : Gazal-Bel.

English, Sanskrit and Regional Names:

Eng. : Cowhage, Cowitch;
Sans. : Durabhigraha, Galrabhange;
B. : Akolshi, Alkusa, Bichchoti;
Bo. : Kuhili;
G. : Kanchan, Kivanch;
H. : Goncha, Kaunch, Kivachh;
Kan. : Hassagunigida, Nasukunni;
Mal. : Naykkuurma;
Mar. : Kuhili;
Oriya : Alokushi;
P. : Gunchagaji, Kanaucha, Kawanch, Kunch;
Tam. : Attumabuttar, Punaiyavarai;
Tel. : Pravrishayani, Telladuradagondi.

Distribution : Throughout India.

Description : Climbers. Leaflets membranous. Flowers purple. Pods curved upward at tip, densely clothed with bristles.

Flowering and Fruiting : July-September.

Habitat Ecology : Drier areas. Open slopes.

Material Examined : EBH-HVP-813, 20.09.2004.

Part Used : Seeds.

Folk Uses : 20-30 ml. decoction of seeds thrice daily till cure prescribed for **impotency**. 1 teaspoonful of powdered roasted seeds with 'gur' and ghee taken every morning with milk to provide **strength** to the body. Seeds mixed with fodder given to animals to increase **lactation** (lactagogue).

Active Constituents : Properties attributed to mucadine, mucunadinine, mucadininene, pruienidine, mucunin, mucunadine, nicotine, L. dopa-glutarthiane, lecithin, gallic acid, a glucoside and indole alkylamines.

Biological Activity : Plant extr. shows marked hypotensive and hypocholestrolaemic effect in normal rats.

Uses in Literature : Recorded to be useful as an anthelmintic, aphrodisiac, astringent, diuretic (vet.), edible (pods), lactagogue, local stimulant, mild resicant, nervine tonic, psychoactive plant, purgative, stimulant, vermifuge (vet.); and for boils (vet.), bone fracture, cough, diarrhoea, digestion, fencing, fertility, impotency, leucorrhoea, madness, nervous disorder (vet.), night dreams, menstruation, scabies (vet.), scorpion sting, seminal weakness, snake bite, spermatorrhoea, syphilis, ulcer (vet.), urinary complaints, weakness, worm (vet.) and wound (vet.) **(Ambasta, 1986; Aminuddin & Girach, 1991; Borthakur**

& Sarma, 1996; Gill & Nyawuame, 1994; Girach *et al.*, 1997; Jain, 1991; Jain & Patole, 2001; Jha & Varma, 1996; Jha *et al.*, 1991; Kurian, 1995; Mohanty *et al.*, 1998; Punjani, 1998; Radhakrishnan *et al.*, 1996; Rama Rao & Henry, 1996; Rana *et al.*, 1996; Sahu, 1996; Sarangi & Sahu, 2004; Satapathy & Brahmam, 1996; Singh, 1991; Singh & Pandey, 1996; Singh & Zaheer, 1996; Sinha, 1996; Siwakoti & Verma, 1996; Srivastava *et al.*, 2003; Watt, 1972).

Major Ayurvedic Formulations : Agastya-haritaki, Aswagandha-churna, Badampak, Brimhana-gutika, Dhatupaustic-churna, Kamdev-ghrita, Kwanch-churna, Kwanch-pak, L-dopa, Masabaladi-pachan, Musli-pak, Sperman tablets, Tentax-forte, Vajikasana-ghrita, Vanari-gutika.

Trade : Seeds sold as 'Kaunch beej'.

130. *Murraya koenigii* Spreng. (Pl. 43B)

Syn. : *Bergera koenigii* Linn.

Family : Rutaceae.

Vern. : Gandhla.

English, Sanskrit and Regional Names:

Eng. : Curry Leaf Tree;
Sans : Alakavhaya, Chhandighna, Kadarya, Kalasaka, Krishtna-nimba, Nimba-pattram, Raman, Surabhi, Surabhi-nimba;
Ass. : Bishahari, Narasingha;
B. : Barsungha, Kariaphyll;
Bo. : Goranimb, Karripat, Karrinim;
G. : Kadhi, Limbdo;
H. : Barasunga, Gandhela, Harri, Karipatha, Katnim;
Kan. : Kari-bevu;
Mal. : Karea-pela, Kariveppilei;
Mar. : Jhirang, Karapacha, Kudianim;
Oriya : Barsan, Basan-go, Bhursunga;
P. : Bowala, Gandanim, Gandalu;
Tam. : Karuvembu, Karuvepallai;
Tel. : Karepeka, Kari-vepa, Karivepaku.

Distribution : Throughout India.

Description : A small strong smelling tree. Leaves imparipinnate. Flowers white, in many-flowered peduncled corymbs. Berry ovoid, apiculate, black rough with glands, two seeded and with whitish pulp. Seeds green, globose.

Flowering and Fruiting : March-June.

Habitat Ecology : Open slopes. Wastelands. Roadsides.

Material Examined : EBH-HVP-780, 22.06.2003.

Parts Used : Leaves. Stem. Fruit.

Folk Uses : Leaves used for **flavouring curries*** and **'chutney'***. Stem considered good for **cleaning teeth***. Fruits **edible**. ½ teaspoon infusion of leaves with 'gur' and onion (*Allium cepa* Linn.) taken with hot water, twice a day, for **stomachache.**

Active Constituents : Properties attributed to mahanimbidine, mahanimbine, isomahanimbine, mahanimbineine, girinimbine, muconicine, cyclomahanimbine, scopoline, carotene, vitamins (leaves), curryangine, curryanine, mukanole, murraycine, murrayacinine, isomurrayazoline, murrayazoline, murrayazolinol, 3 (1, 2-dimethhylallyl) xanthyletin (stem bark); mahanimbine, koenimbine, koenigicine (fruit, leaves) and mahanimbinol (wood).

Biological Activity : Root antiameobic, hypotensive whereas aerial parts antiameobic and spasmolytic. However, koenoline exhibits cytotoxic activity against ICB cell culture. Aerial parts—antiprotozoal and spasmolytic.

Uses in Literature : Known to be used in India as an anthelmintic, carminative, edible (vegetable), febrifuge, laxative, stimulant for eruptions, stomachache, tonic; and for acute abdominal pains, agricultural implements, bleeding, bone fracture, burning pains, common cold, cough, culinary purposes, dandruff, diabetes, diarrhoea, dysentery, fertility (vet.), fever, flavouring curries, grooms as ornamental, hair tonic, jaundice, joint pains, parasites, renal pains, snake bite, swelling in various tract of body, tonic, tuberculosis, vomiting and worms **(Aminuddin & Girach, 1996; Borthakur, 1996a; Chauhan, 1999; Gill *et al.*, 1997; Gogoi & Borthakur, 1996, 2001; Gogoi *et al.*, 2003; Jain, 1991; Kurian, 1995; Lal *et al.*, 1996; Maliya & Singh, 2003; Mohanty & Padhy, 1996; Mohanty *et al.*, 1998; Parmar & Kaushal, 1982; Rama Rao & Henry, 1996; Singh, 1996; Siwakoti & Verma, 1996; Verma & Chauhan, 2006; Watt, 1972).**

131. *Murraya paniculata* (Linn.) Jack. (Pl. 43C)

Syn. : *M. exotica* Linn.
Family : Rutaceae.
Vern. : Gandhla.
English, Sanskrit and Regional Names:

Eng. : Orange Jesmine;
Sans. : Durabhigraha, Galrabhange;
B. : Kamini;
H. : Kamini, marchula;
Kan. : Angarakana gida, Pandry;
Mar. : Pandari, Kunti, Marchulajuti;
Oriya : Banmallika, Harkankali;
Tam. : Konji;
Tel. : Nagagolunga, Karepaku.

Distribution : Throughout India.
Description : An evergreen shrub. Leaves elliptic-lanceolate, shining. Flowers white, fragrant. Fruits red when ripe.
Flowering and Fruiting : June-December.
Habitat Ecology : Open slopes. Wastelands. Roadsides.
Material Examined : EBH-HVP-814, 12.12.2004.
Parts Used : Stem. Fruit.
Folk Uses : Stem used for **scouring teeth***. Fruits **edible*.**
Active Constituents : Indole and murragin.
Uses in Literature : Used so far for agricultural tools, cough, cuts, cosmetics, diarrhoea, dysentery, hysteria, handles, renal pains, rheumatism; and as carminative (**Ambasta, 1986; Kirtikar & Basu, 1935**).

132. *Musa sapientum* Linn. (Pl. 43D)

Syn. : *Musa paradisiaca* Linn.
Family : Musaceae.
Vern. : Kela.
English, Sanskrit and Regional Names:

Eng. : Adam's Fig, Banana, Fig of India, Plantain;
Sans. : Ambusara, Balakapriya, Bhamphala, Mocha, Mous, Nisara, Rambha, Shajratulmouz Sukumara, Tulhtula;
B. : Kela, Khela, Mus;
G. : Kel, Maoz, Mouz;
H. : Amrit, Kach-kula, Kala;
Kan. : Pagnan, Pizang, Vasha, Vazhap, Vellacoi;
Mal. : Ngetpyau, Yakhaing, Ya-thi-lan;
Mar. : Kela;
P. : Kewiro;
Tam. : Anati, Amti, Ariti, Kommuariti, Kadali;
Tel. : Bale, Balenaru.

Distribution : Cultivated throughout India and tropics.
Description : Stem stoloniferous, tall. Leaf stalk stout. Spike drooping; bract ovate, many fid, deciduous. Fruits oblong, trigonous, sweet, tapering at base and apex, yellowish when green.
Flowering and Fruiting : Throughout year.
Habitat Ecology : Planted.
Material Examined : EBH-HVP-808, 25.08.2004.
Parts Used : Fruit. Stem. Leaves.
Folk Uses : Plant considered **sacred*** and used in various religious and ceremonial occasions like **marriage**, **'grih pravesh' (house warming ceremony)** and **death rituals**. Fruits **offered** to appease Lord Satyanaryana; and also **edible.** One banana fruit taken daily to check **dysentery**. Paste of two banana fruits along with sugar and 'elachi' (*Elettaria cardamomum*

Maton) given to cure burning due to **acidity.** Unripe fruits cooked as **vegetable***.

Active Constituents : Principle constituents are 9-19-cyclotriterpene ketone, a teriterpenoid-24 (R)-4α, 14 α, 24 trimethyl 5 α cholesta-8, 25 (27)-dien 3 β-ol, dopamine, dopa, noradrenalin, serotonin and caffein, cinnamic, p-coumaric, ferullic, gallic and protocatechuic acid, compesterol, β-sitosterol, stigasterol, cyclomusalenol, cyclomusalenone, steroids, glycosides, terpenes and organic acid.

Uses in Literature : Used for boils, black spots, burn injuries, cholera, cold, cough, diarrhoea, edible, gastric disorder, jams and jellies, heart pains, headache, hypertension, intestinal disorders, irregular menstrual cycle, leucorrhoea, pooja prasad, pyorrhoea, kidney and bladder stone, snake bites, rheumatism; and as laxative **(Ambasta, 1986; Bhatt *et al.*, 2001; Baruah & Sarma, 1987; Khanna, 2002; Mohanty *et al.*, 1998; Mukherjee *et al.*, 1986; Pandey & Verma, 2002; Rastogi & Mehrotra 1993, 1995a, b; Samati, 2004; Sen *et al.*, 2002; Sen & Behera, 2003; Sharma *et al.*, 2002; Watt, 1972).**

133. *Nerium indicum* Mill. (Pl. 44A)

Syn. : *N. odorum* Soland., *N. odoratum* Lamk., *N. latifolium* Mill.

Family : Apocynaceae.

Vern. : Kaner.

English, Sanskrit and Regional Names:

- **Eng.** : India Oleandre, Sweet Scented Oleander;
- **Sans.** : Asvamaraka, Pratihasa, Vishavrykshanka;
- **Ass.** : Karavi, Karbira, Karvir;
- **B.** : Karabi;
- **Bo.** : Ganira, Kanhera;
- **G.** : Kanera;
- **H.** : Kaner, Karber, Kuruvira;
- **Kan.** : Kanagalu;
- **Kash.** : Gandeela;
- **M.P.** : Arali, Kanave eram;
- **Mal.** : Chuvannarali;
- **Mar.** : Kaneri;
- **Oriya** : Konera, Koniyoro;
- **P.** : Ganhira, Kanhira;
- **Tam.** : Asuvabari;
- **Tel.** : Ettaganneru.

Distribution : N.W. Himalaya at 600-1,500 m. C. & N. India.

Description : Erect glabrous shrubs. Leaves long, pointed, leathery. Flowers pink, fragrant. Fruit paired, cylindric.

Flowering and Fruiting : April-November.

Habitat Ecology : Along Roadsides. Wasteplaces.

Material Examined : EBH-HVP-720, 15.07.2002.
Parts Used : Leaves. Flowers.
Folk Uses : A pinch of dried leaves snuffed to cure **headache** and **bad cold***. Flowers **offered*** to appease deities.
Active Constituents : Leaves possess quercetin, isoquercetin, nerimol, nerifol, oleandrin, ursolic acid, adynerin, a compound similar to rutin, B-D-digitaloside and digitoxigenin.
Biological Activity : Alc. extr. of root, bark, leaf and flowers exhibits antibacterial activity. Oleandrin and oleandrigenin-B-D-diginpyranoside showed cardiopkinetic and diuretic activities.
Uses in Literature : Used so far for blisters, eczema, headache, ringworms, skin diseases, snake bite; and as cardiac tonic, **(Katewa *et al.*, 2001; Kumar & Chauhan, 2005; Verma & Chauhan, 2006).**
Major Ayurvedic Formulations : Karaviradhya-taila, Kasi-sadi-taila, Laghu-visgarbha-taila.

134. *Ocimum basilicum* Linn. (Pl. 44B)

Syn. : *O. pilosum* Willd., *O. album* Linn., *O. minimum* Burm., *O. hispidum* Lamk., *O. menthaefolium* Benth., *O. caryophyllatum* Roxb., *Plectranthus barrelieri* Spreng.
Family : Lamiaceae.
Vern. : Bhabhri.
English, Sanskrit and Regional Names:

Eng. : Basil, Common Basil, Sweet Basil;
Sans. : Ajagendhika, arjakan, Asuvasa, Barbara, Karahi, Surabhi, Tungi, Varvara;
B. : Babui-tulsi, Debunsha;
G. : Damaro, Nasabo;
H. : Bahari, Kalitulsi;
Kan. : Kam-Kasturi, Sajjebiya;
Mal. : Marva, Sabja;
Oriya : Dhala, Zulasi;
P. : Baburi, Rehan;
Tam. : Tirnupatchie, Tirunitru;
Tel. : Rudrajada,Vipudi-patri.

Distribution : Throughout India.
Description : An erect softly hairy strongly-scented annual undershrub. Leaves elliptic-ovate or lanceolate, entire to serrulate. Flowers small, pinkish-white. Nutlets ellipsoid, pitted.
Flowering and Fruiting : June-September.
Habitat Ecology : Commonly cultivated, rarely as an escape in moist slopes.
Material Examined : EBH-HVP-791, 21.09.2003.
Part Used : Leaves.
Folk Uses : Green leaves used for **culinary purposes.** Also 20 ml. of its decoction taken twice a day to cure **gastric** problems and **vomiting***.

Active Constituents : Essential oil from leaves contains methyl cinnamate, 1-linalool, terpinene, macinnamate, methyl chavicol, linalool, cineole, ocimene, borneol, sambulene and safrole.

Biological Activity : Alc. extr. of leaves and flowers exhibited activity against *Micrococcus pyrogenes* var. *aureus* whereas essential oil antibacterial and antifungal.

Uses in Literature : Described earlier to be used as an anthelmintic, antipyretic, carminative, demulcent, diaphoretic, expectorant, nasal douche, stimulant and stomachic; and for alc. intoxication, bowl complaints, cholera, cold, cough, cramp, cosmetics, culinary purposes, diarrhoea, dropsy, dysentery, earache, epilepsy, fever, flavouring, headache, labour complaints, mouth ulcer, pus in ears, ring worms, sinuses, sores, sore throat, snake bite, toothache and wounds **(Ambasta, 1986; Jadhav, 2006; Jain, 1991; Pandey & Verma, 2002; Silori & Rana, 2000).**

Adulterant or Substitute : Seeds used as an adulterant to *Ficus hederacea* Linn. and *Ipomoea nil* (Linn.) Roth.

135. *Ocimum sanctum* Linn. (Pl. 44 C)

Syn. : *O. monachorium* Linn. Mant., *O. tenuiflorum* Linn., *O. inodorum* Burm.

Family : Lamiaceae.

Vern. : Baramasi Tulsi, Ram Tulsi.

English, Sanskrit and Regional Names:

Eng. : Holy Basil, Monk's Basil, Rough Basil, Sacred basil;
Sans. : Ajaka, Arjaka, Amrita, Bharati, Gauri, Laxmi, Madhavi, Punya, Surabhi, Tulasi;
B. : Bantulsi, Tulsi;
Bo. : Tulasa, Tulas-icha-jhada;
G. : Tulsi;
H. : Hural, Kal-tulsi, Tulai, Tulshi;
Kan. : Krishna-tulsi, Nallu-tirtta, Niella-tirtua;
Mal. : Lum;
Mar. : Talasi;
P. : Tulas, Tulasa;
Tam. : Gaggera-chettu, Krishna-tulasi, Tulasi;
Tel. : Tulashi-gida.

Distribution : Throughout India.

Description : Erect, branched, aromatic hairy perennial herbs. Leaves elliptic-oblanceolate, entire, pubescent on both sides. Flowers greenish, pale yellow and in terminal racemes. Nutlets broad-ellipsoid, smooth, yellow with small black marking.

Flowering and Fruiting : July-November.

Habitat Ecology : Cultivated.

Material Examined : EBH-HVP-733, 11.11.2001.

Parts Used : Leaves. Whole Plant.

Folk Uses : Plant considered **sacred*** and worshipped in 'Panch-Bhikham'. Newly stitched clothes are kept overnight covering the plant before being given to children for wearing as it is believed to kill all germs and considered **antiseptic***. Leaves form one of the constituents of **holy water** and believed to possess **curative*** properties. Leaf juice used as nasal drops for curing **cold**, and also as an **expectorant.** Poulitice of leaves applied for suppuration of **boils*.** Decoction of inflorescence good for **cough**. Mixture of 3 g. infusion of leaves along with 5 g. honey and 3 g. ginger (*Zingiber officinale* Rosc.) prescribed twice a day for **asthma, cough** and **fever**. Chewing of leaves good to arrest **foul smell** of mouth.

Active Constituents : Essential oil contains carracrol, caryophyllene, nerol and camphene, camphor, b-caryophyllene, caryophyllene, camphene, eugenol, methyleclavicol, a pinene, volatile oil and ascorbic acid.

Biological Activity : Hypoglycaemic and spasmolytic activities confirmed.

Uses in Literature : Known to be used as antibacterial, antiseptic, contraceptive, expectorant, insecticidal, stomachic; and for black patches, boils, bronchitis, burning of feet, asthma, conjuctivitis, common cold, cough, cuts, dandruff, earache, eczema, enhancing effect of wine, fertility, flavouring, foul breath, fever, gastric trouble, headache, hepatitis, jaundice, night blindness, perfumery, rat bite, respiratory troubles, ringworm, scabies, skin diseases, spermatorrhoea, stomachache, sterilization, tuberculosis, ulcers, urinary troubles **(Ambasta, 1986; Borthakur *et al.*, 2004; Gogoi & Borthakur, 2001; Jadhav, 2006; Maya *et al.*, 2003; Kshirsagar & Singh, 2000; Kumar & Chauhan, 2006; Sharma *et al.*, 2002; Sen & Batra, 1997; Sen & Behera, 2003; Sen *et al.*, 2002; Sharma & Singh, 2001).**

Major Ayurvedic Formulations : Tulsiyadi-Kashaya, Honey Tus.

136. *Opuntia dillenii* Haw. (Pl. 44D)

Syn. : *O. stricta* (Haw.) Haw. var. *dillenii*, *Cactus indicus* Roxb.

Family : Cactaceae.

Vern. : Chu, Thor.

English, Sanskrit and Regional Names:

Eng. : Pricklypear, Slipper Thorn;
Sans. : Bhaudagdhika, Kandashaka, Nagaphana, Vidara;
B. : Nagphana, Phenimana, Nugphana;
G. : Chorhathola, Zhorhatheylo;
H. : Hathathoria, Nagphani, Nag-phana;
Kan. : Papaskalli;

Mal. : Nagmullu, Palakakkalli, Poturiyabiju;
Mar. : Chapal, Nagphana, Samar;
Oriya : Nagophenia, Poturiyabiju;
P. : Chittarthohar;
Tam. : Manjarnagadali, Pattamadugalli;
Tel. : Nagajemudu, Naga-dali, Nagamullu.

Distribution : Planted and naturalised.

Description : An evergreen succulent shrub. Stem jointed, densely elongate-bristly, areoles raised, glochidiate. Spines 5-7 per areole, yellow. Leaves deciduous. Flowers bright yellow. Berries pyriform, truncate, depressed apex, purple when ripe. Seeds many.

Flowering and Fruiting : April-May.

Habitat Ecology : Wasteplaces. Dry slopes. Edges of fields.

Material Examined : EBH-HVP-748, 22.05.2002.

Parts Used : Fruits. Plant.

Folk Uses : Fruits **edible** and good for **heart***. Also it is considered a good **hedge plant.** 1-2 g. powdered burnt plant taken with honey for **whooping cough.**

Active Constituents : These are glucosides of isorhamnetin, quercetin, isoquercetin, traces of free flavanols, arabinogalacton and betanin (plant).

Biological Activity : Ether extr. of stem exhibits some antibiotic properties whereas fruit exhibits CNS activity.

Uses in Literature : Known to be used in India as an antipyretic, carminative, diuretic, edible (fruits), expectorant, laxative, refrigerant, stomachic, symbolic in religious; and for asthma, boils, excessive menstruation, eye trouble, fencing, hedge, guinea-worm, leucoderma, ceremonies of 'Santhals' (spines), expelling worms, piles, pimples, swelling, urinary complaints, and whooping cough **(Ambasta, 1986; Chopra *et al.*, 1956; Jain, 1991; Jain & Tarafder, 1970, Kurian, 1995; Nautiyal, 1981; Parmar & Kaushal, 1982; Singh, 1996, Watt, 1972).**

137. *Oroxylum indicum* Vent. (Pl. 45A)

Syn. : *Calosanthes indica* Blume, *Bignonia indica* Linn., *B. pentandra* Lour., *Spathodea indica* Pers.

Family : Bignoniaceae.

Vern. : Tat-palanga.

English, Sanskrit and Regional Names:

Eng. : Indian Trumpet Flower;
Sans. : Adhvantashatrava, Syonakah, Svarnavalkana;
Ass. : Bhatgila, Dingari, Kering, Toguna;
B. : Nasona;
Bo. : Phalphura, Tetu;
G. : Podval;
H. : Assatsauna, Sauna, Sonapatha;
Kan. : Bunepale, Sonepatta, Tigdu;
Mal. : Palakapayani;

	Mar.	: Taitu;
	Oriya	: Phompania;
	Tam.	: Paluiyudaichi, Vangam;
	Tel.	: Dundilamu, Mundakaparnamu.

Distribution : Throughout India ascending up to 1,350 m.

Description : A small tree having very large, tripinnate, opposite leaves. Flowers numerous, fetid, in large terminal raceme, purple. Capsule straight, flat, acute, valves semi-woody. Seeds numerous, winged.

Flowering and Fruiting : June-August.

Habitat Ecology : Forest. Shrubberies.

Material Examined : EBH-HVP-836, 17.06.2004.

Parts Used : Bark. Leaves. Fruits.

Folk Uses : Poultice of bark applied locally for **healing fracture**. For healing **internal injuries***, powdered bark mixed with wheat dough fed to cattle. Also 15 ml. of its decoction along with that of fruit, twice a day, good to cure **constipation** both in humans and livestock. Leaves used as **fodder***.

Active Constituents : Therapeutic properties attributed to baicalein, its 6-glucuronide and 7-glucuronide, prunetin, scuttellarein, oroxylin A, chrysin and scuttellarein-7-rutinoside (leaves, stem bark); glucuronide-oroxidin, synthesis of biacatein-6-glucuronide and 7 glucuronide (seeds).

Biological Activity : Spasmolytic (fruit) and diuretic (bark) activities +ve.

Uses in Literature : Known so far as anodyne, antidote to snake bite, antiseptic, appetizer, astringent, black dye, carminative, hepatic stimulant, ingredient of 'Chyawanaprasha', lactation, psychoactive plant, purgative (seed), stomachic, tonic, vermic, vegetable (corolla young shoots, unripe fruits); and for affection of brain, blood dysentery, body pain, bone fracture (vet.), dysentery, cracks in nipples, dysentery (vet.), diseases of chest, indigestion, joint pains, headache, liver ailments, oedema, pain, piles, pimples, rutting (vet.), scabies, scorpion sting, small pox, sore back of herds and bullocks, sores of nails and soles (vet.), tanning, tuberculosis, ulcers, urticaria, vitality and wounds (vet.) (**Ambasta, 1986; Borthakur, 1996b; Borthakur & Sarma, 1996; Brahmam & Saxena, 1990; Brahmam *et al.*, 1996; Chhetri, 2005; Chopra *et al.*, 1956; Chowdhery, 1996; Girach, 1992; Girach & Aminuddin, 1995; Jain, 1991; Jain *et al.*, 1994; Kothari & Moorthy, 1996; Mehrotra *et al.*, 1995; Mohanty *et al.*, 1998; Satapathy & Brahmam, 1996; Sen & Behera, 2003; Siwakoti & Verma, 1996; Srivastava *et al.*, 2003; Upadhye *et al.*, 1994; Vishwanathan & Singh, 1996; Watt, 1972**).

Major Ayurvedic Formulations : Amritarishta, Brihatpancamulayadi-kwatha, Chyawanaprasha, Dashmool-ghrita, Dashmul-taila, Mahanarayan-taila, Pushayanug-churna, Shyonaaka-putpale.

138. *Papaver rhoeus* L. (Pl. 45B)

Syn. : *Papaver atropurpureum* Gilib., *Papaver intermedium* Becker.
Family : Papaveraceae.
Vern. : Gulla.
English, Sanskrit and Regional Names:
Eng. : Common poppy, Common Red poppy, Corn poppy, Flanders poppy, Red poppy, Shirley poppy;
Sans. : Rhakta-posta-Vrikshah;
H. : Lala, Lal-Post.
Distribution : Throughout India.
Description : An erect, branched herb with alternate, petiolate leaves which are hispid on both surfaces. Flowers red.
Flowering and Fruiting : May-October.
Habitat Ecology : Cultivated and escaped to fields. Roadsides. Waste grounds.
Material Examined : EBH-HVP-752, 10.10.2002.
Parts Used: Flower. Petals. Leaves. Plant.
Folk Uses : Planted as an **ornamental***. Flowers **offered*** to deities. Extr. of petal and leaves applied externally on **wasp stings***.
Active Constituents : α-tocopherol, mecocyanin chloride, cyanidin derivative (petals), cyanidin-B and pelargonidin-C (flowers); rhoeadine (leaves, flowers, capsules); morphine, thebaine; narcotine, meconic acid (capsules); protopine and coptisine, phenolic and nonphenolic, crystalline and amorphous bases (roots, aerial parts) have been characterised.
Biological Activity : Antioxidant activity and antiradical power have been found to be positive.
Uses in Literature : Used so far as antibiotic, expectorant, sedative; and for cough or hoarseness and colouring medicines and food materials **(Watt, 1972).**
Trade : Anti-Wrinkle Cream.

139. *Parthenium hysterophorus* Linn. (Pl. 45C)

Family : Asteraceae.
Vern. : Congress grass.
English Name : Congress grass.
Distribution : Common weed throughout India.
Description : Profusely branched erect herbs. Leaves dissected. Flowers numerous, white. Achenes compressed.
Flowering and Fruiting : August-September.
Habitat Ecology : Wastelands.
Material Examined : EBH-HVP-759, 15.08.2002.
Parts Used : Root. Plant.
Folk Uses : Poultice of root prepared in ghee recommended for **slipped neck***. Plant used as **fuel***.

Active Constituents : These are bornmylacetate, coronopilin, dihydroisopar-thenin, hysterin, parthenin, α methylene γ lactone sesquiterpene, hexacosanol, myricyl alcohol, galactose, glucose, 6 hydroxy-kaempferol-3-7-dimethyl ether, tetraneurins A, B, C, D, kaempferol, betulin, ursolic acid and a saponin composed of oleanolic acid and glucose, campesterol, β-sitosterol and stigmasterol (plant).

Uses in Literature : Used as an analgesic, emmenagogue, febrifuge and tonic; and for dysentery and nasal blockage **(Ambasta, 1986; Bennet, 1985).**

140. *Pedilanthes tithymaloides* (L.) Poit. (Pl. 45D)

Family : Euphorbiaceae.

Vern. : Nag Bel.

English and Regional Names:

- **Eng.** : Devil backbone, Ribbon cactus, Red bird, Slipper Plant;
- **B.** : Bilati;
- **Bo.** : Vilayti-sher;
- **H.** : Nagdhamni;
- **M.P.** : Nag-phani, Nagdaman.

Distribution : Afghanistan to Kashmir. N. & W. Asia.

Description : Branched succulent bush with zig-zag fleshy stems. Leaves large, ovate, fleshy, green. Flowers red.

Flowering and Fruiting : August-September.

Habitat Ecology : Grown in Indian gardens.

Material Examined : EBH-HVP-795, 2.09.2003.

Part Used : Whole Plant.

Folk Uses : Planted as an **ornamental*** with the belief that it keeps away snakes from the house.

Active Constituents : Azafrin (root).

Uses in Literature : Used so far as emetic; and for boils, latex, leucoderma patches, pimples, warts, venereal diseases **(Ambasta, 1986; Katewa & Sharma, 1998).**

141. *Pennisetum purpureum* Schum. (Pl. 46A)

Family : Poaceae.

Vern. : Chari.

English Names : Napier grass, Elephant grass.

Distribution : Throughout India.

Description : A robust, deep rooted perennial grass with creeping rhizome. Leaves dull green. Spikes dense, erect, long and yellow.

Flowering and Fruiting : September-November.

Habitat Ecology : Cultivated.
Material Examined : EBH-HVP-761, 18.10.2002.
Part Used : Whole Plant.
Folk Uses : Plant considered good **fodder** for livestock.
Uses in Literature : Used so far as fodder, fence; and also for walls of huts, raw material for paper **(Duthie, 1888).**

142. *Phoenix sylvestris* Roxb. (Pl. 46B)

Syn. : *P. acaulis* (L.) Buch. Ham. *ex* Roxb., *Elate sylvestris* Linn., *Katou indel* Ham.
Family : Palmaceae.
Vern. : Khajoor.
English, Sanskrit and Regional Names:

Eng. : Date Sugar Palm, Dwarf Date Palm, Indian Wine Palm, Sugar Palm, Wild Date Palm;
Sans. : Bhumikhajurika, Haluka, Kharju, Kharjura, Kakakarkati, Skandhappala;
B. : Kejur, Kajar, Khjura;
Bo. : Kajar, Khajuri, Khjura, Sendi;
G. : Khjuri, Kharak;
H. : Jangli Khajur, Kejur, Khaji, Salma, Thakil;
Kan. : Ichalu;
Mal. : Inta, Intappana, Kattinta;
Mar. : Boichand, Sendri, Sindikujuri;
Oriya : Khorjura, Khorjuri, Khorjuro;
P. : Khaji, Khajur;
Tam. : Ichambanai, Periyayinju, Madal;
Tel. : Ita, Peddayita.

Distribution : Throughout India.
Description : Trees with stout trunk covered with persistent leaf bases. Leaves long, greyish green, glabrous, pinnate. Flower cluster inclined with spreading branches. Fruits spadix compressed, oblong-ellipsoid, golden orange coloured. Seeds deeply grooved along its whole length.
Flowering and Fruiting : May-June.
Habitat Ecology : Open hill slopes. Rocky slopes.
Material Examined : EBH-HVP-895, 11.06.2005.
Parts Used : Leaves. Fruits.
Folk Uses : Ripe fruits **edible.** Leaves employed for making **brooms, mats, thatching** and **baskets.**
Active Constituents : Active properties due to protein 0.37%, total sugar 11.01%, reducing sugar 0.97%, mineral matter 0.54%, phosphorus 0.10%, calcium trace, iron 0.04%, nicotonic acid, iso nicotonic and hydrazide, organic acid (fresh unfermented sap), amino acids and vitamins.
Biological Activity : Leaves CVS and CNS active.
Uses in Literature : Useful as cardiotonic, constipative, diuretic, refrigerant, beverages, diuretic, edible (buds & fruit); and for anaemia,

asthma, bakery, brooms, bronchitis, cold, diabetes, diarrhoea, dysentery, fever, gum trouble, heart complaints, leprosy, basketry, mats and bags, rope, thatching, toothache, venereal diseases and vomiting (**Ambasta, 1986; Goud & Pullaiah, 1996a; Jain, 1991; Jain *et al.*, 1994; Katewa *et al.*, 2001; Kirtikar & Basu, 1935; Kshirsagar *et al.*, 2003; Mohanty *et al.*, 1998; Patole & Jain, 2002; Ragupathy & Mahadevan, 1991; Rama Rao & Henry, 1996; Ravishankar & Henry, 1992; Sahu, 1996; Singh, 1996; Srivastava *et al.*, 2003; Vishwanathan, 1997; Watt, 1972**).

143. *Phyllanthus urinaria* Linn. (Pl. 46C)

Syn. : *P. leprocarpus* Wight, *P. alatus* Blume, *P. cantoniensis* Hornem., *P. mucronatus* Herb., *P. racemosus* Herb., *P. muricatus* Herb., *P. polyphyllus* Herb., *P. echinatus* Herb.

Family : Euphorbiaceae.

Vern. : Bhumi-Amla.

Sanskrit and Regional Names:

- **Sans.** : Adhyanda, Ajata, Buhuphala, Jhatamala,;
- **B.** : Hazarmani;
- **G.** : Kharsadabonyaanmali;
- **H.** : Hazarmani, Lalbhuinanvalah;
- **Mal.** : Chirukizhukanelli, Chukannakizhanelli;
- **Mar.** : Lalmundajanvalli;
- **Tam.** : Shivappunelli;
- **Tel.** : Ettausirika.

Distribution : Throughout India.

Description : Herbaceous annuals. Leaves oblong, membranous. Flowers yellow. Fruits echinate.

Flowering and Fruiting : August-September.

Habitat Ecology : Moist stony slopes.

Material Examined : EBH-HVP-821, 22.09.2004.

Parts Used : Seeds.

Folk Uses : ½ teaspoon of powdered seeds given once daily for 7 days to cure **stomach disorders.**

Active Constituents : Plant contains astragalin, ellagic, tannin, quercetin, quercitrin, isoquercitrin and rutin (plant).

Uses in Literature : Reported to be described as an appetiser, astringent, diuretic, febrifuge and tonic; and for diarrhoea, dysentery, dyspepsia, gonorrhoea, jaundice, stomach trouble, sleeplessness in children and urinogenital diseases in children (**Ambasta, 1986; Chopra *et al.*, 1956; Jain, 1991; Maya *et al.*, 2003; Sharma, 2003, Uphof, 1968; Watt, 1972**).

Major Ayurvedic Formulations : Chyawanaprasha, Tonali tablets, Livomap tablets and syrup.

Trade : Plant traded as Bhumiamla.

144. *Pinus roxburghii* Sarg. (Pl. 46D)

Syn. : *P. longifolia* Roxb. *ex* Lambert.
Family : Pinaceae.
Vern. : Chil.
English, Sanskrit and Regional Names:
Eng. : Himalayan Long Leaved Pine, Three Leaved Pine;
Sans. : Bhadradaru, Drava, Kshire, Putikashtha, Srivasa;
Ass. : Telia;
B. : Sarala Kashtha;
G. : Sarala Deodara;
H. : Chir, Dhupasarala, Salla, Saral;
Kash. : Salla;
Mal. : Charalam, Putikashtam;
Mar. : Saraladedara;
Oriya : Sorolokatho;
P. : Anandher, Chir, Drabchi, Nashtar, Saral;
Tam. : Simaidevadari, Suruldevadari;
Tel. : Devadaru, Saralu.
Distribution : Afghanistan, Bhutan usually 1,000-2,000 m.
Description : A large gregarious tree with whorled branches and long needle-like leaves. Bark brown, corky furrowed. Seeds winged.
Flowering and Fruiting : March-May.
Habitat Ecology : Forming extensive forest. Roadsides.
Material Examined : EBH-HVP-745, 12.04.2002.
Parts Used : Needles. Wood. Female cone.
Folk Uses : Wood employed for cheap **furniture** and **building** works. Needles with red soil employed for **plastering*** walls of mud houses. Needles also used to make **brooms***. Burning female cones used as **torch light**.
Active Constituents : Various constituents are stilbene, pinosylvin, pinociembrin and pinobanksin (heart wood), vit. C, E and carotene α and β pinene (leaves), Δ 3 carone (turpentine oil), pine oil (needle), β pinene and Δ 3 carone (plants).
Biological Activity : Oil from needles antibacterial against a number of organisms including *Staphyllococcus aureus*.
Uses in Literature : Known for boils, bone fracture, chronic bronchitis, constipation, cracks in soles of feet, edible, skin diseases, snake bite, sprain, swelling, ulcer and urinary complaints, construction purposes, tanning, tea chest, torches, turpenting, typhoid and ulcers; and also as an expectorant, flatulent colic, stimulant (**Ambasta, 1986; Chauhan, 1999; Jain, 1968, 1991; Jain *et al.*, 1994; Sensarma, 1991; Srivastava *et al.*, 2003; Watt, 1972**).
Major Ayurvedic Formulations : Chandanaadi-vati, Rosin, Turpentine.
Trade : Birozaa, Chil, Chir.

145. *Pistacia integerrima* Stewart *ex* Brand. (Pl. 47A)

Syn. : *P. khinjuk* Stew. *ex* Brandis, *P. chinensis* Bunge, *P. chinensis* var. *integerrima* (Stewart) Zohary, *Rhus integerima* Wall., *R. kakrasingee* Royle.

Family : Anacardiaceae.

Vern. : Kakad-singi, Kakarsingi.

English, Sanskrit and Regional Names:

- **Eng.** : Crab's Claw;
- **Sans.** : Chakra, Chakrangi, Chandrasapada, Ghosha, Karkali, Kasavinashini, Mahaghosa, Navanga, Vakra;
- **Ass.** : Kakaisrnghi;
- **B.** : Kakra, Kakrsingi;
- **G.** : Kakra;
- **H.** : Kakra, Kakra singi;
- **Kan.** : Dushtapuchattu, Kakadsasingi, Karkatasrini;
- **Mal.** : Gurgu, Kakkar;
- **Mar.** : Kakra, Kakrasingi;
- **Oriya** : Kakadashringi, Kakadashrungi;
- **P.** : Kakkar, Kakkeram, Kaklashne, Tungu;
- **Tam.** : Karkata-shingi (galls);
- **Tel.** : Kakar-shingi, Karkatashrungi.

Distribution : W. Himalaya. W. Nepal: 600-2,400 m.

Description : Medium-sized glabrous trees. Leaves pinnate. Flowers small, red coloured, in axillary racemes. Drupes globose, glabrous, grey when ripe and wrinkled.

Flowering and Fruiting : March-May.

Habitat Ecology : Open slopes. Grassland. Along Roadsides.

Material Examined : EBH-HVP-777, 12.05.2003.

Parts Used : Leaves. Bacterial galls.

Folk Uses : Burnt bacterial galls mixed with honey good as **expectorant** and for **asthma**. Leaves chewed to cure **toothache*** and also as **fodder.**

Active Constituents : Therapeutic properties attributed to α-pinene, camphene, dl-limonene, cineole, α-terpenol, aromadendrene, β-sitosterol, gummastic, pistacin and pistacinin and two riterpene acids, pistacienonic acid A and B (galls), (drupes).

Biological Activity : Gall oil antibacterial, antiprotozoal and antispasmodic.

Uses in Literature : Known so far as an antidote to scorpion sting, snake venom, expectorant, fodder and tonic; and for dysentery, phthisis, wood for construction and furniture works, ingredient of 'Chyawanaprasha', asthma, cough, fever, furniture, ploughs, respiratory trouble, roping, spinning wheel, tanning, turnery and want of appetite **(Ambasta, 1986; Chauhan, 1999; Gupta, 1981; Jain, 1991; Kapur & Nanda, 1996; Mehrotra *et al.*, 1995; Negi *et al.*, 1985; Watt, 1972).**

Major Ayurvedic Formulations : Ingredient of 'Chyawanaprasha', Sringynadi-churna, Dashmularista Kantakari, Kantakari, Avaleha, Nokufcough, Kasni.

Trade : Kakarsinghi.

146. *Plumbago zeylanica* Linn. (Pl. 47B)

Syn. : *P. auriculata* Blume, *Thela alba* Lour.
Family : Plumbaginaceae.
Vern. : Chitra, Chitrak.
English, Sanskrit and Regional Names:

Eng.	: Ceylon Lead Wort, White Flowered Lead Wort;
Sans.	: Agni, Agnisikha, Analanania, Bhali, Chitraka, Jataveda, Vaishavan-Ara;
B.	: Chita, Chitruk;
Bo.	: Chitra, Chitrack;
G.	: Chitaro, Gholochitro;
H.	: Chita, Chiti, Chitarak, Chitra;
Kan.	: Chitramula, Vahni;
Mal.	: Tumpokotuveli;
Mar.	: Chitraka, Chitramula;
Oriya	: Chitramula, Krishanu;
P.	: Chitrak;
Tam.	: Adigarradi, Tigana, Vellaikkodiveli;
Tel.	: Agnimata, Chitramulamu, Tellachitramulamu.

Distribution : Throughout India ascending to 1,650 m.

Description : A rambling sub-scandent perennial herb. Leaves elliptic ovate, truncate to attenuate, apex apiculate. Flowers white, in long spikes having glandular rachis. Capsule glandular hairy, globose. Seeds solitary.

Flowering and Fruiting : April-November.

Habitat Ecology : Moist slopes. Wastelands. Roadsides.

Material Examined : EBH-HVP-801, 12.11.2004.

Parts Used : Roots. Stem. Leaves.

Folk Uses : Stem used for **scouring teeth***. Extr. of leaves and root mixed with mustard oil applied for relieving **rheumatic pain.**

Active Constituents : Constituents characterised are plumbagin, 3-choloro-plumbagin, droserone, 3, 31-bi plumbagin, zeylanone, iso-zeylanone and ellipitinone, new bi napthaquinone-chitranone with zeylinone, elliptinone drosene and plumbagic acid (roots), isoschinandone, sitosterol, plumbazeylone, a new dihydronaptho-quinone-dihydroserone-tyrosine (4.5), threonine (4.1), alanine (3.5), histidine (3.4), glycine (1.7), methionine (0.5), hydroxyproline (0.02%) (aerial parts).

Biological Activity : Root extr. antibacterial and antimicrobial. Plumbagin (0.75 mg./kg.) active against P 338 lymphocytic leukaemia and at 4 mg./kg. exhibits antibacterial and antifungal activity against wide variety of bacteria and fungi. Bonded plumbagin induces abortifacient, anticoagulant anti-implantation, antiovulatory activities.

Uses in Literature : Known so far as an abortifacient, antiperiodic, appetiser, diuretic, expectorant, vesicant; and for abdominal pain, anti-fertility, aphthae, bodyache, cancer, diarrhoea, dysentery, dyspepsia, eczema, fever, headache, hydrocele, indigestion, leprosy, leucoderma, malaria, muscular pain, paralysis, piles, rheumatism, skin diseases, spleen complaints, stomach disorder, wounds and skin diseases in animals, body pains (roots worn on arm) **(Ambasta, 1986; Balasubramaniam & Prasad, 1996; Bhandary *et al.*, 1996; Borthakur, 1993; Brahmam & Saxena, 1990; Brahmam *et al.*, 1996; Goud & Pullaiah, 1996b; Hembrom, 1996; Hembrom & Goel, 2005; Jain, 1991; Jain & Patole, 2001; Jain & Sharma, 1996; Jain *et al.*, 1994; Jha & Varma, 1996; Manadhar, 1987; Ranjan, 1996; Reddy *et al.*, 1996; Sharma & Borthakur, 2003; Srivastava *et al.*, 2003; Tirkey, 2004; Varma, 1997; Venkataratnam & Venkata Raju, 2004; Verma & Chauhan, 2006; Watt, 1972).**

Major Ayurvedic Formulations : Agnitundi-vati, Chandraprabha-vati, Chitraka, Chitraka-ghrita, Devaradi-kwatha, Mahashankh-vati, Mahamrityunjay-rasa, Navayas-lauha, Panchkola-churna, Sarvajwar-lauha, Sudarshan-churna, Trimalla-churna, Yograj-gugglu.

Trade : Chitrak (roots).

147. *Pogostemon benghalensis* Burm. f (Pl. 47C)

Syn. : *P. plectranthoides* Desf., *Origanum benghalensis* Burm.f., *O. indicum* Roth., *Mentha secunda* Roxb., *M. fruriculosa* Roxb.

Family : Lamiaceae.

Vern. : Dusyan, Kali Basuti.

Regional Names:

- **B.** : Bakoha, Jin, Jui-Lata;
- **Bo.** : Pangla;
- **Gar.** : Lujra;
- **Oriya** : Dumobadottoka, Poksunga;
- **Tel.** : Gondripula, Kasurijang.

Distribution : H.P., E. Nepal up to 1,300 m, W. Himalaya.

Description : Pubescent shrubs. Leaves membranous, irregularly toothed. Flowers whitish purple, crowded in spikes. Nutlets smooth.

Flowering and Fruiting : February-April.

Habitat Ecology : Wastelands. Roadside.

Material Examined : EBH-HVP-705, 11.04.2001.

Parts Used : Roots. Leaves.

Folk Uses : ½ teaspoon of root ash with honey prescribed for 15 days to cure **whooping cough***. Powdered leaves with onion (*Allium cepa* Linn.), 'ajwain' (*Trachyspermum ammi* Linn.) and 'gur' (1 teaspoon, twice daily) cure **stomachache***. Poultice of leaves applied on head to check **headache***.

Active Constituents : 0.07% essential oil (fresh leaves).

Biological Activity : Leaves show antifungal activity against *Helminthos-porium sativum*.

Uses in Literature : Known so far as an antihaemorrhagic, antidote to scorpion sting and snake bite edible (eaten in time of scarcity), febrifuge, green manure, styptic; and for body pains, boils, cuts, detergent, diarrhoea, diarrhoea (vet.), drink, wounds and wounds in animals (**Ambasta, 1986; Brahmam & Saxena, 1990; Chopra *et al.*, 1956; Datt, 1996; Jain, 1991; Kothari & Moorthy, 1996; Mokat & Deokule, 2004; Rama Rao & Henry, 1996; Siwakoti & Verma, 1996).**

148. *Polyalthia longifolia* (Sonn.) Thwaites (Pl. 47D)

Family : Annonaceae.
Vern. : Ashok.
English Names : False Ashoka, Indian Mast Tree, Green Champa.
Distribution : Throughout India.
Description : A tree with pendulous branches and long narrow lanceolate leaves having undulate margins. Flowers star-like pale green. Fruits in cluster of 10-20 and purple or black when ripe.
Flowering and Fruiting : February-April.
Habitat Ecology : Along roadsides.
Material Examined : EBH-HVP-827, 13.04.2004.
Parts Used : Plant. Leaves.
Folk Uses : Plant known for its **ornamental** value and festoons of leaves which are strung across doors along with that of mango on **ceremonial occasions**.
Active Constituents : Active constituents are (3S, 4R)-3, 4, 5-trihydroxypen-tanoic acid-1, 4-lactone (1).
Biological Activity : The ethanolic extr. of stem showed promising antibacterial activity against thirteen Gram-positive and nine Gram-negative organisms.
Uses in Literature : Used so far for barrels, packing cases, pencils, matches and for scaffolding and carriage shafts. Festoons of leaves are often used to make arches or are strung across doors during religious ceremonies, kidney diseases, strengthening of gums, toothache; and as anthelmintic, analgesic, anti-inflammatory, hepatoprotective, antihyperglycaemic and anticancer (**Ambasta, 1986**).

149. *Populus nigra* L. (Pl. 48A)

Syn. : *P. nigra* var. *italica* (Munchn) Koechne.
Family : Salicaceae.
Vern. : Poplar.

English and Regional Names:

Eng. : Black Poplar, Lombardy poplar;
Kash. : Frasht;
P. : Buins, Do, Frast, Kramadi, Pakshubut.

Distribution : N.W. Himalaya.

Description : A large-sized tree with upright branches forming a narrow cylindrical crown.

Flowering and Fruiting : April-July.

Habitat Ecology : Planted along roadsides.

Material Examined : EBH-HVP-749, 10.07.2002.

Parts Used : Wood. Tender shoots.

Folk Uses : Wood used in **building works**, **plywood* industries** and as **fuel***. Tender shoots used as **fodder.**

Active Constituents : Active constituents are salicin, salicortin, d, new phenol, glycoside-nigracin (bark), 2-methylpropyl, 2-methyl-2-propenyl 2-methylbuyl, 3-methylbutyl-, 2-methyl-2-butenyl-, 4-methylpeatyl, 5-methylhexyl and 3,7-dimethyl-2,6-octadienyl caffeates (bud exudates).

Uses in Literature : Reported so far as coacher, depurative, fodder; and for bone fracture, carving, cold, cross-beams, furniture, haemorrhoids, hair growth, matches, poles, small boxes, timber (**Ambasta, 1986; Jain, 1991; Singh *et al.*, 1996a; Watt, 1972**).

150. *Premna latifolia* Roxb. (Pl. 48B)

Syn. : *P. latifolia* var. *mucronata* C.B. Clarke, *P. mucronata* Roxb.

Family : Verbenaceae.

Vern. : Bankar.

Regional Names:

B. : Goharo;
H. : Agnium, Bakarcha, Jhatela;
Mal. : Knappa;
Oriya : Gondhona;
P. : Bankar, Ganhila;
Tam. : Erumaimunai, Munnai, Panchumullai;
Tel. : Koora, Peddanell kure.

Distribution : Temperate Himalaya up to 1,400 m.

Description : A small tree with subcoriaceous leaves and broad-ovate, greenish-white globose, dark purple flowers.

Flowering and Fruiting : July-September.

Habitat Ecology : Common on Roadsides.

Material Examined : EBH-HVP-757, 25.08.2002.

Part Used : Leaves.

Folk Uses : Juice of leaves rubbed as an **emollient** on affected parts. Leaves used as **fodder**.

Active Constituents : Properties attributed to β-sitosterol and its glucoside, 5-

hydroxy-4'-methoxyflavone-7-o-arabinopyranosyl-rhamnoside, two iridoside-7-deoxyloganic acid (bisdesoxydi-hydromonotropein) and geniposidic acid (leaves).

Uses in Literature : So far known as diuretic, firewood, fodder, vegetable (leaves and tender shoots); and for blisters, boils, charcoal, colic, colic (vet.), dropsy, headache, flavouring, gonorrhoea, indigestion, ring worms, syphilis and ulcers **(Ambasta, 1986; Khanna & Kumar 2000; Venkataratnam & Venkata Raju, 2004; Watt, 1972)**.

151. *Prunus amygdalus* Baill. (Pl. 48C)

Syn. : *P. communis* Fritsch, *Amygdalus communis* L., *P. dulcis* (Mill.) D.A. Webb.

Family : Rosaceae.

Vern. : Badam.

English and Regional Names:

- **Eng.** : Almonds;
- **H.** : Badam;
- **Mal.** : Vatam-kotta;
- **Mar.** : Badam;
- **Tam.** : Vadamkottai;
- **Tel.** : Adam vittulu.

Distribution : Throughout India.

Description : A tree with oblong-lanceolate serrate leaves. Flowers solitary, pink. Fruit a drupe. Seed long, oval.

Flowering and Fruiting : April-August.

Habitat Ecology : Planted.

Material Examined : EBH-HVP-896, 22.08.2005.

Parts Used : Fruits. Seeds. Twigs.

Folk Uses : Ripe fruits **edible**. Seeds oil given to check **constipation**. Dried branches used as **fuel wood***.

Uses in Literature : Reported earlier as astringent, demulcent, detergent, diuretic, edible (kernel), laxative, nervine tonic, sores, stimulant, symbolic; and for confectionary, cosmetics, cough, diabetes, peptic ulcer, perfumery, skin eruptions, sore, and turnery **(Ambasta, 1986; Jain, 1991; Shome *et al.*, 1996; Watt, 1972)**.

152. *Prunus domestica* L. (Pl. 48D)

Syn. : *P. insititia* L., *P. communis* Huds. var. *insititia* (L.) Hook. f.

Family : Rosaceae.

Vern. : Aloo Bukhara, Plum.

English and Regional Names:

Eng.	:	Plum Tree, Prum Tree, Common Plum;
B.	:	Alubukhara;
G.	:	Alubukhara;
H.	:	Alubukhara.

Distribution : W. Himalaya.

Description : A cultivated tree without spines. Receptacle hollow. Flowers stalked, pink. Fruits ovoid.

Flowering and Fruiting : April-August.

Habitat Ecology : Planted.

Material Examined : EBH-HVP-833, 12.07.2004.

Parts Used : Fruits. Twigs.

Folk Uses : Ripe fruits **edible**; also employed for making **jams**. Fruit juice prescribed after every two hours to cure **dysentery***. Dried branches used as **fuel wood***.

Active Constituents : Ingredients comprise phloracetophenone, 4-O-methyl-phloracetophenone, and its 6-O-glucoside (I) and frakinol (bark), galacturmic acid, arabinose, galactose, xylose and rhamnose (fruits), naringenin, quercetin, chrysin (heart wood).

Biological Activity : Naringenin possesses significant antiulcer activity in experimentally induced ulcer in pylorus ligated rats (132 mg./kg.).

Uses in Literature : Useful as aperient, astringent, cabinet work, cathartic, cold, demulcent, desert, digestive, edible, laxative, refrigerant; and for confectionary, decoction, fever, nutritious, refreshing, bilious states of the system, inlay and turnery, jams, skeleton of paper match boxes **(Ambasta, 1986; Samati, 2004; Watt, 1972).**

153. *Prunus persica* Benth. & Hook. f. (Pl. 49A)

Syn. : *Amygdalus persica* L., *A. colinus* Wall., *Persica vulgaris* Miller, *P. saligna* Royle.

Family : Rosaceae.

Vern. : Adu, Aru.

English and Regional Names:

Eng.	:	Peach Tree, Nectarine;
H.	:	Aru;
Oriya	:	Pishu;
P.	:	Aru, Bembi, Chirmaru;
Urdu	:	Adud.

Distribution : N.W. Himalaya, cultivated.

Description : Small deciduous trees. Leaves conduplicate. Flowers pink. Fruit downy, succulent.

Flowering and Fruiting : February-May.

Habitat Ecology : Planted.

Material Examined : EBH-HVP-704, 15.02.2001.

Parts Used : Roots. Leaves. Fruit. Gum. Seeds.

Folk Uses : Fruits **edible**. Infusion of leaves and roots good for **cold***. Gum used as an a **adhesive***. Seeds also relished.

Active Constituents : Kaempferol (leaves), 6-o-(β glucoronopyranosyl)-D-galactose, 6-o-(4-o-methyl β-D-glucoronopyranosyl)-d-galactose and 2-o-(β-glucorono pyranosyl), D-mannose, rhamnose, xylose, arabinose and galactose (gum), γ-hexalactone, γ-heptalactone, γ-octalactone, γ-nanolactone, γ-decalactone, ethanol, hexanol, benzylalcohol, acetaldehyde, benzaldehyde, acetic acid, pentanoic acid, hexanoic acid, hexyl formate, methyl acetate and ethylacetate, vit. A, protein 1.2, fat 0.03%, fibre 1.2, C15, Mg 21, oxalic acid, iron 2.4, total P 41, Na 2, K 453.0, Cu 0.06, S 26.0, thiamine 0.02, riboflavin 0.03, nicotinic acid 0.05 and ascorbic acid 6.0 mg./100 g. (fruits).

Biological Activity : Aerial parts diuretic. Alc. extr. of embryo and immature seeds exhibits analgesic and anti-inflammatory activities.

Uses in Literature : Reported earlier as an abortifacient, antiemetic, antiscorbutic, ascaricide, cattle feed, demulcent, diuretic, stomachic and table fruit; and for eczema, headache, scabies, dessert, edible, fertilizer, tonic, vegetable, building purposes, canning, cough, peach brandy, pickling, piles, removing bad smell of mouth and sputum, stomach trouble and whooping cough **(Ambasta, 1986; Jain *et al.*, 1994; Kapur & Singh, 1996; Lal *et al.*, 1996; Parmar & Kaushal, 1982; Samati, 2004; Sharma, 1996; Watt, 1972).**

154. *Psidium guajava* Linn. (Pl. 49B)

Family : Myrtaceae.

Vern. : Amrood.

English, Sanskrit and Regional Names:

Eng. : Common Guava, Guava tree;
Sans. : Aprithaktvacha, Ovidhavija;
Ass. : Madhuriam;
B. : Goaachni, Peyara;
Bo. : Jam, Perala;
G. : Jamrud, Peru;
H. : Amrud, Safed safari;
Kan. : Pera, Koyera;
Mal. : Koyya, Malakkaperu;
Mar. : Jamba;
Oriya : Bodojambo, Julabojamb;
P. : Amrud, Anjirizarod;
Tam. : Segappugnoyya, Vellaikoyya;
Tel. : Ettajama, Tellajama.

Distribution : Widely cultivated throughout India.

Description : Small cultivated trees. Leaves opposite, ovate-oblong. Flowers white. Berries globose, green, pulp yellow with white or pink seed.

Flowering and Fruiting : July-September.

Habitat Ecology : Cultivated.

Material Examined : EBH-HVP-820, 14.08.2004.

Parts Used : Fruits. Leaves.

Folk Uses : Fruits **edible** and considered good for **anaemia** and **constipation.** Leaves used for **scouring teeth** and **tongue** and its paste cures **pyorrhoea***. Chewing of leaves along with 'katha' (*Acacia catechu* Willd.) cures **mouth ulceration**.

Active Constituents : Properties attributed to sesquiguavaene, ellagic acid, ellagitanin, guavin β, guavin A, C and D, galaturonic acid, galactose, arabinose with traces of xylose, rhamnose (fruits), quercetin-3-α arabino pyranoside (leaves).

Biological Activity : Spasmolytic activity of leaves is mainly due to presence of quercetin (+). Gallocatechin exhibits antimutagenic activity against UV-induced mutation in *Escherchia coli*.

Uses in Literature : Known in India as an antiemetic, antinarcotics to overcome the influence of liquor, astringent, coloration material in 'Mithila' paintings, edible, astringent, colic, emetic, laxative, tongue cleaner, tonic; and for carbuncle, cholera, chutneys, cough, cold, diarrhoea, dysentery, headache, inflammation, itches, jam, ketchup, loose motions, malaria, marmalades, mouth wash for swollen gums, piles, spear handle and instrument of various kinds, sprains, stomachache, stones, tanning, toothache, ulcer and wound (vet.) **(Ambasta, 1986; Augustine & Sivadasan, 2004; Bhandary *et al.*, 1996; Bhardwaj & Gakhar, 2003; Borthakur, 1993; Chowdhury *et al.*, 1996; Gill *et al.*, 1993; Girach & Aminuddin, 1995; Gogoi & Borthakur, 2001; Hosagoudar & Henry, 1996c; Jain, 1991; Jain *et al.*, 1991; Jha & Basak, 1994; Lalramnghinglova, 1996; Mohanty *et al.*, 1998; Pandey & Verma, 2002; Samati, 2004; Sen & Behera, 2003; Shome *et al.*, 1996; Singh & Maheshwari, 1992; Siwakoti & Verma, 1996; Tirkey *et al.*, 2003; Upadhyay & Singh, 2005; Verma *et al.*, 1995; Vishwanathan & Singh, 1996; Viswanathan *et al.*, 2001; Watt, 1972).**

155. *Punica granatum* Linn. (Pl. 49C)

Syn. : *P. nana* Linn.

Family : Punicaceae.

Vern. : Daadu, Daardn.

English, Sanskrit and Regional Names:

Eng.	:	Pomegranate Tree;
Sans.	:	Bijapura, Dadima, Dantabijaka, Granat, Pindapushpa, Vrittaphala;
Ass.	:	Dalim;
B.	:	Dalimgachh, Dalimphal;
Bo.	:	Anara, Dalimba;
G.	:	Dadam, Daramphal;
H.	:	Anar, Dhalim, Dharimb;
Kan.	:	Dalimbe, Darrim;
Mal.	:	Pumatalam, Uruyapalam;
Mar.	:	Dalimba;
Oriya	:	Dalim, Dalimba;
P.	:	Anar, Daan, Daru, Daruna;
Tam.	:	Kalumal-adalai, Tusagam;
Tel.	:	Dadimba, Dadimanu, Dalimma, Purvudanimma;
U.P.	:	Anar, Rumman;
Urdu	:	Anamitha.

Distribution : W. Himalaya 700-2,700 m. Throughout India, cultivated.

Description : A small deciduous tree with thorny branches and dark grey bark. Leaves opposite, oblong-lanceolate, glabrous. Flowers sessile, bright red, terminal. Fruits globose, berry. Seeds angled, aril filled with an acidic sweet juice. Rind coriaceous.

Flowering and Fruiting : May-September.

Habitat Ecology : Stony slopes. Roadsides. Edges of agricultural fields.

Material Examined : EBH-HVP-788, 20.07.2003.

Parts Used : Seeds. Fruits. Rind. Plant.

Folk Uses : Dried seeds **'anardana'** used for **'chutney'*** preparation by grinding it with mint leaves and chillies. Fruits **edible** and considered good for **hyperacidity**, **anaemia** and **fatigue**. One teaspoon of powdered rind taken with honey for seven days to cure **cough.** Rind also added in the preparation of 'jimikand' (*Amorphophallus campanulatus* (Roxb.) Blume *ex* Decne.) for removing the **bitterness***. Plant also considered **sacred*** and worshipped by newly wed couple for their long and happy married life.

Active Constituents : Tannin-punicalagin and puncalin, granatin A and B (fruit rind), punicalin, punicalagin, punicafolin, granatins A and B, corilagin, strictinin, 1, 2, 3, 4, 6-Penta-o-galloyl-β-D-glucose, 1, 2, 4, 6-tetra-o-galloyl-β-D-glucose and new hydrolysable tannin 2-o-galloyl-4, 6(δ, I)-galloyl-D-glucose (I) (leaves); cyanidin-3-glucoside, delphindin-3-glucoside, and three, 5-diglucoside (seed coat), elagitannins, punicacorteins A-D, punigluconin, causuariin, punicalin, puicalagin and pullestierine; 28% gallotanic acid together with yellow colouring matter (rind), glucose, fructose, maltose, starch, carotene, thiamine, riboflavin, nicotinic acid, vitamin, pectin and amino acid (edible portion of fruit), elagitannins, punicacorteins A, D, punigluconin, casuarin, casuarinin, punicalin, punicalagin and pullestierine (bark) are the active principles.

Biological Activity : Aerial parts CNS depressant and diuretic whereas flower buds antifungal and antibacterial. Fruit extr. exhibits high activity against *Micrococcus pyogenes* var. *aureus, Escherichia coli*, *Salmonella typhosa*, *S. monterideo*. Aq. extr. of roots completely inhibits activity of *Mycobacterium tuberculosis*.

Uses in Literature : Reported earlier as an anthelmintic, astringent, cardiac, coolant, condiments, edible, expectorant, laxative, refreshing drink, souring agent, stomachic, symbolic and tonic; and for brain affection, bronchitis, cholera, cough, diarrhoea, dysentery, diabetes, emetic, eye complaints, congestion of liver, fertility, intestinal worms, pyorrhoea, pimples, antivomiting, coloration, material in 'Mithila' paintings, good source of vit. C, gastric, jaundice, kidney stones, sore throat, toothache, urinary troubles, uterine disease and wine **(Ambasta, 1986; Anonymous, 1990; Bhardwaj & Gakhar, 2003; Bondya & Sharma, 2003; Chopra *et al.*, 1956, 1983; Dagar & Dagar, 1996; Ghosh *et al.*, 1996; Gogoi & Borthakur, 2001; Jain *et al.*, 1991; Jha & Basak, 1994; Kapur & Singh, 1996; Katewa *et al.*, 2001; Khanna *et al.*, 1996b; Mahato & Mahato, 1996; Mohanty & Padhy, 1996; Mohanty *et al.*, 1998; Pandey & Verma, 2002; Rana *et al.*, 1996; Singh, 1996; Tirkey *et al.*, 2003a, b; Verma & Chauhan, 2006; Viswanathan *et al.*, 2001; Watt, 1972).**

Major Ayurvedic Formulations : Bhaskara-lavana, Dadimastak-churna, Dadimadi-ghrita, Lavanabhaskar-churna, Eladi-churna, Sukramatrka-vati.

Trade : Fruits and dried seeds marketed as 'anardana'.

156. *Putranjiva roxburghii* Wall. (Pl. 49D)

Family : Euphorbiaceae.

Vern. : Putranjan.

English, Sanskrit and Regional Names:

Eng. : Life of the child;
Sans. : Putranjiva;
B. : Putranjiva;
H. : Putrajiva;
Mal. : Pongalam;
Mar. : Jewanputr;
P. : Jiyaputra;
Tam. : Karupale;
Tel. : Kadrajuvi, Putrajivi.

Distribution : Throughout tropical India.

Description : A moderate-sized evergreen tree. Leaves dark green, elliptic-oblong. Flowers yellowish. Drupes tomentose.

Flowering and Fruiting : March-April.

Habitat Ecology : Cultivated in temple courtyard.
Material Examined : EBH-HVP-893, 12.04.2006.
Parts Used : Whole plant. Seeds
Folk Uses : Whole plant considered **sacred** and worshipped for long life of children in the family. Children **play*** with its seeds.
Uses in Literature : Used so far as sacred **(Ambasta, 1986, Sharma, 2003)**.

157. *Pyrus communis* L. (Pl. 50A)

Syn. : *P. pyrifolia* Nakai var. *culta* Nakai., *P. sinensis* Hort. (non L.).
Family : Rosaceae.
Vern. : Nashpati.
English and Regional Names:
Eng. : Pear, Wild Pear;
H. : Naspati;
Kash. : Batang, Kistabahira;
P. : Batank, Naspati, Tangi;
Tam. : Perrikaya;
Tel. : Berikaya;
U.P. : Bagugohsa.
Distribution : Cultivated in N.W. Himalaya.
Description : Trees. Leaves long green. Flowers white. Fruit pyriform, yellow.
Flowering and Fruiting : June-August.
Habitat Ecology : Cultivated.
Material Examined : EBH-HVP-789, 20.06.2003.
Part Used : Fruit.
Folk Uses : Fruits **edible**, also considered good for **dysentery** and **diarrhoea***.
Active Constituents : Fruits rich in lupeol, β-sitosterol, betulin, betulinic acid, hydroguinone and arbutin.
Biological Activity : Lupeol, â-sitosterol, betulin, betulinic acid, hydroguinone and arbutin possess bacteriostatic activity against *Escherchia coli, Salmonella typhii, Shigella flexneri* and *Staphylcoccus aureus*.
Uses in Literature : Useful as an edible, psychoactive; and for acute fever, beverages, carving, engraving purposes, malaria, mathematical and drawing instrument, turning and wines **(Ambasta, 1986; Gill *et al.*, 1993; Jain *et al.*, 1994; Watt, 1972)**.

158. *Pyrus pashia* Ham. (Pl. 50B)

Syn. : *P. variolosa* Wall., *P. verruculosa* Bertol., *P. nepalensis* Herb. Ham. *ex* Hort.
Family : Rosaceae.
Vern. : Kainth, Kaenth.

Regional Names:

Ass.	:	Chalthei, Soh-jhur, Soh-shur;
H.	:	Mehal Mol;
P.	:	Batangi, Kaenth, Keitha, Kiat, Ku, Shehgal, Tang.

Distribution : Kashmir to Bhutan (Temperate Himalaya).

Description : Small deciduous trees. Leaves acuminate. Branches spinescents. Flowers white, in corymbs. Fruits globose, rough due to small white spots.

Flowering and Fruiting : March-May.

Habitat Ecology : Open grassy slopes. Moist and drier areas. Edges of fields.

Material Examined : EBH-HVP-712, 12.03. 2001.

Parts Used : Fruit. Leaves. Stem. Bark.

Folk Uses : Fruits **edible** and its extr. effective against **dandruff*** and **mouth ulcers***. One teaspoon of powdered bark and leaves in 1:2 combination with a pinch of black pepper (*Piper nigrum* Linn.) prescribed for **rheumatic pains*** till relief. Pulp of fruits with black pepper twice a day for three days given to check **dysentery***. Also its pulp applied (two times a day for 3-4 days) for suppuration of **blisters*** on tongue of animals. Plant also used for **grafting** of 'naspati' (*Pyrus communis*). Leaves used as an excellent **fodder**. Wood used as **fuel** and for making **agricultural implements***.

Active Constituents : These are N-triacontane, myricyl alcohol as chloride and sitosterol (leaves), friedelin, β sitosterol (bark).

Biological Activity : Aerial parts anticancerous and CNS depressant.

Uses in Literature : Considered useful as fodder, edible, fuel, bobbins, catarrh (vet.); and for eye complaints, combs, injury (vet.), making mortar and walking sticks **(Ambasta, 1986; Joseph & Kharkonger, 1981; Sinha, 1986).**

159. *Randia dumetorum* (Retz.) Lam. (Pl. 50C)

Syn. : *R. nutans* DC & W. & A., *R. brandisii* Gamble, *R. tomentosa* W. & A. non Blume, *R. longispina* DC & W. & A., *R. floribunda* DC & W. & A.,
R. rotteri W. & A., *R. stipulosa* Miq., *R. spinosa* Blume, *R. malabarica* Wall., *Gardenia nutans* Roxb., *G. spinosa* L., *G. longispina* Roxb., *G. floribunda* Roxb.,
G. dumetorum Retz., *G. stipularis* Rottl. & Willd., *G. glabra* Br., *G. propinqua* Br., *G. latifolia* Schelecht., *Posoqueria dumetorum* Roxb., *P. nutans* Roxb.,
P. longispina Roxb., *P. floribunda* Roxb., *Ceriscus malabaricus* Gaertn., *Catuneregum nutans* (DC.) Tiruv., *Xeromphis spinosa* (Thunb.) Keay.

Family : Rubiaceae.

Vern. : Rara.

English, Sanskrit and Regional Names:

Eng.	:	Common Emetic Nut;
Sans.	:	Bastishodhana, Ramachhardanaka;
Ass.	:	Bahmona, Gural, Mon;
B.	:	Madan;
Bo.	:	Gehala, Gelaphal;
G.	:	Medhola;
H.	:	Arar, Madanphala
Kan.	:	Banegara, Kara, Mangri;
Kash.	:	Kirkala, Kokoa;
Mal.	:	Karalikkaya, Kattumaranna;
Oriya	:	Potova;
P.	:	Mandkolla, mindhal;
Tam.	:	Karadam, Murkka-rai;
Tel.	:	Madanamu, Mangara.

Distribution : Subtropical Himalaya up to 1,200 m.

Description : A small tree armed with strong, straight, opposite decussate spines. Leaves obovate-obtuse, wrinkled, shining above. Flowers creamish-yellow, fragrant, sessile. Fruits ovoid, yellowish-brown. Seeds numerous, compressed, smooth, closely packed in pulp.

Flowering and Fruiting : April-June.

Habitat Ecology : Wasteplaces. Drier areas.

Material Examined : EBH-HVP-721, 14.05.2001.

Parts Used : Roots. Fruits. Seeds.

Folk Uses : One teaspoon of powdered fruit and 'ajwain' (*Trachyspermum ammi* Linn.) taken with 'gur' (jaggery) to cure **stomachache*** (twice a day) and **constipation***. Poultice of fresh root good against **foot infection*** in animals. Also its paste rubbed on boils as an **antiseptic***.

Active Constituents : Isolated constituents are dumetoronins A, B, C, D, E & F, β-D-galactopyran-osyl (1→3) deanolic acid, lupeol and two new triterpenoid glycosides (fruits).

Uses in Literature : Known earlier as an abortifacient, anodyne, anthelmintic, astringent, intensifier, emetic, expectorant, fodder, edible, sacred tree; and wood for walking sticks, boils, bodyache during fever, dysentery, diarrhoea, dyeing, jaundice, pimples, umbrella handles, and soap industry **(Ambasta, 1986; Chauhan, 1999; Chopra *et al.*, 1956; Jain, 1991; Lalramnghinglova, 1996; Singh & Prakash, 1994; Watt, 1972).**

Major Ayurvedic Formulations : Madnaphala-churna, Pipalayadi-churna.

Trade : Madanphal.

160. *Raphanus sativus* Linn. (Pl. 50D)

Syn. : *R. sativus* var. *caudatus* (Linn.) Vilmorin, *R. sativus* var. *mougri* Helm., *R. raphanistrum* subsp. *caudatus* (Linn.) Thell.

Family : Brassicaceae.

Vern. : Mooli.

English and Regional Names:

- **Eng.** : Raddish;
- **Sans.** : Mulaka;
- **B.** : Muli, Mula, Mura, Muri;
- **H.** : Muli, Mula, Mura, Muri;
- **Mar.** : Muli, Mula, Mura, Muri;
- **Tam.** : Malangi;
- **Tel.** : Malangi.

Distribution : Cultivated throughout India.

Description : Tall herbs. Leaves pinnatifid. Flowers pinkish.

Flowering and Fruiting : November-March.

Habitat Ecology : Cultivated.

Material Examined : EBH-HVP-825, 23.10.2004.

Parts Used : Roots. Leaves.

Folk Uses : Fresh roots **eaten** raw and also cooked as **vegetable.** Also roots considered **diuretic** and good for **piles, jaundice*** and **gastrodynia.** Juice of fresh leaves checks **jaundice** and **constipation;** leaves also cooked as 'saag' (a type of vegetable preparation). 'Mooli' **worshipped*** on **'Hoi' festival** for long life of children.

Active Constituents : Root possesses seven new b-carboline derivatives I which are characterised as 1-(2-pyrrolidinethion-3-yl)-1, 2, 3, 4-tetrahydro-B-carbolin-3-carboxylic acid.

Biological Activity : A purified basic protein homologus to non-specific lipid transfer protein from seed showed antifungal activity.

Uses in Literature : Used so far as carminative, diuretic, laxative, vegetable; and for impotency, flatulence, piles, ringworm, reducing acidity and kidney stones **(Ambasta, 1986; Chandra Sekar & Srivastava, 2003; Khanna & Kumar 2000; Kumar & Chauhan, 2005; Samati, 2004).**

161. *Ricinus communis* Linn. (Pl. 51A)

Syn. : *R. inermis* Jacq., *R. lividus* Jacq., *R. speciosus* Burm., *R. spectabilis* Blume, *R. viridis* Willd., *Croton spinosus* Linn.

Family : Euphorbiaceae.

Vern. : Airand, Arind, Erand.

English, Sanskrit and Regional Names:

- **Eng.** : Castor Oil Plant, Palma Christi;
- **Sans.** : Dirghadantaka, Triputiphala, Vyadatvaka;
- **Ass.** : Eri;
- **B.** : Bheranda;
- **Bo.** : Rendi;
- **G.** : Diveligo;
- **H.** : Arand, Arend, Arangi, Erand, Ind, Rand;
- **Kan.** : Avanakka, Chittavanaku;
- **Mar.** : Yorandicha;
- **Oriya** : Bheronta, Chitroka, Tindoko;

P. : Arand, Aneru, Bedanjir, Harnauli;
Tam. : Amanakku, Muttukkottilai, Sittamunuk;
Tel. : Amadam, Chittamudamu, Eramudapu;
U.P. : Rend.

Distribution : Planted and naturalised in Himalayan villages up to 2,400 m.

Description : An evergreen soft wooded shrub. Leaves peltate, large palmately lobed, serrated. Flowers reddish or yellowish, in large terminal branches. Capsules prickly. Seeds shining black with an aril at its base.

Flowering and Fruiting : March-August.

Habitat Ecology : Wastelands. Dry areas. Open slopes. Roadsides.

Material Examined : EBH-HVP-710, 23.03.2001.

Parts Used : Leaves. Seeds.

Folk Uses : Seed paste mixed with butter applied for **skin infection**; massage with seed oil prescribed for **rheumatic pains**. Warm leaves applied as **bandage*** on sprains. Stem used for **scouring teeth***. Two teaspoonful of 'arind oil' with hot milk given to cure **constipation**.

Active Constituents : These are ricinine and 1-methyl-3-cyano-4-methoxy-2-pyridone (seeds, leaves), β-amyrin, stigmasterol, β-sitosterol, ricinine, quercetin, rutin, hyperoside, n-dimethylricinine, 3-o-β-d-xylopyranosides, 3-o-β-d glucopyranosides and 3-o-β-d-rutinosides of kaempferol and quercetin (leaves), lysine, palmitic, stearic, hexadecenoic, oleic, linoleic, linoleni, ricinoleic acids, arachidic, chlorogenic, oleic, palmitic, ricinoleic, stearic and dihydrostearic acid, hexadecanoic, hydrocyanic and uric acid (oil).

Biological Activity : Activities exhibited are antiameobic, anticancer, hypoglycaemic (root, stem), antiviral hypoglycaemic (leaf), diuretic. anticholestatic and hepatoprotective (seed).

Uses in Literature : So far known as an antidote to food poisoning, contraceptive, carminative, fish poison, edible, fodder, psychoactive, purgative, laxative, lubricant, luminent, purgative, spice, symbolic to ward off evil, and vermicide; and for abortion, aching joints, anasarca, boils, burns, cholera, conjunctivitis, constipation, delirium, dropsy, dysentery, earache, gout, headache, heat stroke, hydrocele, increasing milk yield, injury, intestinal worms, irregular menstruation, lumbago, migraine, muscular pain, night blindness, pneumonia, rearing silk worms, rheumatism, sciatica, scorpion bites, skin disease, skin ulcer, sores, swelling, throat problems (vet.), wounds, bone sprain of horses and cough **(Ambasta, 1986; Bhandary *et al.*, 1996; Bhardwaj & Gakhar, 2003; Bhatt *et al.*, 2001; Borthakur, 1993; Hembrom, 1996; Hembrom & Goel, 2005; Jadhav, 2006; Jain, 1968, 1991; Jain & Saklani, 1992; Jha *et al.*, 1991; Kharkonger & Joseph, 1981; Kadel & Jain, 2006; Katewa & Arora, 1997; Khanna *et al.*, 1996a; Kirtikar &**

Basu, 1935; Kumar & Chauhan, 2005; Kumar *et al.*, 2004; Lal, 1996; Mahato & Mahato, 1996; Mokat & Deokule, 2004; Panda, 1996; Pawar & Patil, 2004; Rama Rao & Henry, 1996; Sen & Behera, 2003; Sharma & Borthakur, 2003; Tirkey *et al.*, 2003; Tiwari & Tiwari, 1996; Verma & Chauhan, 2006; Vihari, 1995; Watt, 1972).

Major Ayurvedic Formulations : Balarista, Eranda-pak, Eranda-saptaka-kwatha, Eranda-tel, Erandimubhi-kwatha, Gangharvahastadi-kwatha-churna, Laghuvisgarava-taila, Rasnasaptak-kwatha, Sadanga-guggulu, Ratari-guggulu.

Trade : Erandi, Eranda-bija, Manjeeth (roots). Castor seeds and oil are marketed.

162. *Rosa damascena* Mill. (Pl. 51B)

Family : Rosaceae.

Vern. : Gulab.

English, Sanskrit and Regional Names:

Eng. : Damask rose, Otto of roses;

Sans. : Fasli gulab;

H. : Shataputri.

Distribution : Throughout India.

Description : Perennial shrubs. Stem moderately hooked. Leaf stipulate, compound, imparipinnate. Sweet scented pink flowers in terminal axillary corymbs. Fruit hard achenes.

Flowering and Fruiting : Greater part of the year.

Habitat Ecology : Cultivated.

Material Examined : EBH-HVP-894, 15.11.2006.

Parts Used : Fruit. Flowers.

Folk Uses : 'Gulkhand' prepared from its petals prescribed with milk for **constipation***. Plant considered **sacred*** and flowers offered to deities. Also planted as **ornamental***.

Active Constituents : Active constituents are quercetin, kaempferol and cyaniding (whole plant), cyaniding-3, 5-diglucoside (petals).

Uses in Literature : Used so far as perfume oil, confectionery, drugs, food and beverages **(Ambasta, 1986, Sharma, 2003).**

163. *Rumex hastatus* Don. (Pl. 51C)

Family : Polygonaceae.

Vern. : Maroli.

Regional Names:

Kum. : Amlora, Chulmora;

P. : Amla, Amlora, Ami, Khattimal, Malori-gha.

Distribution : Common in Himalaya from Kashmir to Kumaon.

Description : A bushy shrub. Leaves broadly triangular, stalked. Flowers tiny, greenish-white. Fruits pinkish, wings papery. Nuts brown.

Flowering and Fruiting : November-April.

Habitat Ecology : Stony slopes. Banks. Common on walls of cultivated terraces.

Material Examined : EBH-HVP-739, 12.12.2001.

Part Used : Leaves.

Folk Uses : Leaves **eaten*** raw and as a source of vit. C also used in **culinary** preparations. Poultice of leaves applied on **skin affections***.

Active Constituents : 21-23% tannin (root bark).

Uses in Literature : Used in India for burns, cuts, cooling, chutneys, fever; and also as pot herb **(Ambasta, 1986; Chandra Sekar & Srivastava, 2003)**.

164. *Saccharum arundinaceum* Reitz. (Pl. 51D)

Syn. : *Erianthus arundinaceus* (Retz.) Jesw. *ex* Hyene, *S. cillare* Anders, *S. munja* Roxb., *S. procerum* Roxb.

Family : Poaceae.

Vern. : Sarphar.

Sanskrit and Regional Names:

Sans. : Munja;
B. : Teng;
Mal. : Munja;
P. : Sarkanda;
Tam. : Munji;
Tel. : Adavicheruk.

Distribution : Throughout India.

Description : A gigantic tufted grass. Stem glabrous with very long internodes. Leaf sheath glabrous. Panicle long, pink. Main branches tufed on axis.

Flowering and Fruiting : October-December.

Habitat Ecology : Wastelands. Drier areas. Rocks.

Material Examined : EBH-HVP-768, 21.11.2002.

Parts Used : Whole Plant.

Folk Uses : Plant used for roofing **huts**. Pieces of stem used as **writing pen***. Dried cane used as **fuel***.

Uses in Literature : Used as demulcent, diuretic, refrigerant, aphrodisiac; and for burning sensation, blood troubles, paint brushes, thatched houses, thirst, urinary complaints and after delivery **(Ambasta, 1986)**.

165. *Saccharum officinarum* L. (Pl. 52A)

Syn. : *S. sinense* Roxb.
Family : Poaceae.
Vern. : Ganna, Kamandi.
English, Sanskrit and Regional Names:

Eng. : Sugar Cane, Noble Cane;
Sans. : Adhipatra, Gudakastha, Rasola;
Ass. : Pura-Kuhiyar;
B. : Ak, Kushiar, Puri;
Bo. : Gol, Serdi;
G. : Naisakar, Sheradi, Uns;
H. : Ganna, Ikh, Kumad;
Mal. : Kantarakam, Madhutrianam;
Mar. : Aos, Kabbo;
Oriya : Aku, Gudodaru;
P. : Ganna, Khaud, Paunda;
Tam. : Angarigai, Madudirunam;
Tel. : Arukanupulakranugu, Tellacheraku.

Distribution : Cultivated throughout India.
Description : Tall perennials. Stem solid. Leaves linear, lanceolate, scarbid. Spikelet lanceolate, surrounded by dense silky white hairs.
Flowering and Fruiting : January-April.
Habitat Ecology : Cultivated.
Material Examined : EBH-HVP-766, 22.04.2002.
Part Used : Plant.
Folk Uses : Plant considered **sacred*** and **worshipped** on **'Hoi'** (a holy day when mothers fast for good health and long life of their sons). Cane juice extracted considered good for **jaundice** and considered **refrigerant** and **refreshing;** also used in preparing **'gur'**, **'shakkar'** and **sugar.** Dried cane used as **torches** and **thatching** material for huts.
Active Constituents : Saccharans A-F, neocarlinoside, neoisoschaftoside, orientin, swerti-ajaponin, vicenin-2, p-coumaric, ferulic, caffeic and 3, 4 di-o-methyl caffeic acids (stalk juice) have been isolated.
Biological Activity : Saccharans A, B, C, D, E and F exhibited hypoglycaemic activity in normal and alloxan produced hyperglycaemic rats.
Uses in Literature : Useful as anthelmintic, antidote to arsenical poisoning, antiseptic, diuretic, emollient, fodder, thatching material for huts, laxative, manure, nutritious, refrigerant, sweetener, symbolic in sculpture (bow of Kamdev—God of love made from canes), tonic, torches (dried cane); and for bagasse or fibre, beverages, disorder of bile and gall bladder, hepatitis, jaundice (juice as drink), lactation (vet.), rough cordage, soothing and protecting the alimentary tract, inflammation, strengthening the gum, sugarcane wax, tying, ulcer, wax papers and wounds (vet.) **(Ambasta, 1986; Borthakur, 1996a; Borthakur *et al.*, 2004; Dagar & Dagar, 1996; Gogoi *et al.*,**

2003; Jain, 1991; Kurian, 1995; Lalramnghinglova, 1996; Sahu, 1996; Siwakoti & Verma, 1996; Tirkey *et al.*, 2003a; Watt, 1972).

Major Ayurvedic Formulations : Bala-taila, Navaratnaralamrganka-rasa.

166. *Saccharum spontaneum* Linn. (Pl. 52B)

Family : Poaceae.
Vern. : Kans.
Sanskrit and Regional Names:

Sans. : Kasa;
B. : Kagara, Kanga, Kans, Kas, Kash,;
G. : Kans, Kansado;
H. : Kans, Kas;
Mal. : Nan-nana;
Mar. : Kagara;
P. : Kahi, Kanh, Kans, Sarkara;
Tam. : Nanal;
Tel. : Kakicheraku.

Distribution : Throughout India.
Description : A tall erect grass. Leaves narrowly linear, acuminated, rigid, coriaceous, glabrous. Panicles lanceolate, silky hairy. Rachis slender. Spikelet lanceolated, bearded with spreading silky hair. Glume 4.
Flowering and Fruiting : Greater part of the year.
Habitat Ecology : Drier areas.
Material Examined : EBH-HVP-817, 23.07.2004.
Parts Used : Stem, Whole Plant.
Folk Uses : Plant considered **sacred***. Stem used as **'kalam'*** (writing pens). Also its pieces kept near the dead body of a person for the last **rites***.
Uses in Literature : Known in India as an aphrodisiac, laxative, fodder; and for burning sensation, stranguary, phthisis, urinary troubles, vesical calculi diseases of blood, biliousness, haemorrhagic diatheris, paper manufacturing and thatching **(Ambasta, 1986; Chhetri, 2005).**

167. *Salvia coccinea* Linn. (Pl. 52C)

Syn. : *S. gerardiana* Benth.
Family : Lamiaceae.
Vern. : Phultulsi.
English Name : Sage.
Distribution : More or less cultivated throughout India.

Description : Erect pubescent herbs with scarlet red flowers borne in terminal racemes. Nutlets smooth.

Flowering and Fruiting : June-August.

Habitat Ecology : Dry slopes.

Material Examined : EBH-HVP-785, 16.06.2003.

Part Used : Flowers.

Folk Uses : Flowers used for **offering*** to various deities.

Uses in Literature : Considered useful for lumbago, renal troubles and tubercular cough **(Ambasta, 1986).**

168. *Sapindus mukorossi* Gaertn. (Pl. 52D)

Syn. : *S. abruptus* Lour.

Family : Sapindaceae.

Vern. : Doddae, Doda, Reetha, Ritha.

English, Sanskrit and Regional Names:

Eng. : Soap Nut, Tree of North India;
Sans. : Arishta, Phenila, Urista;
Ass. : Haithaguti;
B. : Chotta-Ritha, Ritha;
Bo. : Kanmar, Ritha;
H. : Aritha, Dodam, Kanmar, Ritha;
Oriya : Ita;
P. : Aritha, Dodan, Haritha (the fruit), Ritha.

Distribution : Cultivated throughout N. India, ascending to 1,650 m.

Description : Large deciduous trees. Leaves long; leaflets opposite, lanceolate, glabrous. Flowers small, pale brown, globose. Fruits drupe, yellow, globose. Seeds black, loosely contained in fleshy pericarp.

Flowering and Fruiting : May-August.

Habitat Ecology : Open slopes and along border of agricultural fields.

Material Examined : EBH-HVP-838, 24.07.2004.

Parts Used : Fruits. Seeds.

Folk Uses : Fruits used for washing **woollens** and **silken** clothes; and as **shampoo** for hair. Powdered fruits also used as a substitute for **soap** for bath.

Active Constituents : Main ingredients are sapindoside V, sapindoside A & B, heedragenin as a common glyconel, sapindoside V, mukoros-side, a new saponin-emarginatoside (fruits), protein (31.87), carbohydrate (2.21) and starch (11.94%) (seeds kernel).

Biological Activity : Saponins show high spermicidal activity in both spot and IPPF tests with human semen, with MED of 0.05 and 0.51, respectively. Saponins from fruits when injected in rabbits lower cholesterols and brings down B.P. by 25%.

Uses in Literature : Reported in India as an anthelmintic, cattle feed, contracep-tive, detergent, emetic, expectorant, febrifuge, fish poison, hair wash, oil seeds, purgative and for charcoal making, chlorosis, epilepsy, polishing jewellery, pneumonia, salvation, sinusitis, washing and bleaching cardamoms (**Ambasta, 1986; Banerjee & Pal, 1996; Borthakur, 1993; Chauhan, 1999; Chhetri, 2005; Chopra *et al.*, 1956; Hajra & Baishya, 1981; Jain, 1991; Jain & Borthakur, 1980; Karnick *et al.*, 1981; Rao, 1996; Rao *et al.*, 1996; Singh, 1999b, Siwakoti & Verma, 1996; Venkataratnam & Venkata Raju, 2004; Vihari, 1995).**

Major Ayurvedic Formulations : Mahalaxadi-tel.

Trade : Dried fruits sold in the local and national markets.

169. *Saraca asoca* (Roxb.) DE Willd. (Pl. 53A)

Syn. : *S. indica* Auct. Non L., *S. arborescens* Burm., *S. minor* Miq., *S. zollingeriana* Miq., *Jonesia asoca* Roxb. *J. pinnata* Willd.

Family : Fabaceae.

Vern. : Ashoka.

English, Sanskrit and Regional Names:

Eng. : Ashok Tree, Ashoka Tree;
Sans. : Karnapuraka;
Ass. : Ashok, Jasundi;
B. : Asok, Asoka;
Bo. : Ashok, Jasundi;
G. : Ashopalava;
H. : Ashok;
Kan. : Aksunkar, Asokadamara, Kenkali;
Mal. : Hemapushpam, Vanjulam;
Mar. : Asoka, Jasundi;
Oriya : Osoka;
Tam. : Asogam, Malaikkarunai;
Tel. : Asokamu, Vanjalamu.

Distribution : W. Ghats; often cultivated.

Description : A large evergreen tree. Leaves paripinnate, leaflets oblong-lanceolate, drooping. Flowers orange-yellow.

Flowering and Fruiting : March-September.

Habitat Ecology : Edges of fields. Open slopes.

Material Examined : EBH-HVP-711, 22.05.2001.

Parts Used : Bark. Plant.

Folk Uses : Plant considered **sacred**. 1/2 teaspoon daily of powdered bark for 15 days prescribed to cure **menorrhagia, leucorrhoea** and **chest infection***.

Active Constituents : Properties attributed to leucopelarogonidin-3-o-β-D glucoside, leucopelarogonodin, leucocyanidin, β-sitosterol (stem bark), β-sitosterol, quercetin, leucocynidin, kaempferol 3-o-B-D glucoside and gallic acid (flowers).

Biological Activity : Flower bud inhibits *Escherichia coli, Salmonella viballerup*,

Vibrio cholerae, Shigella dysenteriae and *Shigella boydii*.

Uses in Literature : Known so far as abode of God (sacred), astringent, blood purifier, diuretic, masticatory, psychoactive plant, substitute of milk in tea, tonic to prevent miscarriage, uterine sedative, haemorrhagic; and for diabetes, edible purposes, heart disease, leucorrhoea, menorrhagia, menstrual irregularity, urinogenital disease, uterine affections, treatment of excessive menstruation, plough, shaft **(Ambasta, 1986; Bhandary *et al.*, 1996; Gogoi & Borthakur, 2001; Jain *et al.*, 1991; Lalramnghinglova, 1996; Maya *et al.*, 2003; Sharma *et al.*, 2002; Siwakoti & Verma, 1996; Srivastava *et al.*, 2003; Watt, 1972)**.

Major Ayurvedic Formulations : Ashokadi-kwatha, Ashoka-vasi, Ashoka-ghrita, Ashoka-rishta.

170. *Sechium edule* Sw. (Pl. 53B)

Family : Cucurbitaceae.

Vern. : Lanku.

English and Regional Names:

Eng. : Chayote, Chow-Chow;
Kan. : Badane, Seema;
Tam. : Kattirikkai, Seeme.

Distribution : Cultivated.

Description : Perennial climbers. Tendril tri- or penta-fid. Leaves hirsute on both sides. Flowers greenish-yellow. Fruits echinate, one seeded.

Flowering and Fruiting : June-September.

Habitat Ecology : Cultivated.

Material Examined : EBH-HVP-792, 22.06.2003.

Part Used : Fruit.

Folk Uses : Fruit cooked as **vegetable** and considered good for **gastro intestinal disorder***.

Active Constituents : These are α-amino γ-ureidobutyric acid (seeds), glucosides of sitosterol and stigmasterol (fruits).

Biological Activity : β-sitosterol and stigmasterol-β-D-glucosides exhibit anti-inflammatory activity in rat paw odema.

Uses in Literature : Used in India for its edible fruits, fibre, fodder, diabetes; and as vegetable **(Ambasta, 1996; Negi & Gaur, 1991; Samati, 2004; Watt, 1972)**.

171. *Sesamum indicum* DC. (Pl. 53C)

Syn. : *S. luteum* Retz., *S. occidentales* Heer & Regel., *S. orientale* Linn.

Family : Pedaliaceae.

English, Sanskrit and Regional Names:

Eng.	:	Gingelly, Sesame;
Sans.	:	Homadhamya, Jattle, Pavitra, Puraphal;
Ass.	:	Tila;
B.	:	Bhadutil, Kalatil, Katil;
Bo.	:	Ashadital, Bariktil, Krishnatil, Tal;
G.	:	Mithutel, Tal, Tel, Til;
H.	:	Bariktel, Gingli, Krishnatel, Mithatel, Til;
Kan.	:	Ellu;
Mal.	:	Bijan;
Mar.	:	Til, Tila, Zilechatil;
Oriya	:	Khasa, Rasi;
P.	:	Kanjad, Til, Tili;
Tam.	:	Ellu, Nuvvulu:yelluchedi;
Tel.	:	Nuvvu, Nuvvula, Pollanuvvulu.

Distribution : Throughout India.

Description : A branched erect hairy annual with linear leaves. Flowers axillary, solitary and light pink. Capsule tetragonal. Seeds numerous.

Flowering and Fruiting : July-September.

Habitat Ecology : Cultivated. Wastelands. Damp places.

Material Examined : EBH-HVP-818, 12.07.2004.

Parts Used : Leaves. Seeds.

Folk Uses : Leaf paste applied externally for **painful swellings***. Seed oil applied externally for **rheumatic pains*** and also used for **cooking** purposes. Infusion of seeds given with butter twice daily till relief for **bleeding piles***. Seeds considered **sacred** and used in performing various religious ceremonies; also constitute an important component of '**havan**'*.

Active Constituents : Active constituents are globulin, thiamine and niacin (seeds), pedaliin (leaves).

Uses in Literature : Used so far as an abortifacient, demulcent, diuretic, emollient, galactagogue, tonic; and for boils, dysentery, eye and spleen complaints, ulcer, kidney and bladder disorder, edible, sacred (**Ambasta, 1986; Bhattacharjee *et al.*, 1980; Chopra *et al.*, 1956; Jain, 1991, Sharma, 2003**).

172. *Solanum erianthum* Don. (Pl. 53D)

Syn. : *S. verbascifolium* Linn., *S. pubescens Roxb.*

Family : Solanaceae.

Vern. : Rada.

English, Sanskrit and Regional Names:

Eng.	:	Potato Tree;
Sans.	:	Gandira, Pathi, Priyamkari, Vidari;
Gar.	:	Asedu;
Kan.	:	Kadusinde, Sow-dangigida;

M.P. : Chichora;
Mal. : Chunta, Erichunta, Malanjunta;
Mar. : Kutri;
P. : Kalamewa, Tiari, Ola;
Tam. : Anaichundai, Malaichundai.

Distribution : Throughout India.

Description : A shrub covered with dense, stellate, tomentum. Flowers greenish-white. Berries globose, yellow.

Flowering and Fruiting : July-October.

Habitat Ecology : Wastelands. Roadsides.

Material Examined : EBH-HVP-797, 14.07.2003.

Part Used : Leaves.

Folk Uses : Poultice of leaves good for healing **wounds***.

Active Constituents : Solasonine (leaves).

Uses in Literature : Described earlier for body pains, inflammations, urinary troubles, vaginal discharge, vertigo and expelling worm, edible purposes and given to horses affected with glanders (**Ambasta, 1986; Banerjee, 1996; Bhargava, 1981; Henry *et al.*, 1996; Hosagoudar & Henry, 1996b; Jain, 1991; Siwakoti & Verma, 1996; Watt, 1972**).

173. *Solanum melongena* L. (Pl. 54A)

Syn. : *S. incarnum* L., *S. insanum* L., *S. zeilanicum* Scop., *S. undulatum* Lamk., *S. trongum* Lamk., *S. longum* Roxb., *S. pseudoundulatum* Blume., *S. esculentum* Dunal., *S. heteracanthum* Dunal., *S. melanocarpum* Dunal., *S. sativum* Dunal., *S. pressum* Dunal., *S. cumingii* Dunal., *S. torrum* var. *inerme* Dalz. & Gibbs.

Family : Solanaceae.

Vern. : Baingan.

English, Sanskrit and Regional Names:

Eng. : Brinjal, Egg Plant;
Sans. : Drighavartaki, Nilakantaka;
Ass. : Jatibengani;
B. : Bartaku, Kalibegun;
Bo. : Waingi;
G. : Rihgni;
H. : Baigun, Baingan;
Kan. : Badanekayi, Doddabadane;
Kash. : Tukmibadanjanurmi, Vangun;
Mal. : Hinguli, Nilavalutina;
Mar. : Vangi;
Oriya : Bartakki;
P. : Bengan, Vataun;
Tam. : Kalliyam, Mullukkattari;
Tel. : Mettavanga, Nidupuvanga.

Distribution : Widely cultivated for its edible fruits in warmer regions of India.

Description : An erect herb closely covered with shoot prickles. Leaves elliptic with rounded lobes and woolly beneath. Flowers solitary blue. Berry elongate, dark purple.

Flowering and Fruiting : June-September.

Habitat Ecology : Cultivated.

Material Examined : EBH-HVP-812, 20.06.2004.

Part Used : Fruit.

Folk Uses : Fruits cooked as **vegetable** and considered good for **heart*** and **gastric problems***. Its infusion applied on hands and feet to stop **sweating***. Ash of fruit mixed with honey applied locally on **piles***.

Active Constituents : Solasodine and diosegin (fruits).

Uses in Literature : Used earlier as an alternative, antiasthmatic, diuretic, laxative, stimulant, tonic, vegetable; and for asthma, fever, dropsy, enlargement of liver, epitaxis (vet.), hydrophobia, inflammation of kidney and bladder, otitis producing uncomfortable pregnancy sleep, skin disease, toothache, ulcers, virulent gonorrhoea (**Ambasta, 1986; Gill *et al.*, 1993; Jain, 1991; Rama Rao & Henry, 1996; Sen & Behera, 2003; Watt, 1972).**

Adulterant or Substitute : Roots used for that of 'sadabahar' (*Catharanthus roseus* G. Don.).

174. *Solanum nigrum* Linn. (Pl. 54B)

Syn. : *S. rubrum* Mill., *S. triangular* Lamk., *A. villsoum* Lamk., *S. incertum* Dunal., *S. nodiflorum* Jacq., *S. uliginosum* Blume., *S. roxburghii* Dunal., *S. fistulosum* Rich., *S. miniatum* Bernh., *S. paludosum* Dunal., *S. erithraeum* Dunal.

Family : Solanaceae.

Vern. : Makoi.

English, Sanskrit and Regional Names:

Eng. : Black Nightshade, Common Nightshade, Garden Nightshade, Hounds Berry, Morelle, Petty Morel;

Sans. : Bahuphala, Bahutikta, Ghanaghana, Jahenephala, Kaka, Rasayani, Saravatik-ata, Sundari, Tiktika, Vayasi;

Ass. : Pichakati;

B. : Ghati, Kamuni;

Bo. : Ghati, Kamnni, Mako;

H. : Gurkamai, Kabaiya, Mekoi;

P. : Kachmach, Kambei, Meko;

Tam. : Manattakkali;

Tel. : Gajuchettu, Kachi, Kakamachi, Kamanchi, Kasaka;

Urdu. : Makoya.

Distribution : Throughout India.

Description : Annual herbs. Leaves ovate-lanceolate, acuminate, glabrous, thin. Flowers white, small. Berry globose, purplish, smooth, shining. Seed discoids, yellow.

Flowering and Fruiting : January-July.

Habitat Ecology : Wastelands. Moist areas.

Material Examined : EBH-HVP-769, 12.02.2003.

Part Used : Fruit.

Folk Uses : Ripe berries good for **fever** and **diarrhoea**; also used as **laxative**.

Active Constituents : Therapeutic properties of the plant due to 12-hydroxy-solasodine, N-methyl solasodine, 23-o-acetyl-12-β-hydroxy solasodine, solanocapsine and solamargine.

Biological Activity : Plant CNS depressant, hypotensive and spasmolytic.

Uses in Literature : Known so far as an antidote to opium toxic, aphrodisiac, antiseptic, diaphoretic, diuretic, emollient, expectorant, laxative, sedative, stomachic, pot herb; and for body swelling, boils, colitis, conjuctivitis, cough, diarrhoea, dysentery, ear complaints, easy expulsion of foetus, eye complaints, fever, facilitate delivery, goitre, heart ailments, hydrogogue, inflammation of scortum, testicles, kidney bladder, jaundice, liver complaints, nostril complaints, piles, psoriasis, rheumatism, ringworm, skin diseases, sores, sprains, swelling throat troubles, ulcers in mouth and urinary complaints (**Ambasta, 1986; Chatterjee & Pakrashi, 1997; Chauhan, 1999; Chopra *et al.*, 1956; Dey, 1973; Dhiman, 1976; Jain, 1991; Khanna & Kumar, 2000; Kurian, 1995; Pawar & Patil, 2004; Upadhyay & Singh, 2005**).

175. *Solanum surattense* Burm. f. (Pl. 54C)

Syn. : *S. armatum* Br., *S. jacquinii* Willd., *S. diffusum* Roxb., *S. virginiamum* Jacq., *S. xanthocarpum* Schrad. & Wendl., *S. maccanni* Sant.

Family : Solanaceae.

Vern. : Choti-Kandayai.

English, Sanskrit and Regional Names:

- **Eng.** : Dwarf Lid Brinjal, Yellow Berried Nightshade;
- **Sans.** : Kantkaree, Nidigadhika;
- **B.** : Kantikari;
- **G.** : Bhoyaringani;
- **H.** : Katai, Kateli, Ringni;
- **Mar.** : Ankranti, Bheji-behun;
- **P.** : Kandyali, Mahori, Warumba;
- **Tam.** : Cundung Katric, Kandankattiri;
- **Tel.** : Nela-mulaka, Pinna-mulaka, Vankuda.

Distribution : Throughout India.

Description : Prickly diffuse perennial herbs. Stem zig-zag with numerous branches. Prickles compressed. Leaves hairy on both sides; armed on midrib with long yellow sharp prickles. Flowers blue, in a few flowered extra-axillary cyme. Berry globose yellow when ripe.

Flowering and Fruiting : April-October.

Habitat Ecology : Dry localities. Roadsides. Wastelands.

Material Examined : EBH-HVP-723, 21.04.2001.

Parts Used : Fruits. Seeds.

Folk Uses : One teaspoon of powdered stem and fruits with water, recommended twice a day till relief to cure **fever**. Smoke of seeds and 'ajwain' (*Trachyspermum ammi* Linn.) inhaled to cure **dental problems**. Seeds kept in nose to check **sneezing***.

Active Constituents : Therapeutic properties attributed to solasodine, solmargine, β-solamargine, solasonine, diosgenin, solasurine, β-sitosterol, stigmasteryl glucoside (fruits), arachidic, linoleic, oleic, palmitic and stearic acids and solanines (fruit oil), isochlorogenic, neochlorogenic, chlorogenic acid and caffeic acid.

Biological Activity : Aq. and alc. extr. of the plant hypotensive.

Uses in Literature : Reported as an appetiser, carminative, diuretic, edible (seeds), laxative; and for asthma, blister, boils, bronchitis, chest pain, cold, cough, diarrhoea, dropsy, ear and eye complaints, fever, gonorrhoea, gum trouble, heart diseases, lumbago, migraine, muscular pain, pain in jaw, paralysis, piles, pyorrhoea, rheumatism, scorpion bite, snake bite, sore throat, sterility in women, swelling, watery eyes and in curries **(Ambasta, 1986; Chopra *et al.*, 1956; Jain, 1991; Kalpesh *et al.*, 2004; Kumar & Chauhan, 2005; Kurian, 1995; Maliya & Singh, 2003; Mohanty *et al.*, 1998; Uphof, 1968).**

Major Ayurvedic Formulations : Chyawanaprasha, Dashamoolarisht, Dasamool-kwatha, Kantakariavaleha, Kantakari-ghrita, Vyaghiharitaki.

Trade : Sold as 'Kantakari' in crude drug markets of India.

176. *Solanum tuberosum* Linn. (Pl. 54D)

Family : Solanaceae.

Vern. : Aloo.

English and Regional Names:

Eng. : Potato;
B. : Alu;
G. : Batata;
H. : Alu;
Kan. : Alugidde, Batate;
Mal. : Kizhangu, Urulam;

Mar. : Batata;
Oriya : Alu;
P. : Alu;
Tam. : Urala kilangu, Wallaraikilangu;
Tel. : Bangaladumpa, Uralagadda.

Distribution : Commonly cultivated.
Description : Herbs with pinnately dissected leaves. Flowers white.
Flowering and Fruiting : May-September.
Habitat Ecology : Cultivated.
Material Examined : EBH-HVP-779, 17.05.2003.
Parts Used : Tuber.
Folk Uses : Tubers relished as **vegetable**. Its paste applied on face to remove **dark circles***. Also tubers boiled eaten on **religious fast**. These are given as '**sidha**' (**sacred offering*** to Brahmin on religious days) with **wheat flour** and **sugar**.
Active Constituents : Isolated constituents are cycloartenol, stigasterol, β-sitosterol, campesterol, cholesterol, a mixture of 4-mono-methyl sterol, chlorogenic, neochlorogenic and iso chlorogenic, 1-caffeoyl-quinic acid and glycosides of quercetin (leaves).
Uses in Literature : So far used as antiscorbutic, aperient, diuretic, galactagogue, antispasmodic; and for burns, food, feed, starch, ethyl alcohol, potato chips, vegetable, medium for microorganisms **(Ambasta, 1986; Samati, 2004).**

177. *Solanum viarum* Dun. (Pl. 55A)

Syn. : *Solanum khasianum* Clarke var. *chatterjeenum* Sen Gupta.
Family : Solanaceae.
Vern. : Kandayai.
Distribution : Greater part of India. Naturalised weed.
Description : Herbs with woody stem. Leaves deeply lobed, hirsute and with straight prickles. Flowers white. Fruits green with white reticulation, yellow when ripe.
Flowering and Fruiting : May-August.
Habitat Ecology : Drier areas. Rocks. Roadsides. Wasteplaces.
Material Examined : EBH-HVP-786, 15.05.2003.
Parts Used : Whole Plant. Fruits. Seeds.
Folk Uses : Freshly prepared extr. of plant mixed with black pepper (*Piper nigrum* Linn.) and honey efficacious in checking **fever*** and **asthma***. Smoke of seeds and 'ajwain' (*Trachyspermum ammi* Linn.) inhaled to cure **dental problems.**
Active Constituents : These are carpestrol, norcarpestrol, solasodine, scopoletin, esculin and esculetin.
Biological Activity : Antiviral, spermicidal (plant), antiarrhythmic (fruits), CNS active and spasmolytic properties exhibited.

Uses in Literature : Considered useful for diarrhoea, menstrual complaints, toothache, worms in teeth, worms, wounds; and as a contraceptive (**Chaudhari** ***et al.*****, 1982; Maheshwari** ***et al.*****, 1981; Mohanty** ***et al.*****, 1998; Verma & Chauhan, 2006).**

178. *Sonchus asper* Hill. (Pl. 55B)

Syn. : *S. ferox* Wall., *S. oleraceus* Wall.
Family : Asteraceae.
Vern. : Bhursalae.
English and Regional Names:
Eng. : Spiny Leaves Sow, Thistle;
H. : Didhi;
Mar. : Mhatara.
Distribution : Throughout India in Himalaya up to 4,000 m.
Description : Milky herbs with clasping stem. Flowers yellow. Achenes compressed. Pappus copious, white.
Flowering and Fruiting : March-April.
Habitat Ecology : Common along roadsides. Wastelands.
Material Examined : EBH-HVP-772, 30.03.2003.
Parts Used : Aerial plant parts.
Folk Uses : ½ teaspoon of powdered aerial plant parts taken twice a day to check **headache***.
Active Constituents : These are α and β lactocerol, ceryl alcohol, mannitol and traces of acetic acid.
Uses in Literature : Considered useful as an emollient, salad (young shoots), vegetable (whole plant); and for boils, cut, injuries and stomach pain (**Anonymous, 1948-1976; Jain & Patole, 2001**).

179. *Spilanthes acmella* Murr. (Pl. 55C)

Syn. : *S. calva* DC., *S. mauritiana* DC.
Family : Asteraceae.
Vern. : Karkara.
Regional Names:
Ass. : Piraza;
Mal. : Kuppa manjel;
Mar. : Akkalkara, Pipulka;
P. : Akarkarha, Pokarmul;
Tel. : Maratimogga, Maratiteega.
Distribution : Throughout India.
Description : Annual herbs. Leaves opposite, stalked, ovate-lanceolate, toothed. Flower heads conical, solitary on long stalk and with yellow flowers.

Flowering and Fruiting : May-August.
Habitat Ecology : Grassy meadows. Moist areas.
Material Examined : EBH-HVP-783, 16.07.2003.
Part Used : Flower.
Folk Uses : Chewing of raw flowers recommended for **mouth ulceration***, **toothache, throat infection** and increasing the flow of **saliva***. Also, poultice of flowers applied to cure **headache***.
Active Constituents : Properties attributed to myricyl alcohol, β-sitosterol its glucosides, stigmasterol glucoside, α and β-amyrin and amyrin acetate, palmitic and stearic acid, a eudesmanolide I (flower heads) and spiloanthol.
Biological Activity : Ether extr. of fresh flowering tops lethal to Anophelin larvae. Spilanthol shows strong sialagogic action, and acts as local anaesthetic.
Uses in Literature : Pungent flowers are chewed for relief in throat infection, paralysis of tongue, stammering of children, caries and inflammation of jaw bones, dysentery, local anaesthesia, toothache, scabies; and as a diuretic, fish poison, lithotropic, silagogue, stimulant **(Ambasta, 1986; Upadhyay & Singh, 2005).**
Major Ayurvedic Formulations : It enters into compound powders, composition of many aphrodisiac pills and lehyams.

180. *Stephania glabra* (Roxb.) Miers. (Pl. 55D)

Syn. : *S. rotuda* Hook. f., *Cissampelos glabra* Roxb., *C. longa* Miers., *C. japonica* Miers., *C. roxburghiana* Miers., *C. glabra* Miers., *C. glandulifera* Miers., *Cocculus glabra* Roxb., *C. roxburghianus* Wall. Cat., *C. finlaysoniansus* Wall., *Clypea wightii* Arn.
Family : Menispermaceae.
Vern. : Bis-khappar.
Regional Names:
Garh. : Gindaru;
U.P. : Gindaru, Purha.
Distribution : Tropical and Temperate Himalaya.
Description : Large climbers. Leaves membranous, glabrous on both sides, pale-glaucescent below. Drupes obovate, compressed.
Flowering and Fruiting : June-August.
Habitat Ecology : Dry places. Open places.
Material Examined : EBH-HVP-811, 28.06.2004.
Parts Used : Roots. Tubers. Fruits.
Folk Uses : Fresh roots mixed with a pinch of rock salt prescribed to cure **flatulence** in cattle. Root paste prepared in mustard oil applied for **inflamed*** parts. Tuber paste applied for **mouth ulceration** both in human beings and livestock. Smoke of fruits cure **inflamed teats*** in animal.

Active Constituents : Tubers rich in gindarine, gindarinine with tetrahydro-palmatine, palmatine, dehydrocorydalmine, palmatrubine, stepharanine, crydalmine, stepholidine.

Biological Activity : Hypoglycaemic, anti-inflammatory, CNS depressant, anticholin esterase and spasmolytic activities found +ve.

Uses in Literature : Useful for asthma, body and headache, diarrhoea, dysentery, eye disease, fever, intestinal complaints, pulmonary tuberculosis, skin diseases, tumour in stomach and veterinary ailments; and as psychoactive plant **(Ambasta, 1986; Chopra *et al.*, 1956; Jain, 1986, 1999; Jain *et al.*, 1991; Mohanty *et al.*, 1998; Shah & Joshi, 1971)**.

181. *Tabernaemontana divaricata* R. Br. *ex* Roem. & Sch. (Pl. 56A)

Syn. : *T. divaricata* Blume, *T. coronaria* R. Br., *Nerium divaricatum* L., *N. coronarium* Jacq., *Jasminum zelanicum* Burm., *Ervatamia coronaria* Stapf.

Familv : Apocynaceae.

Vern. : Tagar.

English, Sanskrit and Regional Names:

Eng. : East India Rosebay;
Sans. : Nandivirksha, Tagara;
Ass. : Tagar;
G. : Tagar;
H. : Chameli, Chandni, Tagar;
Kan. : Kotuhalle, Nantiyavattam;
Mal. : Kuttampale, Takaram;
Mar. : Ananta, Gondetagara, Tagar;
Tam. : Adukkundiyavattai, Nandiyavattam;
Tel. : Gadhitagarpu, Nandivardhamu.

Distribution : Cultivated Throughout India.

Description : An evergreen dichotomously branched, glabrous shrub. Leaves oblong-lanceolate. Flowers white, fragrant.

Flowering and Fruiting : July-October.

Habitat Ecology : Cultivated.

Material Examined : EBH-HVP-826, 22.07.2004.

Parts Used : Flowers. Plant.

Folk Uses : Plant valued for **ornamental** purposes. Flowers **offered*** to deities. 5 ml. decoction of flowers prescribed thrice a day as **anthelmintic**.

Active Constituents : Flowers yield kaempferol, ibogamine, janetine, voaphyllene, voacangine, dregamine, tabernaemonta-nine, voacamine and loroneridine.

Biological Activity : Indole alkaloid (I) inhibits HCl induced ulcer in mice by 48.8%.

Uses in Literature : Used in India as anodyne, anthelmintic, antidote to scorpion, incense, refrigerant; and for boils dental caries, decorative purposes, eye complaints, fever, headache, sores, perfumery **(Ambasta, 1986; Gogoi & Borthakur, 1991; Jain, 1991; Khanna *et al.*, 1996a; Pandey *et al.*, 1996; Rao *et al.*, 1996).**

182. *Tagetes erecta* Linn. (Pl. 56B)

Family : Asteraceae.
Vern. : Ghatak.
English, Sanskrit and Regional Names:

Eng. : Marigold, African Marigold;
Sans. : Gandugai, Sandu, Stahulapushpa;
B. : Genda;
Bo. : Guljaharo, Gul-Jafari, Makhmal;
G. : Chaphul, Rojia;
H. : Genda, Gultera;
Kan. : Chandumallige, Seemes hamantige;
Mal. : Chendumalli;
Mar. : Gulgoto, Rajia-cha-Phul, Zendu;
Oriya : Gendu;
P. : Genda, Gul jafari, Makhmal, Mentok.

Distribution : Naturalized in India; extensively cultivated as a garden plant.

Description : Shrubby herbs. Leaves pinnately dissected, lobes lanceolate, sharply toothed. Ray florets golden-yellow.

Flowering and Fruiting : October-January.

Habitat Ecology : Cultivated.

Material Examined : EBH-HVP-764, 18.10.2002.

Parts Used : Leaves. Flower. Receptacle.

Folk Uses : Flowers considered **sacred** and offered to appease deities. Also its garland hung at the entrance of main door and employed for **decorative** purposes. Infusion of leaves rubbed to subside pain and swelling caused due to **bee or wasp stings***. Receptacle **eaten*** by children after removing the florets.

Active Constituents : Constituents isolated are syringic acid, methyl 3, 5-dihydroxy-4-methoxybenzonate, quercetin (flowers), piperitone (20.0), (E)-β-ocimene (13.1), terpinolene (12.4), limonene (11.0 %) (leaf essential oil) and tagetus oil (flower).

Biological Activity : Essential oil exhibited antimicrobial activity against fungi, Gram +ve and Gram –ve microorganism.

Uses in Literature : Useful in India for boils, bleeding piles, bronchitis, carbuncle, cold, caterpillar bite, eye troubles, laxative, muscular pains, renal trouble, ringworms, scorpion bite, rheumatism; and as carminative, diuretic, emmenagogue, vermifuge, chicken feed **(Ambasta, 1986; Mokat & Deokule, 2004; Sen & Behera, 2003; Upadhye & Kulkarni, 2004).**

183. *Tamarindus indica* Linn. (Pl. 56C)

Syn. : *T. occidentalis* Gaertin., *T. officinalis* Hook.
Family : Fabaceae.
Vern. : Imli.
English, Sanskrit and Regional Names:

Eng. : Tamarind Tree;
Sans. : Amlika, Bhukta, Charitra, Sukha, Tintrini;
Ass. : Tetuli;
B. : Anbli, Tentul;
G. : Ambli, Amli;
H. : Amli, Amlika, Anbli, Imli, Nuli, Teter;
Kan. : Amli, Huli;
Mal. : Amlam, Puli;
Mar. : Chicha, Chinchu;
Oriya : Konya, Tentuli;
P. : Imbli;
Tam. : Amliam, Puli;
Tel. : Chintachettu, Chintapandu, Sintachettu.

Distribution : Most parts of India.
Description : Trees having even pinnate, oblong, glabrous, entire leaves. Flowers yellow with pink striations, in terminal.
Flowering and Fruiting : May-August.
Habitat Ecology : Open fields.
Material Examined : EBH-HVP-810, 16.05.2004.
Parts Used : Fruits. Seeds
Folk Uses : Pulp of fruit used for preparing **'chutney'** and **sauces**. Seeds used by children for playing various **games***.
Active Constituents : Fruit pulp rich in 10% free tartaric acid, 8% acid potassium tartarate, 25-40% invert sugar, tamarindienal, alkaloids, glycosides, tannins, organic acid and amino acids.
Biological Activity : Tamarindienal showed fungicidal activity against *Aspergillus niger* and *Candida albicans* and strong bacterial activity against *Bacillus subtilis, Staphylococcus aureus, Escherichia coli* and *Pseudomonas aeruginosa*.
Uses in Literature : Useful in India as refrigerant, carminative, laxative, anti-scorbutic, fodder, green manure, menorrhoea; and for boils, biliousness, common cold, diarrhoea, dysentery, indigestion, souring agent for sauces, chutney, beverages, dyeing, tanning, polishing and clearing metalware, jams, jellies, kibs, agricultural implements, scorpion stings, tool handles, vomiting, wounds (**Ambasta, 1986; Bhardwaj & Gakhar, 2003; Bondya *et al.*, 2006; Gogoi & Borthakur, 2001; Hembrom & Goel, 2005; Kadel & Jain, 2006; Katewa *et al.*, 2001; Kumar & Jain, 1998; Mohanty *et al.*, 1998; Pandey & Rout, 2006; Ram *et al.*, 2002; Sen & Behera, 2003; Sharma & Singh, 2001; Upadhye & Kulkarni, 2004**).

184. *Taraxacum officinale* Wigg. (Pl. 56D)

Syn. : *T. dens-leonic* Desf., *Leontodon taraxacum* Linn.
Family : Asteraceae.
Vern. : Dudhli, Dulal.
English and Regional Names:
Eng. : Dandelion, Fortune teller, Milk Gowan;
B. : Pitachumki;
G. : Pathardi;
H. : Barau, Dulal, Kanphul;
Kan. : Kaadu, Servent;
Kash. : Hamd;
Mar. : Undarkani;
P. : Baran, Bathur, Dudli, Kanphul, Shamukei.
Distribution : Throughout Temperate Himalaya.
Description : Perennial scapigerous milky herbs with stout rootstocks. Leaves basal. Heads solitary on leafless scapes. Achenes ribbed.
Flowering and Fruiting : March-November.
Habitat Ecology : Grassy meadows. Roadsides. Damp places.
Material Examined : EBH-HVP-767, 24.04.2002.
Parts Used : Leaves.
Folk Uses : ¼ teaspoon of powdered leaves taken twice a day for two days to cure **stomachache***.
Active Constituents : These are taraxin, taraxacin, taraxacerin and inulin, saponin, cerylalcohol, lactuce-roltaraxacin (rhizome and roots).
Biological Activity : Plant exhibits slight antibiotic activity against *Mycobacterium tuberculosis*.
Uses in Literature : Recorded earlier as an aperient, antiscorbutic, blood purifier, diuretic, laxative, liver tonic and vegetable; and for blisters, blood complaints, dislocation of joints, dysentery, gastric ulcers, headache, kidney disorder, liver complaints, vertigo, wounds, preparation of beer and wines **(Ambasta, 1986; Chhetri, 2005; Jain, 1991).**
Adulterant or Substitute : Leaves used as an adulterant and substitute to Henbane leaves and flower tops of *Hyoscyamus niger* Linn.

185. *Tecoma grandiflora* Loisel. (Pl. 57A)

Syn. : *Campsis grandijlora* (Thunb.) K. Schum., *Bignonia grandiflora* Thunb.
Family : Bignoniaceae.
Vern. : Shehnai Phool.
English Name : Chinese Trumpet Creeper.
Distribution : Planted.

Description : A handsome shrub. Leaflets 7-9. Flowers large, orange coloured, in drooping clusters.

Reproductive Cycle : Summer season.

Habitat Ecology : Cultivated.

Material Examined : EBH-HVP-886, 10.05.06.

Parts Used : Flowers. Plant.

Folk Uses : Plant grown for **ornamental*** purposes; and its flowers considered **sacred*** for offerings.

Active Constituents : These are furfural, 5-methylfural, furfurlyl alcohol and 2-acetylfuran (essential oil of flowers).

Uses in Literature : So far used in diabetes **(Ambasta, 1986)**.

186. *Terminalia arjuna* Wt. & Arn. (Pl. 57B)

Syn. : *T. berryi* W. & A., *T. glabra* W.& A., *T. ovalifolia* Rottl., *Pentaptera glabra* Roxb., *P. angustifolia* Roxb.

Family : Combretaceae.

Vern. : Arjun.

English, Sanskrit and Regional Names:

Eng. : Arjun, White Murdah;
Sans. : Arjunah, Dhananjaya, Karaviraka, Savya Sachi;
Ass. : Orjun;
B. : Arjun, Kahu;
Bo. : Arjunasadadra, Jamla;
G. : Arjunasadada, Dhaulasadr;
H. : Arjun, Arjun saadadaa, Kangach, Kowa;
Kan. : Maddi;
Mal. : Vellamarutu;
Mar. : Arjunladada, Saumadat;
Oriya : Orjuna, Ponoyosahajo;
P. : Arjan, Jumla;
Tam. : Attumarudu, Kulamatrudu;
Tel. : Kakubhamu, Tittumaddi;
U.P. : Arjan, Kaucha.

Distribution : Commonly along riverian areas in the sub-Himalayan tract.

Description : A large tree with huge buttressed trunk. Leaves glabrous oblong, elliptic. Flowers green, in panicle spikes. Fruit penta-winged, wings straight.

Flowering and Fruiting : May-September.

Habitat Ecology : Roadsides. Open slopes.

Material Examined : EBH-HVP-830, 11.06.2004.

Parts Used : Bark. Wood.

Folk Uses : One teaspoon of powdered bark prescribed twice a day as **cardiotonic***. Its decoction (twice a day till relief) good for **heart palpitation.** Wood used for **agricultural implements** and **building works.**

Active Constituents : Therapeutic properties attributed to cevasidine, β-sitosterol, friedelin, methyl deanolate and gallic, ellegic and arjunic acid, (+) catehol, (+) gallocatenol, epicateenol, epigalco, catechol, arjunolone, arjunglucoside III, arjunoside I and II, arjunflucoside I, arjunetin, leucoganidin, terminoic acid and arjunin, arjungenin.

Biological Activity : Stem bark abortifacient.

Uses in Literature : Reported as an antidote to poison, antidysenteric, astringent, blood purifier, detergent, expectorant, hair wash, psychoactive plant, purgative, sacred, sex stimulant; and for agricultural implements, anaemia, anthrax, blood diseases, blood dysentery, blood pressure, boats and canoes, boils, carbuncle, chest pain and heart palpitation, cardiac complaints, colic pain, cuts and wounds, earache, fever, fractures, gastric troubles, heart burn, high blood pressure, house construction, impaction, leucorrhoea, menstrual disorder, musical instrument, painful urination, plywood manufacture, pneumonia, sores, tanning, tumours, ulcers, water traps, water trough, wheels of country carts, wounds (**Ambasta, 1986; Chauhan, 1999; Datt, 1996; Gogoi & Borthakur, 2001; Hembrom, 1991; Hosagoudar & Henry, 1996a; Jain, 1968, 1991; Jain *et al.*, 1994; Khanna, 2002; Kothari & Moorthy, 1996; Kumar & Jain, 1998; Kumar & Pullaiah, 1998; Kurian, 1995; Lalramnghinglova, 1996; Masih, 2003; Rama Rao & Henry, 1996; Reddy *et al.*, 1998; Satapathy & Brahmam, 1996; Sen & Behera, 2003; Sharma & Sood, 1997; Singh & Prakash, 1994; Srivastava *et al.*, 2003; Watt, 1972**).

Major Ayurvedic Formulations : Arjuna-ghrita, Arjunakasivapalla, Arjunarishta, Arjunatwal-churna, Arjuna-kwatha, Arvindaasava, Devadarishta, Hridayarnavaasa, Kakubhaadi-churna, Nagarjunabhra-rasa, Parthadyarishta.

Trade : Arjun.

187. *Terminalia bellirica* (Gaertn.) Roxb. (Pl. 57C)

Syn. : *Myrobalamus bellirica* Gaertn.

Family : Combretaceae.

Vern. : Behra, Bahera.

English, Sanskrit and Regional Names:

Eng. : Bastard Myrobalan, Beddanuts, Belleric Myrobalan;
Sans. : Anilaghnaka, Kaliyugalaya, Karshaphala;
Ass. : Banri, Hulluch;
B. : Bahera, Buhura;
Bo. : Buhudda, Bherda, Yella;
G. : Bahedamunjad, Bahedo;
H. : Bahera, Bhairach, Sagona;
Kan. : Tare-kai, Shanti Kayi;

Kash.	:	Babelo, Balali;
Mal.	:	Tanni, Tusham;
Oriya	:	Horinda, Jonghihorida;
P.	:	Bahira, Bayrah;
Tam.	:	Akkam, Kalitturuman;
Tel.	:	Karakkaya, Nallakaroka.

Distribution : Throughout the plains. Sub-Himalayan tract.

Description : Large deciduous trees with ashy grey bark. Leaves alternate, ovovate-elliptic, coriaceous. Flowers yellow, in spike. Fruit long, ovoid, velvety.

Flowering and Fruiting : May-September.

Habitat Ecology : Open slopes. Roadsides.

Material Examined : EBH-HVP-773, 15.06.2002.

Parts Used : Fruit. Seed.

Folk Uses : 'Triphla' powder obtained after grinding 'harar' (*Terminalia chebula* Retz.): 'behra' (*Terminalia bellirica* (Gaertn.) Roxb.): 'amla' (*Emblica officinalis* Gaertn.) in the ratio of 1:2:3 considered good **appetizer*, laxative, liver tonic*** and for **piles*.** If taken on empty stomach with lukewarm water it cures **bronchitis*** and **gastrointestinal problems.**

Active Constituents : Fruit pulp comprises hexahydroxydiphenic acid ester, β-sitosterol, gallic acid, ellegic acid, ethyl gallate, galloyl glucose, chebulegic acid, and cardioglycoside bellericanin.

Biological Activity : Anticancerous activity +ve.

Uses in Literature : Known in India as anthelmintic, antiseptic, aperient, astringent, brain tonic, colic, diuretic, expectorant, fish poison, fodder best for milching cows, ornamental around temple yard, important constituent of Triphala, laxative, narcotic, psychoactive plant, vegetable and writing inks; and for acidity, agricultural tools, asthma, black hairs, boards, boats, carts, cholera, coffee boxes, cold, colic pain, constipation, diabetics, diarrhoea (vet.), dropsy, enlargement of spleen, fever, foot and mouth diseases, gastric complaints, headache, hoarseness, inflammation, inflammatory disease of eyes, jaundice, leprosy, liver complaints, muscular pain, piles, plough, skin disease, slate frames, small crafts, snake bite, soap manufacture, stomach pain, tannin, throat trouble, traditional dwelling, turning, ulcer (vet.), vigour, vitality, vomiting and weakness **(Ambasta, 1986; Aminuddin & Girach, 1993; Borthakur, 1996b; Borthakur & Gogoi, 1994; Chauhan, 1999; Khanna, 2002; Kothari & Moorthy, 1996; Kumar & Pullaiah, 2003; Lalramnghinglova, 1996; Masih, 2003; Mishra *et al.*, 1997; Mohanty *et al.*, 1998; Nazarudeen *et al.*, 1996; Patel *et al.*, 2003; Patil & Merat, 2003; Prasad *et al.*, 1996; Pullaiah & Kumar, 1996; Rama Rao & Henry, 1996; Rao, 1996; Sharma & Borthakur, 2003; Sharma & Singh, 2001; Sharma & Sood, 1997; Singh, 1999b;**

Siwakoti & Verma, 1996; Tirkey *et al.*, 2003; Venkataratnam & Venkata Raju, 2004; Verma *et al.*, 1995; Watt, 1972).

Major Ayurvedic Formulations : Agnitundi-vati, Arogyavardhini-vati, Bibhitak-taila, Chandraprabha-vati, Gokshuvadi-guggulu, Herbotone, Karanjadiyoga, Khadiraista, Kutajavaleha, Lavangudi-vati, Navayasloaha, Unarnava-mandur, Sanjivani-vati, Triphadi-taila, Triphala-churna, Triphala-kwatha, Vidangadi-louha.

Trade : Baehra, Belleric myrobalam.

188. *Terminalia chebula* Retz. (Pl. 57D)

Syn. : *T. aruta* Ham., *T. reticulata* Roth., *Myrobalanus chebula* Gaertn., *Embryogonia arborea* Teys & Binn.

Family : Combretaceae.

Vern. : Harar.

English, Sanskrit and Regional Names:

Eng. : The Chebulic or Black Myrobalan;
Sans. : Haritaki, Harra;
Ass. : Haritaki, Silikha;
B. : Hirda, Haritaki;
G. : Hardo;
H. : Harna, Har, Harar, Harra;
Kan. : Hirda;
Mal. : Kadukka;
Mar. : Hirao, Hirda;
Oriya : Karedha, Haridra;
P. : Har, Harar;
Tam. : Kadakai, Tadukkai;
Tel. : Karaka;
U.P. : Harr.

Distribution : Throughout India.

Description : A moderate-sized tree with glabrous, alternate, elliptic-oblong leaves having acute rounded base. Flowers sessile, yellow, with an offensive smell. Drupes pendulous, ellipsoid, glabrous, when dry yellowish-green. Stone bony, very thick, obscurely angled.

Flowering and Fruiting : April-June.

Habitat Ecology : Open dry slopes. Along agricultural fields.

Material Examined : EBH-HVP-754, 12.04.2002.

Part Used : Fruit.

Folk Uses : Powdered fruits constitute one of the important constituents of '**Triphala**' powder. ½ teaspoon of 'harar murabba' (fruits preserved in sugary syrup) with milk at night is a good **laxative**. The mixture of 'harar' (*Terminalia chebula* Retz.), 'behra' (*Terminalia bellirica* (Gaertn.) Roxb.), 'amla' (*Emblica officinalis* Gaertn.) and 'mehendi' (*Lawsonia alba* Lam.)

considered good **dye** for greying hair. ½ teaspoon of powdered fried harar with honey given twice a day to cure **cough**. Also, 1 teaspoon of powdered 'harar' and 'saunf' (*Foeniculum vulgare* Mill.) fried in ghee taken once daily with hot water to check **dysentery.** The fruit in combination with almond, Belliric myrobolan and Indian gooseberry given to cure **diabetes.**

Active Constituents : Fruits rich in ellagitannin, terchebulin, punicalagin, terflavin A, B, C and D, shilimic acid, gallic triacontanoic and palmitic acid, β-sitosterol, daucosterol, triethyl ester of chebulic acid and ethyl ester of gallic acid, arjungenin, terminoic acid, arjunolic acid and terchebin.

Biological Activity : Antibacterial and spasmolytic activity positive.

Uses in Literature : Used so far as an alternative, astringent, blood purifier, cardiac tonic, colic, dye for hairs, gargles in inflammation of mucous membrane of mouth, fodder, important constituent of Triphala, ornamental, psychoactive plant, sacred for making ink; and for acidity, stomach pain, agricultural tools, asthma, boards, boats, body pains, carts, cholera, coffee boxes, cold, constipation, diabetes, diarrhoea, dropsy, enlargement of spleen, fever, gastric complaints, headache, hoarseness, inflammation, inflammatory disease of eyes, jaundice, leprosy, liver complaints, muscular pain, piles, plough, skin disease, slate frames, small crafts, snake bite, soap manufacture, sweating, tannin, throat trouble, turning, ulcer (vet.), ulceration of gums, vaginal discharge, vigour and vitality, vomiting, worms, wounds **(Ambasta, 1986; Aminuddin & Girach, 1993; Borthakur, 1996a, b; Borthakur & Gogoi, 1994; Chowdhery, 1996; Girach, 1992; Gogoi & Borthakur, 1991, 2001; Hosaoudar & Henry, 1996a; Jain, 1986; Jain & Patole, 2001; Jain *et al.*, 1994; Khanna & Kumar 2000; Kumar, 1996; Kshirsagar & Singh, 2000; Lalramnghinglova, 1996; Masih, 2003; Mohanty & Padhy, 1996; Mohanty *et al.*, 1998; Rama Rao & Henry, 1996; Sarangi & Sahu, 2004; Sharma & Borthakur, 2003; Sharma & Sood, 1997; Srivastava *et al.*, 2003; Tirkey *et al.*, 2003; Verma & Chauhan, 2006; Verma *et al.*, 1995; Watt, 1972).**

Major Ayurvedic Formulations : Abhaya-lavana, Abhaya-modak, Abhayarishta, Agastya-haritaki-rasayana, Amrita-haritaki Avaleha, Brahma-rasayana, Chandraprabha-vati, Dantiharitaki, Gokshuraadi-gugglu, Haritaki-churna, Karanjaadi-yoga, Medigatone, Panchraska-churna, Sanjivani-vati, Sudarshan-churna, Triphala, Triphala-churna, Triphaladi-taila, Triphala-ghrita.

Trade : Myrobalam, Chebulic-myrobalam (tree and fruit), Harada, Jangli Harad (for small variety).

189. *Thevetia neriifolia* Juss. *ex* Steud. (Pl. 58A)

Syn. : *T. peruviana* (Pers.) K. Schum., *Cerbera peruviana* Pers., *Cascabela thevetia* (L.) Lipp.

Family : Apocynaceae.

Vern. : Kaner.

English, Sanskrit and Regional Names:

Eng. : Bastard Oleander, Exile Oleander, Lucky Nut Tree, Yellow Oleander;
Sans. : Ashvamaraka, Nakharatlia, Sidhapushpa;
B. : China-Karab, Kokilphul;
Bo. : Pilakaner, Zardakunel;
G. : Pilakanir;
H. : Pila, Pila kananir;
Kan. : Kadukasi; Kanogalu;
Mal. : Pachchaarali;
Mar. : Pivalakanhera;
Oriya : Konyar-phul;
Tam. : Pachaiyalari, Tiruvachippu;
Tel. : Pila Kaner, Pit Kaner.

Distribution : Throughout India.

Description : Shrubs. Leaves linear-lanceolate. Flowers yellow. Drupes angular, poisonous.

Flowering and Fruiting : July-September.

Habitat Ecology : Forest. Shrubberies.

Material Examined : EBH-HVP-829, 17.07.2004.

Parts Used : Leaves. Plant.

Folk Uses : Powdered leaves with honey (1 teaspoonful thrice daily till cure) prescribed as **blood purifier***. Planted in temple courtyards as an **ornamental**.

Active Constituents : Important constituents of fruits are arabinose, galactose, glues, galacturonic acid and xylose.

Biological Activity : Cardiotonic and isotonic effects and musculotropic activity confirmed.

Uses in Literature : Useful as abortifacient, aeronarcotic poison, alexetric, cardiotonic against pentobarbital induced heart failure, emetic, laxative, ornamental (beads), poison (vet.), vegetable; and for carotene disease, dropsy, impotency, skin disease, snake bite, swelling (vet.), swelling of foot, toothache, tumours and worms **(Ambasta, 1986; Bhandary *et al.*, 1996; Chauhan, 1999; Huidrom, 1996; Jain, 1991; Jha *et al.*, 1996; Kurian, 1995; Mishra *et al.*, 1996a; Pandey & Verma, 2002; Saini, 1996a; Sharma & Sood, 1997; Singh & Pandey, 1996; Siwakoti & Verma, 1996; Watt, 1972).**

Major Ayurvedic Formulations : Used in traditional preparation and its glucosides have been found to have cardiotonic activity.

Warning : All parts of the plant contain cardiac glycosides and considered poisonous particularly seed kernels and milky juice from all parts of plant.

190. *Thuja orientalis* L. (Pl. 58B)

Syn. : *Biota orientalis* (L.) Endl.
Family : Cupressaceae.
Vern. : Morpankh.
English and Regional Names:
Eng. : Oriental Arbor-Vitae;
H. : Mayur Pankh, Morphankhi.
Distribution : Cultivated.
Description : An aromatic evergreen, gymnospermous shrub. Leaves scale-like, overlapping. Cone compressed, pyriform.
Flowering and Fruiting : July-October.
Habitat Ecology : Cultivated.
Material Examined : EBH-HVP-734, 17.07.2001.
Parts Used : Leaves.
Folk Uses : Planted for **ornamental*** purposes. Leaves kept in books by children as '**vidya patri**'* for gift of knowledge from goddess Saraswati.
Active Constituents : Pimeric and isopiameric acid (Leaves).
Biological Activity : Quercitin exhibits haemostatic activity in mice.
Uses in Literature : Known in India for barrels and casks, dysentery, fencing posts, furniture, house building and tannin; and also as antipyretic, tonic, diuretic, psychoactive plant **(Ambasta, 1986; Jain *et al.*, 1994; Singh & Prakash, 1994).**

191. *Tinospora cordifolia* (Willd.) Miers (Pl. 58C)

Syn. : *T. palminervis* Miers, *Cocculus. cordifloius* DC., *C. convolvulaceus* DC., *C. verrucosus* Wall., *Menispermum cordifolium* Willd.
Family : Menispermaceae.
Vern. : Galon, Gulja, Giloy.
English, Sanskrit and Regional Names:
Eng. : Gulancha Tinospora, Heart:leaved Mvonseed;
Sans. : Amrita, Amritalata, Chakrangi, Dhira, Jwarana-shini, Guluchi, Guduchi, Jirantiha, Madhuparnika, Nirajara, Tantri, Vara, Vishalya;
Ass. : Amarlata, Gulancha, Siddhilata;
B. : Gadancha, Giloe, Gulancha;
Bo. : Ambarrel, Gharol, Giroli, Guloe, Gulwel;
G. : Gado, Galo, Gulo;
H. : Giloe, Gulancha, Gulbel, Gurach, Gurcha;
Kan. : Amrita balli, Amruta balli;
Kash. : Amrita Gilo;
Mal. : Amrytu, Peyamrytam, Sittamrytu;
Mar. : Ambarvel, Gharol, Giroli, Gulavela, Gulvel;
Oriya : Gulancha, Gulochi;
P. : Batindu, Gilogularich, Garum, Gilo;

Tam. : Amudam, Amridavalli, Kaippuchindil, Kunale;
Tel. : Duyutige, Guduchi, Jivatika, Somida;
U.P. : Gurch.

Distribution : Throughout tropical India.

Description : A large glabrous climber. Stem succulent corky. Branches sending down slender pendulous fleshy roots. Leaves membranous, roundish, deeply cordate. Flowers yellowish-green.

Flowering and Fruiting : May-July.

Habitat Ecology : Wastelands. Common in hedges.

Material Examined : EBH-HVP-758, 18.05.2002.

Parts Used : Stem. Leaves.

Folk Uses : Stem and leaves mixed with fodder given to improve **lactation*** in animals. Pieces of dried stem used in **'havan'*** for purification of surroundings. ½ teaspoon of powdered stem given twice a day for 15 days to cure **abdominal diseases, congestion of liver** and **persistent fever**. 15-20 ml. of decoction of fresh stem considered good against **rheumatic*** ailments (twice daily for 20 days).

Active Constituents : Therapeutic properties attributed to giloin, giloinin, gilosterol and berberine (stem bark), tinosporin, tinosporon, tinosporol, tinosporic acid, tinosporoside, cordifolide and a furanoid diterpene identical with columbin, clerodane, clerodane furanoditerpene I, II, III, IV (plant).

Biological Activity : Anti-inflammatory, analgesic, anticancer, antipyretic, antiviral, CNS depressant, and hypoglycaemic activity found +ve.

Uses in Literature : Recorded in literature as an antiperiodic, antipyretic, antidote to snake bites, aphrodisiac, colic (vet.), expectorant, fodder, psychoactive plant, febrifuge and tonic; and for anthrax, asthma, bone fracture, cough, debility, diarrhoea, diptheria, dysentery, fever, gonorrhoea, headache, jaundice, malaria, nocturnal emission, piles, pimples, poultry diseases, pulmonary tuberculosis, sexual strength, urinary diseases, venereal complaints, colic and stomach ailments in animal; and also for fibre for tying bundle, ingredient of 'Chyawanaprasha', anthrax (vet.), chronic gonorrhoea, gastric disorder, headache, indigestion (vet.), liver disorder, malaria, reducing bile production, ulcer (vet.), urinary disease, vein injuries, venereal complaints and visceral obstruction **(Ambasta, 1986; Chopra *et al.*, 1956; Goud & Pullaiah, 1996b; Jain & Saklani, 1992; Jain, 1968, 1991; Jain *et al.*, 1994; Jha *et al.*, 1991; Katewa & Arora, 1997; Kshirsagar *et al.*, 2003; Lalramnghinglova, 1996; Mahato & Mahato, 1996; Mehrotra *et al.*, 1995; Pandey & Verma, 2003; Patil & Merat, 2003; Kumar & Pullaiah, 1998; Reddy *et al.*, 1996, 1998; Sen & Behera, 2003; Silori & Rana, 2000; Singh & Pandey, 1996; Srivastava *et al.*, 2003; Verma & Chauhan,**

2006; Verma *et al.*, 1995; Watt, 1972).

Major Ayurvedic Formulations : Amritaashtaka-kwatha, Amritasava, Chandraprabha-vati, Giloyasatva, Guduchyadi-kwatha, Kaishore-guggulu, Rasna-panchak-kwatha, Sanjivani-vati, Sudarshan-churna.

Trade : Giloy or Gulancha (Dried stem).

192. *Trifolium alexandrinum* L. (Pl. 58D)

Family : Fabaceae.
Vern. : Barseem.
English Names : Barseem, Egyptian Clover.
Distribution : Cultivated for green manure.
Description : Annual herbs. Leaves trifoliate. Flowers clustered in dense head.
Flowering and Fruiting : March-June.
Habitat Ecology : Cultivated for green manure.
Material Examined : EBH-HVP-839, 15.04.2004.
Parts Used : Whole Plant.
Folk Uses : Whole plant used as **fodder.**
Active Constituents : Contains protein, calcium and phosphorus.
Uses in Literature : Known so far as green fodder, silage, hay, pasture **(Ambasta, 1986, Khanna, 2000).**

193. *Trigonella foenum-graecum* Linn. (Pl. 59A)

Family : Fabaceae.
English, Sanskrit and Regional Names:
Eng. : Fenugreek, Greek Hayes;
Sans. : Methi;
B. : Methi;
G. : Methi;
H. : Methi, Muth;
Kan. : Menthya;
P. : Methi;
Tam. : Vendayam;
Tel. : Mentulle.
Distribution : Kashmir-Punjab, Upper Gangetic plains.
Description : An erect herb. Leaflets shortly serrate from the middle upward. Flowers yellow. Pod glabrous, curved.
Flowering and Fruiting : May-August.
Habitat Ecology : Cultivated.
Material Examined : EBH-HVP-865, 10.08.2005.
Parts Used : Leaves. Seed.
Folk Uses : 5-10 g. powdered seeds taken daily with cold water as an easy

remedy for **diabetes*** and if taken with hot water it cures **gastric problems**. Leaves used in preparation of **'saag'** which is relished along with 'makki-ki-roti' (a thick 'chappatti' prepared from maize floor) by one and all alike.

Active Constituents : Active principles are due to β-sitosterol, kaemferol, quercetin, saponins-graecumins A, B, C, D, E, F and (leaves); tigogenin, neotigogenin, diosgenin, yamogenin, gitogenin and 4-hydroxy isoleucine, vitexin, iso-vitexin, saponaeretin, homoorientin, vicenin-1 and vicenin-2 (seeds).

Uses in Literature : Known earlier for acne, baldness, boils, enhance lactation (vet.), gastric trouble, gonorrhoea, odema, piles, rheumatism, sores, syphilis and ulcers; and as an aphrodisiac, carminative, emollient in intestinal inflammation, fodder, galctagogue, tonic, pot herb and spice **(Ambasta, 1986; Chopra *et al.*, 1956; Jain, 1991; Kumar *et al.*, 2004; Sen & Behera, 2003; Uphof, 1968).**

Adulterant or Substitute : Seeds considered a good alternative source of *Dioscorea deltoidea* Wall. *ex* Kuntz.

194. *Triticum aestivum* Linn. (Pl. 59B)

Syn. : *T. sativum* Lam., *T. vulgare* Vill.

Family : Poaceae.

Vern. : Gehun.

English, Sanskrit and Regional Names:

Eng. : Common Wheat, Bread Wheat;
Sans. : Godhuma, Mahagodhuma, Madhuli, Mihsuki;
B. : Gam, Giun, Gom;
G. : Ghavum, Govum, Gawn;
H. : Gehun, Giun, Gandham, Kanak;
Kan. : Godhi;
Mal. : Gendum, Godamba, Kotanpam;
Mar. : Gahun, Gahung;
P. : Gehun, Kanak;
Tam. : Godumai, Godumbayarisi;
Tel. : Goodhumalu.

Distribution : Throughout India.

Description : Tall cultivated herbs. Leaves broadly linear. Stem clasping. Spikelets awned.

Flowering and Fruiting : November-April.

Habitat Ecology : Cultivated.

Material Examined : EBH-HVP-740, 14.11.2001.

Parts Used : Grain. Whole plant.

Folk Uses : Plant considered **sacred***. Its grain boiled **eaten** or flour used for making 'chapatis' and '**prasadam for offering'*** to deities on every festive and religious occasion. Wheat straw used as **fodder** and **bedding** for cattle to rest comfortably.

Active Constituents : Gluten (grain).

Uses in Literature : So far used for anaemia, bedding for cattle, cooling, fattening, gastric problems (vet.), increase appetite, chapatis, loaf bread, biscuit, alcohol, malted wheat; and as laxative, tonic **(Ambasta, 1986; Kumar *et al.*, 2004).**

195. *Verbascum thapsus* Linn. (Pl. 59C)

Syn. : *V. indicum* Wall., *V. blattaria* Schmidt.

Family : Scrophulariaceae.

Vern. : Ban tambakhoo.

English and Regional Names:

Eng. : Common Mullein, Cow's Lungwort;
H. : Gidar tamaku;
Kan. : Kaduhogesooppina gida;
P. : Gidar tamaku, Gurganna, Karathri, Phasruk, Vulr.

Distribution : Temperate Himalaya. Kashmir-Bhutan.

Description : Erect herbs densely clothed with soft, yellow grey stellate hairs. Stem robust winged with prolonged leaf bases. Leaves entire, radical; lower stalked, ovate, upper sessile, oblong-lanceolate. Flowers sessile, yellow. Capsule tomentose, ovoid.

Flowering and Fruiting : January-April.

Habitat Ecology : Wastelands. Open slopes. Rocky slopes. Cultivated areas.

Material Examined : EBH-HVP-799, 15.01.2004.

Parts Used : Aerial plant parts. Leaves.

Folk Uses : Poultice of hot leaves good for **joint pains***. Powdered aerial plants parts mixed with dough given to cure **constipation*** in animals.

Active Constituents : Isolated principles are rotenone and coumarin (leaves) and crocetin (flowers).

Uses in Literature : Already known as an aphrodisiac, demulcent in pectoral complaints, emollient, febrifuge, fish poison, heart stimulant and symbolic to keep evils away; and for asthma, cough, diarrhoea, inflammation, leucorrhoea, pulmonary disease of cattle, cramps and migraine **(Ambasta, 1986; Chandra Sekar & Srivastava, 2003; Chopra *et al.*, 1956; Jain, 1991).**

Adulterant or Substitute : Dry leaves substituted for that of tobacco leaves.

196. *Vitex negundo* L. (Pl. 59D)

Syn. : *V. bicolor* Willd., *V. arborea* Desf., *V. paniculata* Lamk., *V. incisa* Lamk.

Family : Verbenaceae.

Vern. : Bana.

English, Sanskrit and Regional Names:

Eng. : Common Chaste tree, Indian Privet, Nigandi;
Sans. : Indrani, Nilapushpa, Shepkali, Svetasuraso;
Ass. : Aggla-chita, Pasutia;
B. : Nishinda, Samalu;
Bo. : Katri, Lingur, Nirgur, Shiwari;
G. : Nagaol, Nigod, Nirgari;
H. : Mewri, Panikisambhalu, Nirgori, Nirgunda, Sanbhalu, Shiwali, Sindhuca, Sinduari;
Kan. : Lakkigida, Nakkilu, Nekki;
M.P. : Nagad;
Mal. : Indrani, Nochi, Vellanochi, Vennochi;
Mar. : Lingur, Nirguda, Nirgundi, Nirgur;
Oriya : Begundia, Indrani;
P. : Bankahu, Moraum;
Tam. : Nirkkundi, Nochi, Siduvaram, Tiributamerirran, Vennocho;
Tel. : Nellavavili, Sinduvaramu, Valili;
U.P. : Tarvan, Kali-Nirgundi.

Distribution : Throughout India.

Description : A large shrub or small tree. Leaves foliolate, opposite, long-stalked. Leaflets lanceolate-acute. Flowers bluish-purplish, in branched cymes. Drupes black when ripe.

Flowering and Fruiting : March-June.

Habitat Ecology : Wastelands. Roadsides. Dry areas.

Material Examined : EBH-HVP-798, 16.03.2003.

Parts Used : Leaves, Buds.

Folk Uses : Inhalation of steam emanating from hot water containing its leaves is good for **pains** due to **inflammation of joints**. Its smoke repels **insects**. Buds with 'gur' (jaggery) given to cure **stomach disorders**. A garland of its branches and leaves is hung on main door to ward off **communicable diseases***. Stem used for **scouring teeth*** and also chewed for healing **mouth sores**.

Active Constituents : Various constituents are iridoid glycosides, cucubin, agnuside, nishindaside, negundoside (leaves), vanitin and as mixture of unidentified alkaloids (seeds).

Biological Activity : Extr. of leaves exhibits anticancerous activity against *Ehrlich ascites* tumour cells. Also, leaf possesses anti-inflammatory, analgesic and CNS depressant activity.

Uses in Literature : Reported earlier as an abortifacient, astringent, cephalic, diuretic, expectorant, insect repellent, sterilizer, stomachic, tonic, symbolic to keep evils away, basketry, vermifuge; and for anasarca, blister, boils, bone fracture, bodyache, cold, colic, cough, diarrhoea, dropsy, dysentery, earache, eczema, eye complaints, fever, foetid discharge, foot diseases, gonorrhoea, epididymitis, giddiness, gout, gum trouble, headache, inflammation, itching, leprosy, liver disorder, mental disorder, paralysis, piles, pneumonia, post-natal debility, reducing sex desire, rheumatism, scabies, sciatica, skin diseases, sores, spleen enlargement, swelling, syphilis, testies disorder, toothache, ulcers, veterinary ailments, waist pain and wounds (**Ambasta, 1986; Aminuddin & Girach, 1996; Bhandary *et al.*, 1996; Bhatt *et al.*, 2001; Borthakur *et al.*, 2004; Chhetri, 2005; Geetha *et al.*, 1996b; Gogoi & Borthakur, 2001; Gogoi *et al.*, 2003; Hembrom & Goel, 2005; Jain *et al.*, 1994; Katewa & Arora, 1997; Khanna *et al.*, 1969; Kshirsagar & Singh, 2000; Lal *et al.*, 1996; Mahato & Mahato, 1996; Painuli & Maheswari, 1996; Pandey & Rout, 2006; Ragupathy & Mahadevan, 1991; Ram *et al.*, 2002; Rama Rao & Henry, 1996; Reddy *et al.*, 1998; Saini, 1996a; Sharma & Sood, 1997; Singh *et al.*, 1996b, Srivastava *et al.*, 2003, Watt, 1972**).

Major Ayurvedic Formulations : Himacolin cream, Nirgundi-ghrita, Nirgundi-kwatha, Pilex, Rumalayas (cream and tablets), Varnasodhana-taila.

Trade : Nirgundi-phal, Samhaalu-phal.

197. *Vitis jacquemontii* Parker (Pl. 60A)

Syn. : *Ampelocissus jacquemontii* (Parker) Raizada

Family : Vitaceae.

Vern. : Gedar dhakh.

Distribution : Sub-tropical regions of India.

Description : A stout woody climber with branched tendrils. Leaves ovate-cordate toothed, woolly hairy beneath. Flowers in leaf opposed dense cluster.

Flowering and Fruiting : June-August.

Habitat Ecology : Forests. Along Roadsides.

Material Examined : EBH-HVP-872, 25.08.2005.

Part Used : Fruit.

Folk Uses* : Wild grapes **edible*** and considered as **laxative*.**

Uses in Literature : There is no mention of its use in literature.

198. *Wendlandia heynei* (Roem. & Schult.) Sant & Merch. (Pl. 60B)

Syn. : *W. exserta* (Roxb.) DC., *W. cinerea* Wall., *Rondeletia exserta* Roxb., *R. cinerea* Wall., *R. thrysiflora* Roth., *R. orissensis* Roth., *R. heynei* Roem. & Sch.

Family : Rubiaceae.

Vern. : Panseera.

Regional Names:

Ass. : Kadam;

B. : Mimri, Tilki;

	H.	: Birsa, Chaulai, Chila, Chilkiya, Tila, Tilki;
	Kan.	: Tilai;
	Mar.	: Padgira;
	P.	: Chilkiya, Pansara, Pudhara;

Distribution : Abundant in dry forest of Tropical Himalaya ascending to 1,350 m.

Description : A small tree with terete branches. Leaves coriaceous. Flowers white, fragrant, sessile in panicles. Capsule globose, many seeded.

Flowering and Fruiting : April-June.

Habitat Ecology : Wasteplaces. Edges of cultivated fields. Slopes. Dry forests.

Material Examined : EBH-HVP-840, 15.04.2004.

Parts Used : Leaves. Twigs.

Folk Uses : Twigs used as **tooth brush***. Leaves used as **fodder** and its juice helps in **expelling leeches*** from nostrils of livestock.

Uses in Literature : Used as fodder; and for skin diseases, snake bites, urinary complaints, agricultural implements and building works (**Ambasta, 1986; Jain, 1991; Murthy *et al.*, 1986; Saxena, 1986; Watt, 1972**).

199. *Woodfordia floribunda* Salisb. (Pl. 60C)

Syn. : *W. fruticulosa* (L.), Kurz., *W. tomentosa* Bedd., *Lythrum fruticosum* L., *Grislea tomentosa* Roxb., *G. uniflora* Rich., *G. punctuata* Ham., *G. micropetala* Hocsht. *ex* Steud.

Family : Lythraceae.

Vern. : Dhai, Dhavae, Dhavi, Dhaun.

English, Sanskrit and Regional Names:

Eng.	: Fire Flame Bush, Red Bell Bush, Shiranjitea Woodfordia;
Sans.	: Agnijvala, Bahupusphika, Dhataki, Reyya-manu, Rodhrapushpini;
Ass.	: Eda, Era;
B.	: Dhadki, Dhavanji, Dawa;
Bo.	: Dhavri, Dowari;
G.	: Dhawadina;
H.	: Dawi, Dhai Dhaura, Thawi;
Kan.	: Dhataaki, Huvvu, Tamrapushpi;
Kash.	: Thavi;
Mal.	: Tattivipu, Tatire;
Mar.	: Dhaiphal, Dhazatichi, Phulsatti;
Oriya	: Datubiphala, Horibadi;
P.	: Dawi, Gulbahar, Khurd;
Tam.	: Thatarijargi, Velakkai;
Tel.	: Gaddapisniki, Gaji-Godari, Kusumamu Reyyamanu.

Distribution : Throughout India.

Description : Evergreen shrubs having opposite, lanceolate, pointed leaves. Flowers bright red, tubular in axillary cluster. Capsule included.

Flowering and Fruiting : March-June.
Habitat Ecology : Light forests. Open slopes.
Material Examined : EBH-HVP-770, 20.03.2003.
Parts Used : Flowers. Wood.
Folk Uses : Application of flower poultice on forehead relieves **headache***. Also, 1 teaspoon decoction of flowers, twice a day for 7-10 days regulate **liver disorders.** Wood used as **fuel***.
Active Constituents : Properties attributed to cyaniding-3, 5-diglucoside, pelargonidin-3, 5 diglucoside, myricetin-3-diglucoside, polystachoside, quercetin-3-rhamnoside, naringenin-7-glucoside, kaempferol-3-glucoside, chrysophanol-8-D-β-glucopyranoside (flowers), a new C-glucoside-norbergenin (stem), ellagic acid, polystachoside, myricetin-3-galactoside and pelargonidine-3, 5-diglucoside (leaves, flowers).
Biological Activity : Flowers abortifacient and CNS depressant.
Uses in Literature : Useful earlier as an astringent, febrifuge, stimulant, red dye, cooling drink, fuel wood, tonic; and for bowl complaints, bleeding in pregnancy and menstruation, bone fracture, burns, cancer, carbuncle, cholera, cooling, conception, conjunctivitis, cough, complaints of liver, diarrhoea, dysentery, edible purposes, injuries, loss of appetite in pregnancy, menorrhoea, miscarriage, muscular pain, nausea, night blindness, rheumatism, seminal weakness, skin diseases, uterus pain, vomiting, wounds and sores, ulcers in animals and eaten raw or cooked **(Ambasta, 1986; Aminuddin & Girach, 1996; Borthakur, 1993; Chhetri, 2005; Hembrom & Goel, 2005; Jain, 1991; Kothari & Moorthy, 1996; Kulkarni & Kumbhojkar, 1996; Masih, 2003; Mohanty *et al.*, 1998; Pandey & Rout, 2006; Patil, 2004; Patole & Jain, 2002; Rama Rao & Henry, 1996; Sarangi & Sahu, 2004; Singh, 1999b; Verma & Chauhan, 2006; Watt, 1972).**
Major Ayurvedic Formulations : Arvindasava, Brihatgangadhara-churna, Dhatakyadi-taila, Gangadhar-churna, Kutajarishta, Laghugangadhar-churna, Pushayanug-churna.
Trade : Dhaataki-phul, Dhai phool.

200. *Xanthium strumarium* Linn. (Pl. 60D)

Syn. : *X. indicum* DC., *X. roxburghii* Discolor, *X. orientale* Blume.
Family : Asteraceae.
Vern. : Bhakhadda.
English, Sanskrit and Regional Names:
Eng. : Bur-weed, Cockle bur;
Sans. : Arishta, Chanda, Hara, Pitapushpi;
Ass. : Agara;

B. : Banokra;
Bo. : Shankeshvara;
Guj. : Gadriyun;
Hin. : Banokra, Chhotgokhru;
Kash. : Lanetsuru;
Mal. : Buah anjang;
Mar. : Dumundi, Sankeshwara;
P. : Chirru, Gudal;
Tam. : Marlumutta;
Tel. : Marulamatangi, Parsvapu.

Distribution : Throughout the hotter parts of India.

Description : Coarse annual, unarmed herbs. Leaves 3-lobed and greenish-white. Heads in terminal and axillary racemes. Achenes clothed with strong hooked spines.

Flowering and Fruiting : September-November.

Habitat Ecology : Open slopes. Wastelands. Roadsides.

Material Examined : EBH-HVP-736, 19.09.2001.

Part Used : Seeds.

Folk Use : Seeds eaten raw to cure **headache**.

Active Constituents : Active properties are due to sesquiterpene lactone-xanthinin, its stereoisomers-xanthinim and xanthatin.

Biological Activity : Root hypoglycaemic.

Uses in Literature : Used in India as a bitter tonic, demulcent, diaphoretic, diuretic, drink, emollient, pot-herb, sedative and sudorific; and for boils, cancer wounds, cooling, eye diseases, fencing, headache, herpes, inflammatory swellings, killing maggots, leucorrhoea, malaria, menorrhoea, night blindness, piles, rheumatism, ringworm, scrofula, snake-bite toothache, ulcers and urinary complaints **(Ambasta, 1986; Jain, 1991; Kirtikar & Basu, 1935; Patil & Merat, 2003; Punjani, 1998; Venkataratnam & Venkata Raju, 2004)**.

201. *Zanthoxylum armatum* DC. (Pl. 61A)

Syn. : *Z. alatum* Roxb., *Z. planispinum* Sieb & Zucc., *Z. hostile* Wall.

Family : Rutaceae.

Vern. : Tirmer.

English, Sanskrit and Regional Names:

Eng. : Wing leaf prickly ash;
Sans. : Andhaka, Dhira, Gandhalu, Mahamuni, Sanuja, Sauravanaja, Saurabh, Sugandhi, Tumburn, Tumbaru;
B. : Gaira, Nepali-dhamia, Tambul, Tun;
G. : Tejabala, Tejbal;
H. : Darmar, Tejbal, Tirmur, Tezmal, Timbur, Tumru;
Kan. : Jimmi, Tejovanti, Tejpatri, Tumbura, Tumburudra;
Mal. : Thumboonal, Thunbooni, Valiyavaluzhavam;
Mar. : Tejabal;
Oriya : Tundopoda;

P. : Tezbal, Timmal, Timru, Kababa;
Tam. : Thejyovathi;
Tel. : Tejovathi.

Distribution : Hot Valleys of subtropical Himalaya.

Description : A shrub or small tree with corky bark and numerous long straight spines on branchlet and leaf stalk. Leaves pinnate. Flowers yellow in short lateral clusters.

Flowering and Fruiting : April-June.

Habitat Ecology : Open slopes. Dry areas.

Material Examined : EBH-HVP-831, 17.04.2004.

Parts Used : Bark. Twigs. Seed.

Folk Uses : Gargle of bark decoction cures **mouth ulcer***. Twigs used for **scouring teeth**. Paste of seeds and bark mixed with a pinch of rock salt and black pepper (*Piper nigrum* Linn.) considered good for **dental problems**.

Active Ccnstituents : Properties attributed to dictamnine (stem-bark), 1.5% essential oil (wartara oil), a source of linalool (fruit), ketone free fraction of essential oil (from leaves) linalyl acetate, sesquiterpene hydrocarbon and tricabane, β sitosterol, pinoresionol, dimethyl ether a mixture of aerial parts (aerial parts).

Biological Activity : Stem exhibits hypoglycaemic and spasmolytic properties. Essential oil from fruits shows antifungal activity.

Uses in Literature : Known to be used as an anthelmintic, carminative, deodorant, disinfectant, fish poison, spice, stomachic and tonic; and for cough, cholera, dental troubles, diarrhoea, fever, flatulence, eczema, itching, leucoderma, madness, piles, rheumatism, small pox, snake bite and indigestion, scouring teeth, toothache, making walking sticks and clubs, religious ceremonies, rheumatism, soap making, warding off house flies, house building **(Ambasta, 1986; Ansari, 1991; Borthakur & Sarma, 1996; Chauhan, 1999; Chhetri, 2005; Chowdhery, 1996; Jain, 1991; Kapur & Singh, 1996; Kirtikar & Basu, 1935; Lal *et al.*, 1996; Manandhar, 1996; Mohanty *et al.*, 1998; Negi & Gaur, 1991; Ravisankar & Henry, 1992; Siwakoti & Verma, 1996; Verma & Chauhan, 2006; Watt, 1972).**

Major Ayurvedic Formulations : Brahmi-vati, Dantamanjan, Hingodi-taila, Kalaka-churna (Lepa), Tumbarwadi-churna.

Trade : Timroo or Tejbel (dried fruits), Tumbru.

202. *Zea mays* L. (Pl. 61B)

Family : Poaceae.

English, Sanskrit and Regional Names:

Eng. : Maize;
Sans. : Yavanala;

B.	:	Bhutta;
Bo.	:	Makai;
H.	:	Makai;
Kan.	:	Mekkejola;
Mal.	:	Cholam;
P.	:	Makai;
Tam.	:	Malsbasholam;
Tel.	:	Mokhajanna.

Distribution : Cultivated.

Description : Tall stout annual herbs. Leaves large. Glumes membranous, lodicules 2. Stamen 3. Female spikelet lodicules zero. Grains sub-globose.

Flowering and Fruiting : July-August.

Habitat Ecology : Planted.

Material Examined : EBH-HVP-883, 26.07.2005.

Parts Used : Grain. Stem. Cobs.

Folk Uses : Grain eaten roasted as popcorns and its flour used to make **'makki-ki-roti'** which is relished with 'saag' prepared from mustard leaves. Stem used as **fuel** and **fodder.** Cobs used as **fuel*.**

Active Constituents : These are 3 galactoside-cyanidol coumarate (corn pigment) and a-b-D-glucoside from roots.

Biological Activity : Plant-nephro protector.

Uses in Literature : Used as astringent, diuretic, fodder, food; and for antifertility, diabetes, corn flakes, popcorns **(Ambasta, 1986; Hembrom & Goel, 2005; Kadel & Jain, 2006; Samati, 2004).**

203. *Zingiber officinale* Rosc. (Pl. 61C)

Syn. : *Amomum zingiber* Linn.

Family : Zingiberaceae.

Vern. : Adrak.

English, Sanskrit and Regional Names:

Eng.	:	Ginger;
Sans.	:	Anupama, Ardraka, Ardrashaka, Kandara, Machhaka, Mahija, Mulaja;
B.	:	Ada;
H.	:	Ada, Adraka;
Mal.	:	Andrakam, Inchi;
Mar.	:	Ale;
Tam.	:	Allam, Inji;
Tel.	:	Allamu, Sonthi.

Distribution : Cultivated in hot and humid areas of S. India, Assam, H.P.

Description : Herbs. Leaves linear-lanceolate, sessile. Flowers in spike. Fruit capsule. Seeds globose, arillate.

Flowering and Fruiting : July-October.

Habitat Ecology : Cultivated.

Material Examined : EBH-HVP-800, 16.07.2004.

Part Used : Rhizome.

Folk Uses : One teaspoon of rhizome juice prescribed with honey twice a day till cure for **cough** and **asthma***. Rhizome also used as **condiments** and for **pickling.** Pieces of rhizome added to tea for **flavouring*** and to cooked vegetables for checking **flatulence**. Also, 5-10 g. of powdered rhizome fried in ghee prescribed with milk for **rheumatic pains***.

Active Constituents : Pungent activities attributed to gingerol, shagols, dihydrogingerol, hexahydrocurcumin, gingerdiols, zingerone (rhizome), 4 new diarylheptenones-gingerenones A, B, C and isogingerenone-B, two more diarylheptanoids, meso-3, 5-diacetoxy-1, 7-bis (4-hydroxy-3, 5-dimethoxyphenyl)-7-(4-hydroxy-3-methoxyphenyl heptane).

Biological Activity : Gingerenone A exhibited anticoccidium activity *in vitro* in cultural kidney cells of chicken and strong antifungal effect against *Pyricularia oryzae*.

Uses in Literature : Used for arthritis, indigestion (vet.), sore throat, throatache, cough, flatulence, soup, snakebite and insect, vomiting; and also as carminative, stimulant, vegetable **(Ambasta, 1986; Gogoi & Borthakur, 2001; Kumar *et al.*, 2004; Samati, 2004; Sharma & Singh, 2001).**

204. *Zizyphus jujuba* Mill. (Pl. 61D)

Syn. : *Z. mauritiana* Herb., *Z. sororia* Scult., *Z. trinervis* Roth., *Z. sativa* Gaertn DC., *Z. vulgaris* Lamk., *Rhamnus jujuba* L.

Family : Rhamnaceae.

Vern. : Ber.

English, Sanskrit and Regional Names:

Eng. : Chinese date, Indian Cherry, Indian jujube, Indian plum;
Sans. : Ajapriya, Badari, Dridhabija, Koli, Prithukoli, Rajakoli, Svaccha, Tanubija;
B. : Bogri;
Bo. : Bhurmi, Bordi;
G. : Ber, Bŏreli, Boyedi;
H. : Baer, Ber, Beri;
Kan. : Elanji, Yalachi;
Mal. : Badaram, Badari, Perinutulali;
Mar. : Baher, Bor;
Oriya : Barkoli, Bodokoli;
P. : Ber, Beri, Unab;
Tam. : Attiram, Ilandai, Iratti, Koli, Kulvali, Sivagam;
Tel. : Badari, Badaramu, Gangaregu, Karkhanduvu;
Urdu : Ber.

Distribution : Throughout India.

Description : A small glabrous bushy shrub. Branches divaricate, zig-zag. Leaves elliptic, spinous dentate. Flowers greenish, in axillary cymes. Drupes globose, glabrous, red when ripe.

Flowering and Fruiting : June-August.
Habitat Ecology : Wastelands. Dry areas.
Material Examined : EBH-HVP-787, 18.07.2003.
Parts Used : Leaves. Fruit.
Folk Uses : Fruits **edible**. Leaves used as **cattle feed***. Twigs used as **fuel***. Fruits also considered **sacred** and offered to appease Lord Shiva on the day of 'Mahashivratri'.
Active Constituents : Fruits rich in fatty acids, phytosterol, organic acid, reducing sugar, pectic mucilage, catechols, tannins, catechins, p-coumaraylates of alphitolic acid, gunosine-3'-5' monophosphate, saponins I, II, III and jujubuside B.
Biological Activity : Aerial parts hypotensive.
Uses in Literature : So far used as an anodyne, antiemetic, astringent, blood purifier, colic, diaphoretic, edible, febrifuge, purgative stomachic, sedative and tonic (vet.); and for bleeding from nose, cholera, cooling, constipation, diarrhoea, dysentery, conjunctivitis, fall of hair, fencing, fever, foot and mouth diseases, headache, inflammation, odema, rheumatism, scorpion stings, sores, spleen complaints, ulcer, wounds, whooping cough, yolk galls. Fruits and plant symbolically worshipping (**Bhatt *et al.*, 2001; Chatterjee & Pakrashi, 1997; Gogoi & Borthakur, 1991; Jain, 1991; Kadel & Jain, 2006; Kumar & Chauhan, 2006; Kumar *et al.*, 2004; Lalramnghinglova, 1996; Mohanty & Padhy, 1996; Mohanty *et al.*, 1998; Punjani, 1998; Ragupathy & Mahedevan, 1991; Ram *et al.*, 2002; Rama Rao & Henry, 1996; Sahu, 1996; Sharma *et al.* 2002; Shome *et al.*, 1996; Shukla & Verma, 1996; Sikarwar & Kaushik, 1992; Singh, 1996; Srivastava *et al.*, 2003; Upadhye & Kulkarni, 2004**).
Major Ayurvedic Formulations : Madhusdudikshay.

EPILOGUE

Altogether 570 ethnic uses of 204 plant spp belonging to 165 genera under 69 families have been reported for the rural communities of Pragpur [medicinal – 153 spp representing 29% of the plants used ethnobotanically, edible (fruits, vegetables, culinary purposes, condiments, spices, flavouring, pickle, chutney) – 81 spp, sacred – 48 spp, fodder – 47 spp, miscellaneous purposes (games, house building, musical instruments, plates, bowls, roofing, torch light, plastering walls, thatching, water container, ladder, mats, measure of weight, shuttering purposes, writing pen, etc.) – 40 spp, fuel – 24 spp, veterinary ailments – 20 spp, ornamentals – 17 spp, scouring teeth – 16 spp, agricultural implements – 12 spp, fencing – 9 spp, furniture – 8 spp, bites (scorpion, wasp, insects) – 7 spp, dye – 6 spp, death rituals – 5 spp, fibre (cordage, chicks, basketry, stuffing) – 5 spp, facial cosmetics – 4 spp, massage – 4 spp, bedding for animals ('pattri') – 4 spp, gum (adhesive) – 3 spp, insect repellent – 3 spp, magico-religious beliefs – 3 spp, hair care – 3 spp, eye sight – 1 sp, chewing gum – 1 sp, expelling leeches – 1 sp, grafting purposes – 1 sp, incense – 1 sp, rearing silkworms – 1 sp, scrubber – 1 sp, shampoo and detergent – 1 sp, souring – 1 sp, suppressing hunger – 1 sp, etc.] (Figs. 6, 7). The overall diversity comprises 65 herbs, 50 shrubs, 74 trees and 15 climbers of which 79 are cultivated (Fig. 3; Table XIII) and the remaining 125 spp occur in wild state. These include mostly the dicotyledonous taxa (175 spp in 142 genera) with predominance of Fabaceae (18 spp), followed by Solanaceae (10 spp), Asteraceae (9 spp), Cucurbitaceae (9 spp), Rutaceae (9 spp), Euphorbiaceae (8 spp), Moraceae (8 spp), Rosaceae (8 spp), Apocynaceae (7 spp), Lamiaceae (7 spp), Myrtaceae (5 spp), Amaranthaceae (4 spp), Meliaceae (4 spp), Verbenaceae (4 spp) (Table VI), followed by monocots (27 spp in 21 genera) and gymnosperms (2 spp in 2 genera). Among monocotyledons, Poaceae with 7 spp predominate in uses, followed by Liliaceae (5 spp), Araceae (3 spp), Zingiberaceae (3 spp), Bambusaceae (2 spp), Dioscoreaceae (2 spp), and the rest of families, viz., Agavaceae, Cannaceae, Punicaceae, Rhamnaceae, and Salicaceae are represented by one sp each. Likewise, the relative percentage in taxa for predominant dicotyledonous families varies from 9.3 (Fabaceae) to 1.9 (Verbenaceae) whereas for monocots it is 3.3 (Poaceae) to 2.4 (Liliaceae) (Table VI). The most frequently used plant parts are: leaves (104) representing 35% of the plant parts used, fruits (69), whole plant (44), seed (42), flower (22), stem (18), roots (17), bark (12), twigs (9), tuber (5), rhizome (3), latex (3), rind (2), etc.; aerial parts mostly used for their sustenance in comparison to underground ones (Figs. 4, 5). As many as 147 plant spp are used singly and 57 plants in combination with others. It is needless to say that rural communities of Pragpur have deep traditional wisdom about the uses of plants available in the region for their sustenance, health-care and veterinary needs, and using resources judiciously since time immemorial.

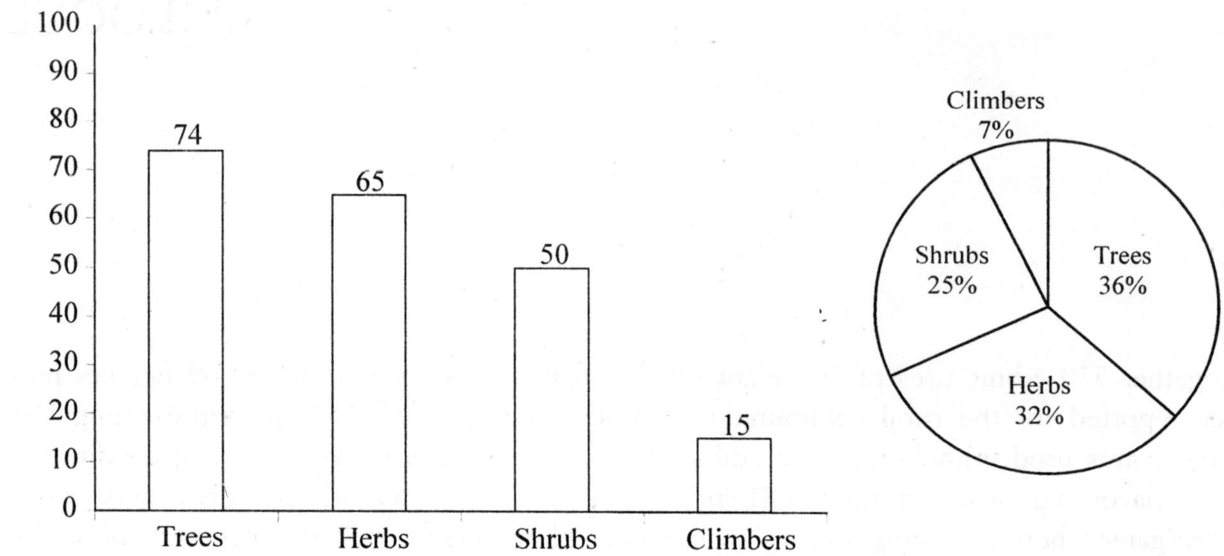

FIG. 3: *Habit-wise classification of ethnobotanical plants of Pragpur and their percentage, respectively.*

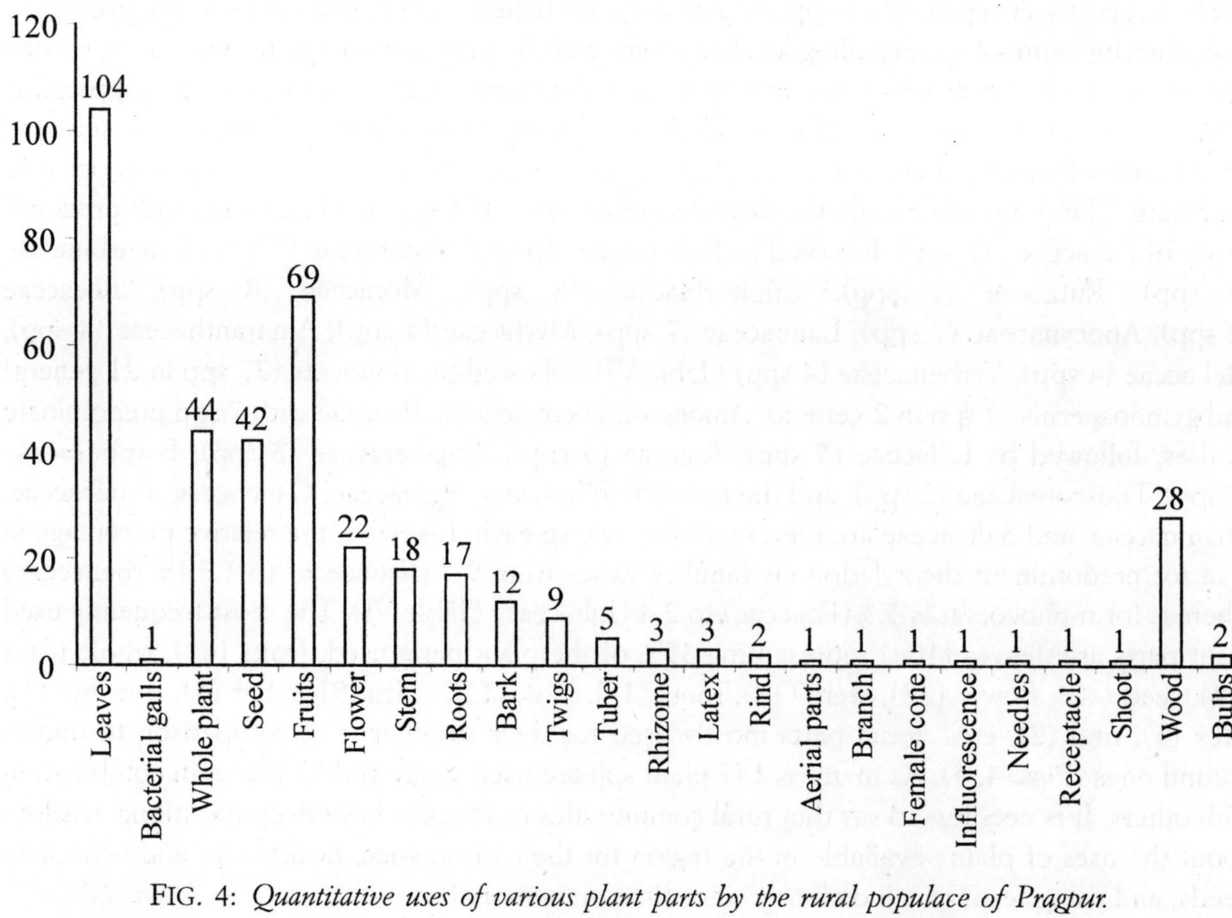

FIG. 4: *Quantitative uses of various plant parts by the rural populace of Pragpur.*

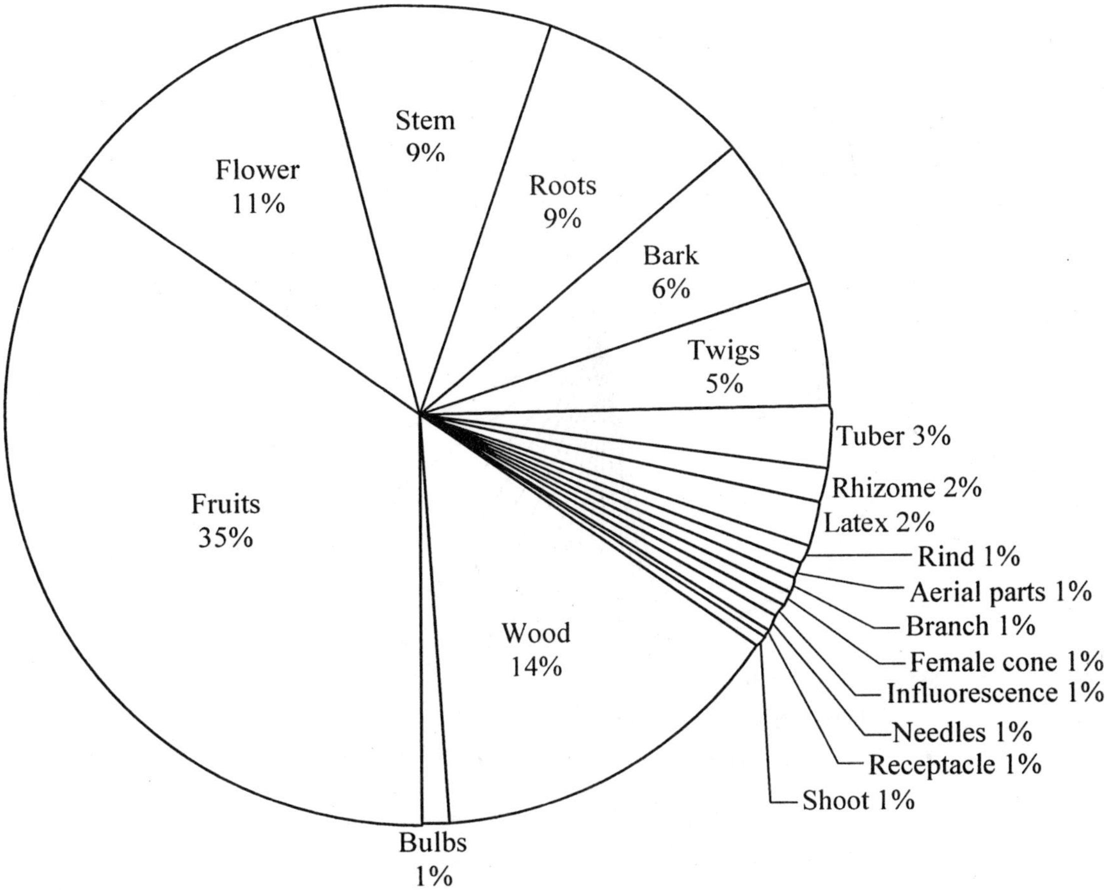

FIG. 5: *Per cent uses of various plant parts by the rural populace of Pragpur.*

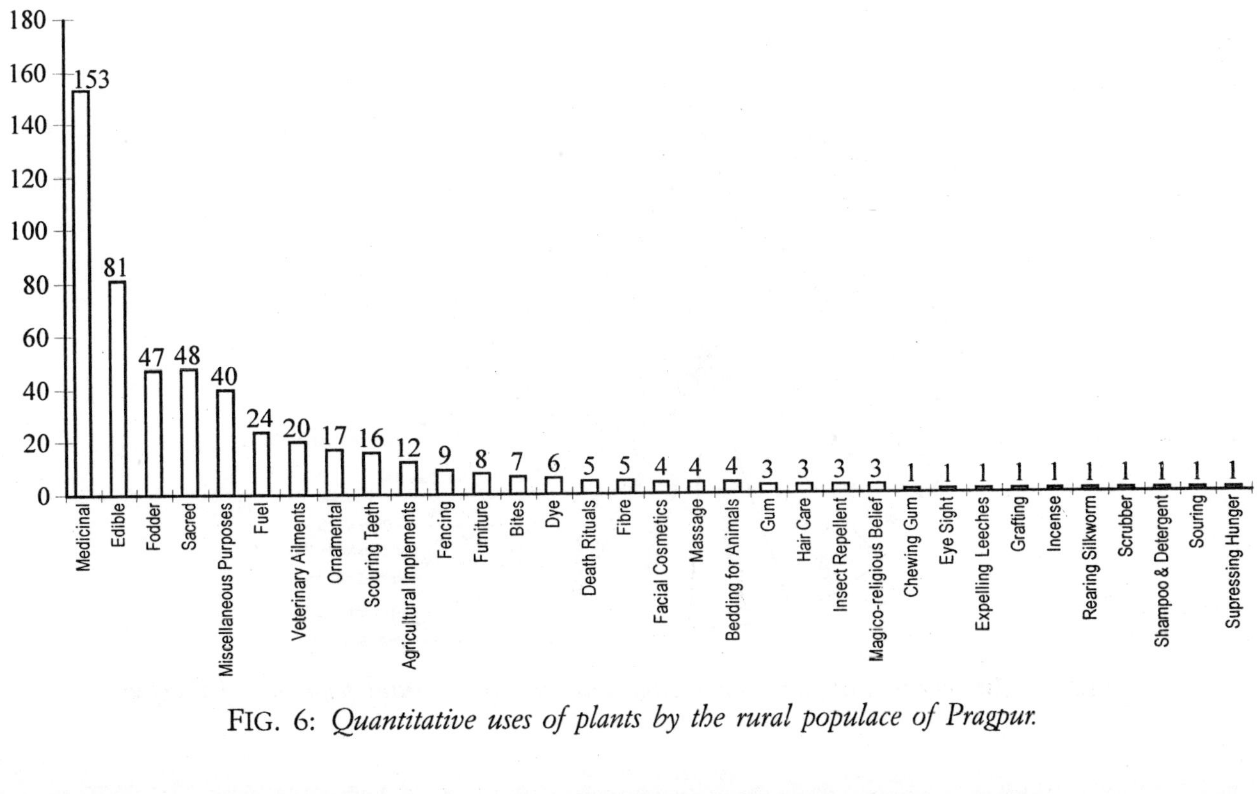

FIG. 6: *Quantitative uses of plants by the rural populace of Pragpur.*

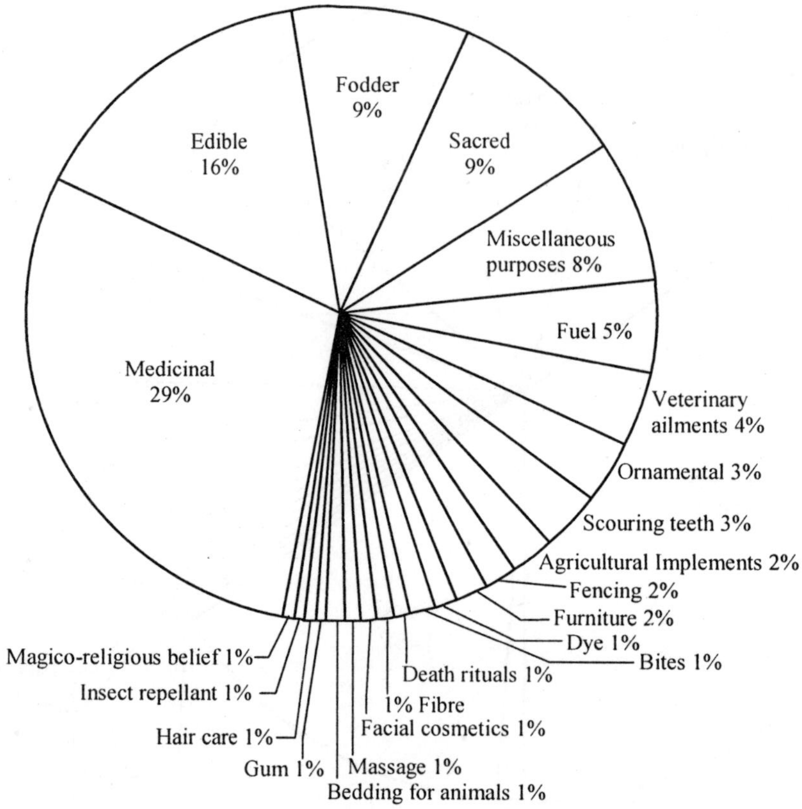

FIG. 7: *Per cent uses of plants by the rural populace of Pragpur.*

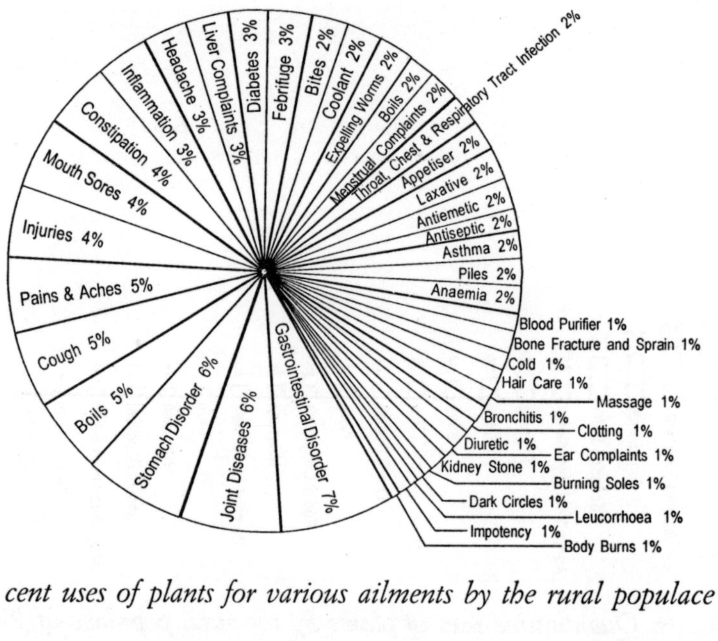

FIG. 8: *Per cent uses of plants for various ailments by the rural populace of Pragpur.*

TABLE VI
PREDOMINANT ETHNOBOTANICAL IMPORTANT FAMILIES OF PRAGPUR

Family	*Species of Ethnic Use*	*Percentage in the Taxa*
Fabaceae	18	9.3
Solanaceae	10	4.9
Asteraceae	9	4.7
Cucurbitaceae	9	4.7
Rutaceae	9	4.7
Euphorbiaceae	8	3.8
Moraceae	8	3.8
Rosaceae	8	3.8
Apocynaceae	7	3.3
Lamiaceae	7	3.3
Poaceae	7	3.3
Liliaceae	5	2.4
Myrtaceae	5	2.4
Amaranthaceae	4	1.9
Meliaceae	4	1.9
Verbenaceae	4	1.9

It is evident from Table VII that as many as 153 spp belonging to 129 genera are used by the inhabitants and traditional healers against 61 types of ailments affecting the populace of the area. Further analysis of the data, however, reveals that the most prevalent ailment in the region as indicated by the use of maximum number of plants for cure happens to be gastrointestinal disorders (Fig. 8) 23 spp—a major health problems in the developing world that alone accounts in India (Bilochi & Dutta, 1993), followed by that of joint diseases (20 spp), stomach disorders (20 spp), boils (15 spp), cough (15 spp), pains and aches (15 spp), injuries (14 spp), dental problems (14 spp), mouth sores and ulcers (14 spp), skin ailments (14 spp), constipation (13 spp), inflammation (10 spp), headache (9 spp), liver complaints (9 spp), diabetes (8 spp), febrifuge (8 spp), bites (7 spp), coolant (7 spp), expelling worms (7 spp), heart ailments (6 spp), menstrual complaints (6 spp), throat, chest and respiratory tract infection (6 spp), appetiser (6 spp), laxative (6 spp), antiemetic (6 spp), antiseptic (5 spp), asthma (5 spp), piles (5 spp), anaemia (5 spp), blood purifier (4 spp), bone fracture and sprains (4 spp), cold (4 spp), hair-care (4 spp), body massage (4 spp), bronchitis (3 spp), clotting (3 spp), diuretic (3 spp), ear complaints (3 spp), kidney stones (3 spp), etc. For other ailments like burning soles, dark circles, impotency, leucorrhoea, body burns, body heat, amenorrhoea, blisters on tongue, communicable diseases, flow of saliva, hernia, neuralgic disorders, nose bleeding, rabies, sex disease, slipped neck, sweating, etc. relatively lesser number of plants have been reported; indicating less prevalence of these diseases in the area. For alleviating these complaints, various plant parts used in the order of preference are: leaves (64 spp), fruits (48 spp), seed (30 spp), roots (15 spp), whole plant (15 spp), flowers (11 spp), bark (9 spp), stem (4 spp), wood (3 spp), rhizome (3 spp), tubers (3 spp), buds (2 spp), bulbs (2 spp), latex (2 spp), spine (1 sp), twig (1 sp), galls (1 sp) and rind (1 sp). Of the total 56 families used, the top ranking families are: Fabaceae (13 spp), Solanaceae (10 spp), Asteraceae (8 spp), Cucurbitaceae (8 spp), Rosaceae (7 spp), Apocynaceae (7 spp), Euphorbiaceae (6 spp), Lamiaceae (6 spp), Liliaceae (5 spp), Moraceae (5 spp), Rutaceae (5 spp), Verbenaceae (4 spp), Amaranthaceae (3 spp), Anacardiaceae (3 spp), Apiaceae (3 spp), Brassicaceae (3 spp), Combretaceae (3 spp), Lythraceae (3 spp), Meliaceae (3 spp), Menispermaceae (3 spp), Myrtaceae (3 spp), Zingiberaceae (3 spp), Araceae (2 spp), Malvaceae (2 spp), Papaveraceae (2 spp), Poaceae (2 spp), Sapindaceae (2 spp) and rest of 29 families with one sp each. By and large, the inhabitants have good knowledge of herbal drugs and preferred these to modern system of medicine as these are considered cheap, safe and with no side effects; and administered orally either singly or in combination with others. More importantly, all those species (86 spp) which have been marked with asterisk against their names listed for their uses against various ailments (Table VII) have been reported for the first time in India. For identical uses of some of the other plants of the studied area with those of ethnic societies elsewhere (Jain, 1991) may not be a mere coincidence but provides greater credibility to their potential for wider usage, a view also held by Jain & Saklani (1992). Moreover, detailed phytochemical and pharmaceutical analysis of the above plants for finding out the phytoconstituents may prove worthwhile as perspective medicinal plants for wider human welfare, and ultimately help in integrating traditional with modern system of medicine.

Table VII

Plants Used for Various Ailments

Ailment	*Plant Species*
Anaemia	*Annona squamosa, Carissa inermis, Carissa spinarum, Psidium guajava, Punica granatum*
Antidote for bites (Scorpion, Wasp Stings, Insect bite)	*Boerhaavia diffusa**, *Cannabis sativa**, *Catharanthus roseus, Martynia annua, Papaver rhoeus, Tagetes erecta, Verbascum thapsus*
Antiemetic	*Asparagus adscendens**, *Citrus medica, Eugenia jambolana, Ocimum basilicum, Mentha piperita, Mentha spicata*
Antilithic	*Bryophyllum caylcinum**, *Chenopodium album*
Antiseptic	*Amorphophallus campanulatus**, *Euphorbia royleana, Lannea coromandelica, Ocimum sanctum**, *Randia dumetorum*
Appetiser	*Amaranthus spinosus**, *Amaranthus viridis**, *Emblica officinalis, Eriobotrya japonica**, *Lycopersicum esculentum, Terminalia bellirica**
Asthma	*Datura innoxia, Ocimum sanctum, Pistacia integerrima, Solanum viarum**, *Zingiber officinale**
Astringent	*Centella asiatica**
Bandage	*Ricinus communis**
Blood purifier	*Melia azadirachta, Momordica charantia, Mallotus philippinensis, Thevetia neriifolia**
Blood sugar (diabetes)	*Catharanthus roseus, Carica papaya, Coccinia grandis, Eugenia jambolana, Momordica charantia, Momordica dioica**, *Terminalia chebula, Trigonella foenum-graecum**
Body heat	*Holarrhena antidysenterica**
Boils	*Allium cepa, Aloe vera, Artocarpus lakoocha**, *Bombax ceiba, Bryophyllum calycinum, Cinnamomum camphora, Coriandrum sativum**, *Euphorbia royleana, Ficus benghalensis, Cynodon dactylon**, *Hibiscus rosa-sinensis, Holarrhena antidysenterica, Melia azedirachta, Ocimum sanctum**, *Randia dumetorum*
Bone fracture, sprains	*Artemisia japonica**, *Dodonaea viscose**, *Oroxylum indicum, Ricinus communis*
Bronchitis	*Celastrus paniculata**, *Elettaria cardamomum, Terminalia bellirica**

Ailment	*Plant Species*
Burning sensation (soles)	*Annona squamosa, Lawsonia alba*
Burns	*Aloe vera, Aloe maculata**
Clotting	*Cynodon dactylon, Cuscuta reflexa*, Cassia occidentalis*
Cold	*Datura innoxia*, Nerium indicum*, Ocimum sanctum, Prunus persica**
Communicable diseases	*Vitex negundo**
Constipation	*Cassia fistula, Carica papaya, Chenopodium album*, Cissampelos pareira, Emblica officinalis, Mallotus philippinensis, Oroxylum indicum, Prunus amygdalus, Psidium guajava, Randia dumetorum*, Raphanus sativus, Ricinus communis, Rosa damascena**
Coolant (refrigerant)	*Aegle marmelos, Aloe vera, Centella asiatica, Emblica officinalis*, Foeniculum vulgare*, Morus alba, Saccharum officinarum*
Cough, Expectorant	*Abrus precatorius, Achyranthes bidentata, Acorus calamus, Adhatoda vasica, Argemone maxicana, Celastrus paniculata, Datura innoxia, Emblica officinalis, Ocimum sanctum, Opuntia dillenii, Pistacia integerrima, Pogostemon benghalensis*, Punica granatum, Terminalia chebula, Zingiber officinale*
Dark circles	*Cucumis sativus, Solanum tuberosum**
Delivery	*Achyranthes bidentata*
Dental problems (toothache, gum diseases, pyorrhoea, etc.)	*Accacia catechu, Argemone maxicana, Curcuma domestica*, Duranta repens*, Jatropha curcas*, Lawsonia alba*, Ocimum sanctum, Pistacia integerrima*, Psidium guajava, Spilanthes acmella*, Solanum surattense, Solanum viarum, Xanthium strumarium, Zanthxylum armatum*
Diuretic	*Allium cepa, Boerhaavia diffusa, Raphanus sativus*
Ear complaints (earache)	*Allium cepa, Allium sativum, Colebrookia oppositifolia**
Eczema	*Carica papaya**
Expelling worms	*Allium cepa, Chenopodium album, Cucurbita hispida, Lycopersicum esculentum*, Mallotus philippinensis, Momordica charantia, Tabernae-montana divaricata*
Eyesight	*Cynodon dactylon**
Febrifuge (fever)	*Adhatoda vasica, Cissampelos pareira, Fumaria indica, Ocimum sanctum, Solanum nigrum, Solanum surattense, Solanum viarum, Tinospora cordifolia*
Flow of saliva	*Spilanthes acmella**
Gastrointestinal disorders (loose motion, dysentery, indigestion, diarrhoea)	*Aegle marmelos, Artocarpus integrifolia*, Bauhinia vahlii, Bauhinia variegata, Cassia fistula*, Cassia occidentalis, Citrus medica, Cordia dichotoma, Cuscuta reflexa, Elettaria cardamomum, Mentha spicata, Musa sapientum, Ocimum basilicum, Prunus domestica*, Pyrus communis*, Pyrus pashia, Raphanus sativus, Sechium edule, Solanum melongena*, Solanum nigrum, Terminalia chebula, Terminalia bellirica, Trigonella foenum-graecum*
Hair-care (baldness, dandruff, hair-tonic)	*Datura innoxia*, Emblica officinalis, Melia azedarach*, Pyrus pashia**
Headache	*Eucalyptus umbellata, Jasminum sambac, Mentha piperita*, Nerium indicum, Pogostemon benghalensis*, Sonchus asper*, Spilanthes acmella*, Woodfordia floribunda*, Xanthium strumarium*
Heart ailments	*Lagenaria siceraria*, Litchi chinensis*, Luffa aegyptica*, Opuntia dillenii*, Solanum melongena*, Terminalia arjuna*
Hernia	*Dendrocalamus hamiltonii*
Impotency	*Malvaviscus conzatii, Mucuna pruriens*
Injuries (bruises, cut, internal injuries, haemorrhages)	*Artocarpus lakoocha*, Bryophyllum calycinum, Curcuma domestica*, Cuscuta reflexa*, Cynodon dactylon, Ficus glomerata, Helianthus annuus, Jatropha curcas, Lantana camara, Lannea coromandelica, Lawsonia alba, Mimosa pudica, Oroxylum indicum*, Solanum erianthum*
Joint diseases (arthritis, knee pain, rheumatism, inflammation of joints, rheumatic pains)	*Acorus calamus, Allium sativum*, Aloe vera, Boerhaavia diffusa, Brassica campestris*, Cannabis sativa*, Chenopodium album, Curcuma domestica*, Cuscuta reflexa, Cynodon dactylon, Datura innoxia, Dioscorea deltoidea,*

Ailment	Plant Species
	*Plumbago zeylanica, Pyrus pashia**, *Ricinus communis, Sesamum indicum, Tinospora cordifolia**, *Verbascum thapsus**, *Vitex negundo, Zingiber officinale**
Kidney stones	*Celosia argentea**, *Cucumis sativus**, *Momordica dioica**
Laxative	*Cassia fistula, Jatropha curcas, Solanum nigrum, Terminalia bellirica, Terminalia chebula, Vitis jacquemontii*
Leucorrhoea	*Cissampelos pareira**, *Saraca asoca*
Liver complaints (congestion, hepatic complaints, jaundice)	*Aloe vera, Boerhaavia diffusa, Butea monosperma, Emblica officinalis**, *Carica papaya, Raphanus sativus**, *Saccharum officinarum, Tinospora cordifolia, Woodfordia floribunda**
Massage	*Acorus calamus**, *Aloe vera, Cuscuta reflexa, Ricinus communis*
Menstrual complaints (excessive bleeding, pains, amenorrhoea, menorrhagia, irregular cycle)	*Aloe maculata**, *Amaranthus viridis**, *Asparagus adscendens**, *Butea monosperma, Holarrhena antidysenterica, Saraca asoca*
Mouth sores, Ulcer	*Acacia catechu, Acacia nilotica, Carissa spinarum**, *Cassia occidentalis**, *Cestrum nocturnum**, *Cordia dichotoma, Flacourtia ramontchi* *, *Jatropha curcas**, *Psidium guajava, Pyrus pashia**, *Spilanthes acmella**, *Stephania glabra, Vitex negundo, Zanthoxylum armatum**
Neuralagic disorders	*Litchi chinensis*
Nose bleeding	*Cynodon dactylon*
Pains and aches (backache, bodyache, muscular pains)	*Acorus calamus, Allium sativum, Aloe vera, Boerhaavia diffusa, Butea monosperma, Datura innoxia, Eruca sativa**, *Eucalyptus umbellata, Lagerstroemia indica**, *Nerium indicum, Pogostemon benghalensis, Sonchus asper, Spilanthes acmella, Verbascum thapsus, Vitex negundo Woodfordia floribunda*
Piles	*Allium cepa, Raphanus sativus, Sesamum indicum**, *Solanum melongena**, *Terminalia bellirica**
Rabies	*Capsicum annuum*
Sex disease	*Bombax ceiba*
Skin ailments (cracked skin, skin eruptions, skin rashes dermatitis, emollients, itch)	*Abrus precatorius, Artemisia japonica, Capsicum annuum**, *Catharanthus roseus, Cinnamomum camphora, Celastrus paniculata, Curcuma domestica, Helianthus annuus, Jatropha curcas**, *Premna latifolia, Mallotus philippinensis, Melia azedirachta, Momordica charantia, Rumex hastatus*
Slipped neck	*Parthenium hysterophorus*
Stomach disorders (abdominal pains, acidity, dyspepsia, flatulence, stomachache)	*Acorus calamus**, *Allium sativum**, *Aloe vera**, *Asparagus adscendens**, *Bombax ceiba, Canna indica, Catharanthus roseus, Coriandrum sativum**, *Elettaria cardamomum**, *Mallotus philippinensis, Mentha piperita, Mentha spicata, Murraya koenigii, Musa sapientum, Phyllanthus urinaria, Pogostemon benghalensis**, *Punica granatum, Randia dumetorum**, *Stephania glabra, Taraxacum officnale**, *Vitex negundo, Zingiber officinale*
Sweating	*Solanum melongena**
Swelling (inflammation)	*Artemisia japonica**, *Canna indica**, *Cannabis sativus**, *Cassia occidentalis**, *Cuscuta reflexa, Cynodon dactylon, Sesamum indicum**, *Stephania glabra**, *Tagetus erecta, Vitex negundo*
Throat, chest and Respiratory tract infection (tonsils, sore throat)	*Acacia catechu, Achyranthes bidentata**, *Allium sativum**, *Elettaria cardamomum**, *Martynia annua**, *Saraca asoca, Spilanthes acmella*

*New Records.

It has been specially observed that for their beauty and oral health-care needs, the inhabitants use a wide range of plant spp available in their vicinity (Tables VII, XI). These are: *Cicer arietinum, Citrus aurantium, Cucumis sativus, Curcuma domestica* for facial cosmetics; *Juglans regia* as substitute for lip-stick; *Luffa aegyptica* as body scrubber; *Sapindus mukorossi* as shampoo and detergent; *Acorus calamus, Aloe vera, Cuscuta reflexa, Ricinus communis* for massage; *Datura innoxia, Emblica officinalis, Melia azedarach, Pyrus pashia* for hair-care, *Cucumis sativus, Solanum tuberosum* for dark circles; *Acacia catechu, Argemone maxicana, Curcuma domestica, Duranta repens, Jatropha curcas, Lawsonia alba, Ocimum sanctum, Pistacia integerrima, Psidium guajava, Spilanthes acmella, Solanum surattense, Solanum viarum, Xanthium strumarium, Zanthxylum armatum* for dental caries, pyorrhoea, bleeding gums; *Acacia nilotica, Citrus limon, C. jambhiri, C. medica, C. grandis, Jatropha curcas, Juglans regia, Melia azadirachata, Murraya koenigii, M. paniculata, Plumbago zeylanica, Psidium guajava, Ricinus communis, Vitex negundo, Wendlandia heynei* and *Zanthoxylum armatum* as brush for scouring teeth, suggesting useful properties in such plants. Amongst these, the use of *Acorus calamus, Cicer arietinum, Citrus aurantium, C. limon, C. medica, Cucumis sativus, Curcuma domestica, Datura innoxia*, *Duranta repens, Jatropha curcas, Lawsonia alba, Melia azedarach, Pistacia integerrima, Plumbago zeylanica, Pyrus pashia, Solanum tuberosum* and *Spilanthes acmella* have been recorded for the first time. This information is a great natural gift from these people and holds considerable scope for popularization to be developed as safe beauty aids for modern civilization.

A wide range of plants are being used as tonics (Table VIII), and prescribed for several days for the desired effect. Evidently, these are: *Amaranthus spinosus* (source of iron), *A. viridis* (source of iron), *Annona squamosa* (anaemia), *Carica papaya* (cardiac), *Carissa inermis* (anaemia), *C. spinarum* (anaemia), *Centella asiatica* (memory), *Citrus aurantium* (vitamin C), *Cucumis sativus* (refreshing drink), *Emblica officinalis* (memory, vitamin C), *Mucuna pruriens* (strength), *Psidium guajava* (anaemia), *Punica granatum* (anaemia, fatigue), *Rumex hastatus* (vitamin C), *Saccharum officinarum* (refreshing drink), *Terminalia arjuna* (cardiac) and *T. bellirica* (liver). Earlier, the use of *Amaranthus spinosus, A. viridis* (Amaranthaceae), *Cucumis sativus* (Cucurbitaceae) and *Emblica officinalis* (Euphorbiaceae) as tonics were not recorded (Jain, 1991) and being reported for the first time from this area, thereby indicating that traditional herbal medicament is still very much alive in this part of Himalaya.

TABLE VIII
TONIC PLANTS OF PRAGPUR

Plant Species	*Curative Property*	*Plant Species*	*Curative Property*
*Amaranthus spinosus**	Source of iron	*Emblica officinalis**	Memory, Vitamin C
*Amaranthus viridis**	Source of iron	*Mucuna pruriens*	Strength
Annona squamosa	Anaemia	*Psidium guajava*	Anaemia
Carica papaya	Cardiac	*Punica granatum*	Anaemia, Fatigue
Carissa inermis	Anaemia	*Rumex hastatus*	Vitamin C
Carissa spinarum	Anaemia	*Saccharum officinarum*	Refreshing drink
Centella asiatica	Memory	*Terminalia arjuna*	Cardiac
Citrus aurantium	Vitamin C	*Terminalia bellirica*	Liver
*Cucumis sativus**	Refreshing drink		

*New Records.

Of the estimated diversity of 3,900 spp of wild edible plants in India (Anonymous, 1994), the present study identifies 36 genera and 45 spp of edible fruits, 22 genera and 22 spp for culinary preparations, 20 genera and 24 spp as pot herbs and vegetables, 9 genera and 12 spp for pickle purposes, 9 genera and 10 spp as condiments, spices and flavourings, 7 genera and 7 spp for chutney preparations; belonging to 81 spp under 65 genera of 31 families (Table IX). Overall, this corresponds to 2.08% of total wild edibles reported from India. Of these, fruits of 72 species, leaves of 22 species with most of the spp eaten as 'saags' which are uncommon elsewhere, seeds of 10 spp, rhizome of 5 spp, buds of 4 spp, tender shoots of 3 spp, tubers of 3 spp, flowers of 2 spp, and bulbs of 1 sp, whole plant of 1 sp and receptacle of 1 sp are consumed by the rural communities of Pragpur either raw, roasted, cooked and/or boiled or as pickle, condiments, spices, flavourings, chutney and for culinary preparations with maximum diversity of supplementary foods represented by Rutaceae (9 spp), followed by Cucurbitaceae (8 spp), Rosaceae (6 spp), Fabaceae (5 spp), Moraceae (5 spp), Solanaceae (4 spp), Lamiaceae (3 spp), Zingiberaceae (3 spp), Amaranthaceae (2 spp), Apiaceae (2 spp), Apocynaceae (2 spp), Araceae (2 spp), Asteraceae (2 spp), Liliaceae (2 spp), Myrtaceae (2 spp), Sapotaceae (2 spp), etc. Among the genera *Citrus* (5 spp) showed highest species diversity, followed by *Prunus* (3 spp), *Amaranthus* (2 spp), *Artocarpus* (2 spp), *Carissa* (2 spp), *Cucurbita* (2 spp), *Ficus* (2 spp), *Mentha* (2 spp), *Momordica* (2 spp), *Murraya* (2 spp), *Pyrus* (2 spp), *Solanum* (2 spp), and rest of the genera represented by 1 sp each. The findings also revealed 15 new records of food plants which could be utilized as alternative food or in terms of scarcity for allaying hunger. These are *Holoptelea integrifolia* (seeds), *Punica granatum* (fruits), *Rumex hastatus* (leaves), *Tagetes erecta* (receptacle), *Amorphophallus campanulatus* (rhizome), *Musa sapientum* (unripe fruit), *Bambusa nutans* (tender shoot), *Citrus limon* (leaves), *Bauhinia variegata* (buds, flowers), *Colocasia esculenta* (rhizome), *Cucurbita hispida* (fruits), *Curcuma domestica* (leaves), *Juglans regia* (fruits), *Mentha piperita* (leaves), and *Zingiber officinale* (rhizome). It is a fact realized by everyone that due to our expanding needs and increased pressure of human population, the per capita availability of land, water and forest is shrinking. And this challenge of sustainable food security can be met only by conserving genetic wealth and utilizing it effectively. In the light of this, more importantly, the resources of wild edible plants, especially fruits and vegetables of this region is very rich and would form an alternative resource base of perspective food plants and promising types for domestication that can solve the food problem likely to be faced in the near future.

Table IX
Edible Plants of Pragpur

Plant Species	*Edible Part/s*	*Plant Species*	*Edible Part/s*
Edible (Raw, Boiled/Roasted)		*Citrus limon*	Fruits
Achras zapota	Fruits	*Citrus grandis*	Fruits
Aegle marmelos	Fruits	*Colocasia esculenta*	All parts
Agave americana	Tender shoots	*Cucumis sativus*	Fruit
Annona squamosa	Fruits	*Cucurbita pepo*	Seed
Artocarpus integrifolia	Fruits	*Eriobotrya japonica*	Fruits
Bryophyllum caylcinum	Leaves	*Eugenia jambolana*	Fruits
Carica papaya	Fruits	*Ficus carica*	Fruits
Carissa inermis	Fruits	*Ficus roxburghii*	Fruits
Carissa spinarum	Fruits	*Flacourtia ramontchi*	Fruits
Cicer arietinum	Seeds	*Fragaria chiloensis*	Fruits
Citrus aurantium	Fruits	*Holoptelea integrifolia**	Seeds
Citrus jambhiri	Fruits	*Juglans regia*	Fruits

Plant Species	*Edible Part/s*
Litchi chinensis	Fruits
Mangifera indica	Fruits
Morus alba	Fruits
Murraya paniculata	Fruits
Murraya koenigii	Fruits
Musa sapientum	Fruits
Opuntia dillenii	Fruits
Phoenix sylvestris	Fruits
Prunus amygdalus	Fruits
Prunus domestica	Fruits
Prunus persica	Fruits
Psidium guajava	Fruits
*Punica granatum**	Seeds
Pyrus pashia	Fruits
Pyrus communis	Fruits
*Rumex hastatus**	Leaves
*Tagetes erecta**	Receptacle
Vitis jacquemontii	Fruits
Zea mays	Fruits (Grains)
Zizyphus jujuba	Fruits
Vegetables and Pot Herbs	
Amaranthus viridis	Leaves
Amaranthus spinarum	Leaves
*Amorphophallus campanulatus**	Rhizome
Artocarpus integrifolia	Fruits
Bauhinia variegata	Buds
Brassica campestris	Leaves
Chenopodium album	Leaves, Tendershoot
Coccinia grandis	Fruits
Colocasia esculenta	Rhizome, Leaves
Cordia dichotoma	Fruits
Cucurbita hispida	Fruits
Cucurbita pepo	Fruits
Dioscorea bulbifera	Tuber
Lagenaria siceraria	Fruit
Luffa aegyptica	Fruit
Lycopersicum esculentum	Fruit
Momordica charantia	Fruit
Momordica dioica	Fruit
*Musa sapientum**	Fruit (Unripe)
Raphanus sativus	Fruit
Sechium edule	Fruit
Solanum melongena	Fruit
Solanum tuberosum	Fruit
Pickle	
*Amorphophallus campanulatus**	Rhizome
Artocarpus lakoocha	Fruits
Artocarpus integrifolia	Fruits
Bauhinia variegata	Buds
*Bambusa nutans**	Tender shoot
Carissa inermis	Fruits
Cassia occidentalis	Seed
Citrus jambhiri	Fruits
Citrus limon	Fruits
Citrus medica	Fruits
Cordia dichotoma	Fruits
Zingiber officinale	Rhizome
Condiments, Spices, Flavourings	
Allium cepa	Bulb
*Citrus limon**	Leaves
Coriandrum sativum	Leaves
Elettaria cardamomum	Fruit
Foeniculum vulgare	Fruit
Mentha piperita	Leaves
Mentha spicata	Leaves
Murraya koenigii	Leaves
Ocimum basilicum	Leaves
Zingiber officinale	Rhizome
Culinary Preparations	
*Bauhinia variegata**	Buds, Flowers
Capsicum annuum	Fruits
Cannabis sativa	Leaves
Cassia occidentalis	Seed
Cicer arietinum	Seed
*Colocasia esculenta**	Rhizome
Cucumis sativus	Fruits
*Cucurbita hispida**	Fruits
*Curcuma domestica**	Leaves
*Juglans regia**	Fruits
Lagenaria siceraria	Fruits
Luffa aegyptiaca	Fruits
Lycopersicum esculentum	Fruits
Mangifera indica	Fruits
Mentha piperita *	Leaves
Prunus persica	Fruits
*Punica granatum**	Seed
Mentha spicata	Leaves
Ocimum basilicum	Leaves
Rumex hastatus	Leaves
Tamarindus indica	Fruits
*Zingiber officinale**	Rhizome
Chutney	
Bauhinia variegata	Bud, Flower
Cassia occidentalis	Seed
Mangifera indica	Fruits
Mentha spicata	Leaves
Murraya koenigii	Leaves
Punica granatum	Seed
Tamarindus indica	Fruits

*New Edible Species/Uses.

Despite the invasion of modern culture, the local populace of heritage village is deeply religious as evident from the use of 48 plant spp belonging to 42 genera and 30 families for offerings, various religious rituals and ceremonies (Table X). Of these, the uses of 12 plants, viz., *Aloe vera* (Liliaceae), *Asparagus adscendens* (Liliaceae), *Bambusa nutans* (Bambusaceae), *Cannabis sativa* (Canabinaceae), *Cicer arietinum* (Fabaceae), *Cucurbita pepo* (Cucurbitaceae), *Holarrhena antidysenterica* (Apocynaceae), *Lagerstroemia indica* (Lythraceae), *Nerium indicum* (Apocynaceae), *Papaver rhoeus* (Papaveraceae), *Raphanus sativus* (Brassicaceae), *Tabernaemontana divaricata* (Apocynaceae) are considered new and documented for the first time (Sood *et al.*, 2005d). Besides plants such as *Capsicum annuum, Citrus limon, Datura innoxia* are employed for magico-religious purposes and likewise, *Aegle marmelos, Bambusa nutans, Emblica officinalis, Musa sapientum* and *Saccharum spontaneum* are considered for various rituals of the departed soul so that it rests peacefully in the heaven. Dried fragrant leaves of *Cestrum nocturnum* are burnt as incense. As a matter of fact, most of the sacred plants (present study) are also endowed with peculiar medicinal, culinary and other properties that account for their reverence and conservation by the local communities for the benefit of the present and future generations.

TABLE X
SACRED PLANTS OF PRAGPUR

Family	*Plant Species*
Poaceae	*Cynodon dactylon, Saccharum officinarum, Saccharum spontaneum, Triticum aestivum*
Apocynaceae	*Catharanthus roseus, Holarrhena antidysenterica**, *Nerium indicum**, *Tabernaemontana divaricata**
Fabaceae	*Butea monosperma, Cicer arietinum**, *Saraca asoca*
Malvaceae	*Hibiscus rosa-sinensis, Gossypium arboreum, Malvaviscus conzatii*
Moraceae	*Ficus benghalensis, Ficus glomerata, Ficus religiosa*
Euphorbiaceae	*Emblica officinalis, Putranjiva roxburghii*
Lamiaceae	*Ocimum sanctum, Salvia coccinea*
Liliaceae	*Aloe vera**, *Asparagus adscendens**
Oleaceae	*Jasminum multiflorum, Jasminum sambac*
Solanaceae	*Cestrum nocturnum, Solanum tuberosum*
Anacardiaceae	*Mangifera indica*
Asteraceae	*Tagetes erecta*
Bambusaceae	*Bambusa nutans**
Bignoniaceae	*Tecoma grandiflora*
Brassicaceae	*Raphanus sativus**
Cannaceae	*Canna indica*
Cucurbitaceae	*Cucurbita hispida, Cucurbita pepo**
Canabinaceae	*Cannabis sativa**
Cupressaceae	*Thuja orientalis*
Lythraceae	*Lagerstroemia indica**
Lauraceae	*Cinnamomum camphora*
Meliaceae	*Mirabilis jalapa*
Menispermaceae	*Tinospora cordifolia*
Musaceae	*Musa sapientum*
Paparveraceae	*Papaver rhoeus**
Pedaliaceae	*Sesamum indicum*
Punicaceae	*Punica granatum*
Rosaceae	*Rosa damascena*
Rhamnaceae	*Zizyphus jujuba*
Zingiberaceae	*Curcuma domestica*

*New Records of Sacred Plants.

As is true elsewhere, the economy is mainly based on agriculture. For their agricultural operations, the various implements got fabricated from *Accacia nilotica, Albizia lebbeck, Cassia fistula, Cedrela toona**, *Dalbergia sissoo, Eugenia jambolana, Flacourtia ramontchi, Grewia oppositifolia, Mallotus philippinensis**, *Melia azedarach**, *Pyrus pashia** and *Terminalia arjuna*. Injuries while handling the tools and implements are unavoidable and for their quick healing, plants such as *Artocarpus lakoocha**, *Bryophyllum calycinum, Curcuma domestica**, *Cuscuta reflexa**, *Cynodon dactylon, Ficus glomerata, Helianthus annuus, Jatropha curcas, Lantana camara, Lannea coromandelica, Lawsonia alba, Mimosa pudica, Oroxylum indicum**, *Ricinus communis** and *Solanum erianthum* are being used. Likewise for relief from inflamed joints and pain, plants viz., *Acorus calamus, Allium sativum**, *Aloe vera, Boerhaavia diffusa, Brassica campestris**, *Cannabis sativa**, *Chenopodium album, Curcuma domestica**, *Cuscuta reflexa, Cynodon dactylon, Datura innoxia, Dioscorea deltoidea, Plumbago zeylanica, Pyrus pashia**, *Ricinus communis, Sesamum indicum, Tinospora cordifolia**, *Verbascum thapsus**, *Vitex negundo* and *Zingiber officinale** are employed. Moreover, for other agricultural related activities like rearing silkworms and grafting, plants such as *Morus alba* and *Pyrus pashia* are valued, respectively. For animal-care activities like providing bedding ('pattri') for livestocks, *Holoptelea integrifolia, Eugenia cerasoides**, *Mallotus philippinensis, Triticum aestivum* are considered whereas *Ageratum conyzoides, Eucalyptus umbellata** and *Vitex negundo* are used as insect repellent. Besides, 6 spp are used as dye as considered safe over synthetic ones (*Cucurbita hispida, Emblica officinalis, Juglans regia, Lawsonia alba, Mallotus philippinensis, Terminalia chebula*), 3 spp as gum (*Aegle marmelos, Cordia dichotoma**, *Prunus persica**), 1 sp for souring purposes (*Citrus jambhiri*), 8 spp for furniture (*Albizia lebbeck, Cedrela toona, Dalbergia sissoo, Eucalyptus umbellata**, *Grewia oppositifolia**, *Mangifera indica, Melia azedarach, Pinus roxburghii*), 17 spp as ornamentals (*Bougainvillea spectabilis, Catharanthus roseus**, *Callistemon lanceolatus**, *Grevillea robusta, Hibiscus rosa-sinensis**, *Juglans regia, Lagerstroemia indica, Mimosa pudica**, *Papaver rhoeus**, *Pedilanthes tithymaloides**, *Polyalthia longifolia, Mimosa pudica, Rosa damascena**, *Tabernaemontana divaricata, Tecoma grandiflora**, *Thevetia neriifolia, Thuja orientalis*) and 40 spp belonging to 39 genera for various purposes like building works, ladder, mats, musical instruments, plastering of walls, measure of weight, bowls, plates, thatching, games, torch light, water container and writing pen (*Abrus precatorius, Agave americana, Aloe maculata**, *Amaranthus viridis**, *Asparagus adscendens**, *Bambusa nutans**, *Bauhinia vahlii, Bombax ceiba**, *Carissa spinarum, Cassia fistula, Cucurbita pepo, Dalbergia sissoo, Dendrocalamus hamiltonii, Celastrus paniculatus, Eucalyptus umbellata, Eugenia jambolana,** *Ficus glomerata, Flacourtia ramontchi, Holarrhena antidysenterica, Lantana camara, Mimosa pudica**, *Parthenium hysterophorus, Phoenix sylvestris, Pinus roxburghii, Polyalthia longifolia, Populus nigra**, *Prunus domestica, Punica granatum**, *Putranjiva roxburghii**, *Pyrus pashia, Saccharum arundinaceum**, *Saccharum officinarum, Sapindus mukorossi, Sesamum indicum, Tagetes erecta, Tamarindus indica, Terminalia arjuna, Thuja orientalis, Woodfordia floribunda, Zizyphus jujuba*). Of these, the uses of spp with asterisk against their names have been hitherto unrecorded and form the first record of its type for this region (Table XI).

Nine spp such as *Bambusa nutans, Bougainvillea spectabilis**, *Carissa spinarum**, *Duranta repens**, *Euphorbia royleana**, *Ipomoea cairica**, *I. carnea**, *Lantana camara, Opuntia dillenii* are being used by the local populace to raise both homogenous and heterogenous fences in their cropland due to their characteristics like occurrence of thorns, spines, prickles and profuse branching so as to counter damage to crops by any external agents, thereby indicating deep traditional lore of biological fences implanted in their sustainable farming system. Of these, the use of *Bougainvillea spectabilis, Carissa spinarum, Duranta repens, Euphorbia royleana, Ipomoea cairica, I. carnea* are new additions to the existing record of bio-fences. The present investigation also lends support to the view-point of Punjani (1998) that the plants constituting bio-fences strengthen mud boundaries of the field due to their strong soil binding roots (Table XI).

Other important features of the present study are the use of latex of *Ficus benghalensis* as a substitute for chewing gum and *Achyranthes bidentata* for suppressing hunger (Table XI). This interesting information has a lot of scope for its exploitation especially while imparting training to our soldiers for mountain and jungle warfare as it would help them to survive during real combat situations when no food is available.

TABLE XI
PLANTS USED FOR MISCELLANEOUS PURPOSES

Use	*Plant Species*
Agricultural implements	*Accacia nilotica, Albizia lebbeck, Cassia fistula, Cedrela toona*, Dalbergia sissoo, Eugenia jambalana, Flacourtia ramontchi, Grewia oppositifolia, Mallotus Philippinensis*, Melia azedarach*, Pyrus pasia*, Treminalia arjuna*
Bedding for animals (Pattri)	*Holoptelea integrifolia, Eugenia cerasoides*, Mallotus philippinensis, Triticum aestivum*
Chewing Gum	*Ficus benghalensis**
Death rituals	*Aegle marmelos*, Bambusa nutans, Emblica officinalis*, Musa sapientum, Saccharum spontaneum*
Dye	*Cucurbita hispida, Emblica officinalis, Juglans regia, Lawsonia alba, Mallotus philippinensis, Terminalia chebula*
Facial cosmetics	*Cicer arietinum*, Citrus aurantium*, Cucumis sativus, Curcuma domestica*
Fencing	*Bambusa nutans, Bougainvillea spectabilis*, Carissa spinarum*, Duranta repens*, Euphorbia royleana*, Ipomoea cairica*, Ipomoea carnea*, Lantana camara, Opuntia dillenii*
Furniture	*Albizia lebbeck, Cedrela toona, Dalbergia sissoo, Eucalyptus umbellata*, Grewia oppositifolia*, Mangifera indica, Melia azedarach, Pinus roxburghii*
Grafting	*Pyrus pashia*
Gum (adhesive)	*Aegle marmelos, Cordia dichotoma*, Prunas persica**
Incense	*Cestrum nocturnum*
Insect repellent	*Ageratum conyzoides, Eucalyptus umbellata*, Vitex negundo*
Lice	*Anona squamosa*
Magico-religious belief	*Capsicum annuum*, Citrus limon*, Datura innoxia**
Miscellaneous Purposes (bowls, building works, games, house building, ladder, mats, measure of wt., musical instruments plates, roofing, plastering walls, shuttering purposes, thatching, torch light, water container, writing pen etc.)	*Abrus precatorius, Agave americana, Aloe maculate*, Amaranthus viridis*, Asparagus adscendens*, Bambusa nutans*, Bauhinia vahlii, Bombax ceiba*, Carissa spinarum, Cassia fistula, Cucurbita pepo,Dalbergia sissoo, Dendrocalamus hamiltonii, Celastrus paniculatus,Eucalyptus umbellata, Eugenia jambalana,* Ficus glomerata, Flacourtia ramontchi, Holarrhena antidysenterica, Lantana camara, Mimosa pudica*, Parthenium hysterophorus, Phoenix sylvestris, Pinus roxburghii, Polyalthia longifolia, Populus nigra*, Prunus domestica, Punica granatum*, Putranjiva roxburghii*, Pyrus pashia, Saccharum arundinaceum*, Saccharum officinarum, Sapindus mukorossi, Sesamum indicum, Tagetes erecta, Tamarindus indica, Terminalia arjuna, Thuja orientalis, Woodfordia floribunda, Zizyphus jujuba*
Ornamentals	*Bougainvillea spectabilis, Catharanthus roseus*, Callistemon lanceolatus*,*

	*Grevillea robusta, Hibiscus rosa-sinensis**, Juglans regia, Lagerstroemia indica, Mimosa pudica**, Papaver rhoeas**, Pedilanthes tithymaloides**, Polyalthia longifolia, Mimosa pudica, Rosa damascena**, Tabernaemontana divaricata, Tecoma grandiflora**, Thevetia neriifolia, Thuja orientalis*
Rearing silkworms	*Morus alba*
Reddening lips	*Juglans regia*
Scrubber	*Luffa aegyptica*
Shampoo and detergent	*Sapindus mukorossi*
Souring	*Citrus jambhiri*
Suppressing hunger	*Achyranthes bidentata**
Tooth brush (scouring teeth)	*Acacia nilotica, Citrus limon**, Citrus jambhiri, Citrus medica**, Citrus grandis, Jatropha curcas, Juglans regia, Melia azadirachata, Murraya koenigii, Murraya paniculata**, Plumbago zeylanica**, Psidium guajava, Ricinus communis, Vitex negundo, Wendlandia heynei, Zanthoxylum alatum*

*New Records/Species.

Like most of the other ethnic societies in India (Arora & Mehra, 1977; Brierly, 1985; Padoch & de Jong, 1991; Ruddle, 1974), the inhabitants in this part of Himalaya maintain their home gardens or household gardens as an age old agroecosystem with great importance for subsistence and as well a repository of tree species, grasses and herbs, medicinal plants, landraces, cultivars and spp of cultivated crops having a significant role in their economy as cash resources. As many as 79 plant spp are grown by them (Table XII), around their dwellings for various purposes which is surprisingly diverse in plant spp with predominance of Cucurbitaceae (7 spp), followed by Rosaceae (6 spp), Rutaceae (6 spp), Fabaceae (5 spp), Lamiaceae (5 spp), Poaceae (4 spp), Solanaceae (4), Liliaceae (3 spp), Zingiberaceae (3 spp), Annonaceae (2 spp), Apiaceae (2 spp), Araceae (2 spp), Moraceae (2 spp), Myrtaceae (2 spp), Sapindaceae (2 spp), etc. Not only this reflects their knowledge about environmental diversity and biodiversity but also their skill in the maximum use of available land for growing diverse plant spp on utilitarian basis every year. Documentation of such promising plants is important for ensuring sustainable utilization of our vast potential resources and in maintaining germplasm collections for scientific purposes.

TABLE XII
PLANTS CULTIVATED BY RURAL POPULACE OF PRAGPUR

Botanical Name	*Family*	*Botanical Name*	*Family*
Achras zapota	Sapotaceae	*Capsicum annuum*	Solanaceae
Allium cepa	Liliaceae	*Carica papaya*	Caricaceae
Allium sativum	Liliaceae	*Celosia argentea*	Amaranthaceae
Aloe vera	Liliaceae	*Cicer arietinum*	Fabaceae
Amorphophallus campanulatus	Araceae	*Cinnamomum camphora*	Lauraceae
Annona squamosa	Annonaceae	*Citrus aurantium*	Rutaceae
Artocarpus integrifolia	Moraceae	*Citrus grandis*	Rutaceae
Artocarpus lakoocha	Moraceae	*Citrus jambhiri*	Rutaceae
Bauhinia variegata	Fabaceae	*Citrus limon*	Rutaceae
Bougainvillea spectabilis	Nyctaginaceae	*Citrus medica*	Rutaceae
Brassica campestris	Brassicaceae	*Colocasia esculenta*	Araceae
Callistemon lanceolatus	Myrtaceae	*Cordia dichotoma*	Cordiaceae
Canna indica	Cannaceae	*Coriandrum sativum*	Apiaceae

(*Contd.*)

Botanical Name	Family	Botanical Name	Family
Cucumis sativus	Cucurbitaceae	*Polyalthia longifolia*	Annonaceae
Cucurbita hispida	Cucurbitaceae	*Populus nigra*	Salicaceae
Cucurbita pepo	Cucurbitaceae	*Prunus amygdalus*	Rosaceae
Curcuma domestica	Zingiberaceae	*Prunus domestica*	Rosaceae
Elettaria cardamomum	Zingiberaceae	*Prunus persica*	Rosaceae
Eruca sativa	Zingiberaceae	*Psidium guajava*	Myrtaceae
Foeniculum vulgare	Apiaceae	*Pyrus communis*	Rosaceae
Fragaria chiloensis	Rosaceae	*Pyrus pashia*	Rosaceae
Gossypium arboreum	Malvaceae	*Raphanus sativus*	Rutaceae
Grevillea robusta	Proteaceae	*Rosa damascena*	Rosaceae
Helianthus annuus	Asteraceae	*Saccharum officinarum*	Poaceae
Hibiscus rosa-sinensis	Malvaceae	*Salvia coccinea*	Lamiaceae
Juglans regia	Juglandaceae	*Saraca asoca*	Fabaceae
Lagenaria siceraria	Cucurbitaceae	*Sechium edule*	Cucurbitaceae
Lagerstroemia indica	Lythraceae	*Sesamum indicum*	Pedaliaceae
Litchi chinensis	Sapindaceae	*Solanum melongena*	Solanaceae
Luffa aegyptica	Cucurbitaceae	*Solanum tuberosum*	Solanaceae
Lycopersicum esculentum	Solanaceae	*Tabernaemontana divaricata*	Apocynaceae
Mentha piperita	Lamiaceae	*Tagetes erecta*	Asteraceae
Mentha spicata	Solanaceae	*Tecoma grandiflora*	Bignoniaceae
Momordica charantia	Cucurbitaceae	*Thuja orientalis*	Cupressaceae
Musa sapientum	Musaceae	*Trifolium alexandrianum*	Fabaceae
Ocimum basilicum	Lamiaceae	*Trigonella foenum-graecum*	Fabaceae
Ocimum sanctum	Lamiaceae	*Triticum aestivum*	Poaceae
Papaver rhoeus	Papaveraceae	*Zingiber officinale*	Zingiberaceae
Pedilanthes tithymaloides	Euphorbiaceae	*Zea mays*	Poaceae
Pennisetum purpureum	Poaceae		

To supplement the fodder requirements, a total of 47 spp under 40 genera and 26 families are being widely used, indicating rich diversity in forages held by the people in this part of North-West Himalaya (Table XIII). Some of the commonly employed forage genera are: *Ficus* (4 spp), *Amaranthus* (2 spp), *Bauhinia* (2 spp), *Carissa* (2 spp), *Eugenia* (2 spp), etc. with diversity predominantly belonging to Fabaceae (7 spp), Moraceae (5 spp), Poaceae (4 spp), Apocynaceae (3 spp), Amaranthaceae (2 spp), Bambusaceae (2 spp), Brassicaceae (2 spp), Meliaceae (2 spp), Myrtaceae (2 spp), Rosaceae (2 spp), etc. As many as 10 fodder spp, such as *Bauhinia vahlii, Ficus religiosa, Amaranthus spinosus, A. viridis, Bambusa nutans, Eugenia cerasoides, Eriobotrya japonica, Oroxylum indicum, Fumaria indica* and *Gossypium hirsutum* are new additions to the list of 400 forage spp documented for our country (Anonymous, 1994), and their importance as invaluable genetic stocks cannot be underestimated.

For their fibre needs, the plants belonging to families Agavaceae (*Agave americana*), Bombacaceae (*Bombax ceiba*), Bambusaceae (*Bambusa nutans*), Malvaceae (*Gossypium arboreum*), Moraceae (*Morus alba*) and Tiliaceae (*Grewia oppositifolia*) are mainly used (Table XIV). The resultant information is of great significance for achieving the goal of economic benefits for the local people by setting several cottage and rural industries in the region.

Fuel wood requirements are largely met from 24 plant spp belonging to 17 families with predominance of representatives from Fabaceae (4 spp), followed by Rosaceae (3 spp), Bambusaceae (2 spp) and rest of the families having one sp each (Table XV). Plants namely:

TABLE XIII
FODDER PLANTS OF PRAGPUR

Family	*Plant Species*
Fabaceae	*Albizia lebbeck, Bauhinia vahlii*, Bauhinia variegata, Cicer arietinum, Dalbergia sissoo, Leucaena leucocephala, Trifolium alexandrianum*
Moraceae	*Ficus carica, Ficus glomerata, Ficus religiosa*, Ficus roxburghii, Morus alba*
Poaceae	*Cynodon dactylon, Triticum aestivum, Pennisetum purpureum, Zea mays*
Apocynaceae	*Adhatoda vasica, Carissa spinarum, Carissa inermis*
Amaranthaceae	*Amaranthus spinosus*, Amaranthus viridis**
Bambusaceae	*Bambusa nutans*, Dendrocalamus hamiltonii*
Brassicaceae	*Brassica campestris, Eruca sativa*
Meliaceae	*Cedrela toona, Melia azedarach*
Myrtaceae	*Eugenia cerasoides*, Eugenia cuminii*
Rosaceae	*Eriobotrya japonica*, Pyrus pashia*
Agavaceae	*Agave americana*
Anacardiaceae	*Pistacia integerrima*
Bignoniaceae	*Oroxylum indicum**
Chenopodiaceae	*Chenopodium album*
Convolvulaceae	*Ipomoea cairica*
Euphorbiaceae	*Mallotus philippinensis*
Flacourtiaceae	*Flacourtia ramontchi*
Fumariaceae	*Fumaria indica**
Malvaceae	*Gossypium hirsutum**
Rhamnaceae	*Zizyphus jujuba*
Rubiaceae	*Wendlandia heynei*
Salicaceae	*Populus nigra*
Sapindaceae	*Celastrus paniculatus*
Tiliaceae	*Grewia oppositifolia*
Ulmaceae	*Holoptelea integrifolia*
Verbenaceae	*Premna latifolia*

*New Records of Fodder Plants.

TABLE XIV
FIBRE (CORDAGE, BASKETRY, STUFFING) YIELDING PLANTS OF PRAGPUR

Family	*Plant Species*
Agavaceae	*Agave americana*
Bombacaceae	*Bombax ceiba*
Bambusaceae	*Bambusa nutans*
Malvaceae	*Gossypium arboreum*
Moraceae	*Morus alba*
Tiliaceae	*Grewia oppositifolia*

Carissa spinarum, Parthenium hysterophorus, Bambusa nutans, Dendrocalamus hamiltonii, Mallotus philippinensis, Bauhinia vahlii, Cassia fistula, Colebrookia oppositifolia, Eucalyptus umbellata, Saccharum arundinaceum, Zea mays, Prunus amygdalus, Populus nigra and *Lantana camara* are new records of fuel wood plants. Interestingly, the locals never extract them in unsustainable manner as only dead or dried branches, twigs and/or plants are being used for the said purpose.

The study also highlights the use of 20 plant spp belonging to 15 families for a wide array of

TABLE XV
FUEL WOOD PLANTS OF PRAGPUR

Family	*Plant Species*
Apocynaceae	*Carissa spinarum**
Asteraceae	*Parthenium hysterophorus**
Bambusaceae	*Bambusa nutans**, *Dendrocalamus hamiltonii**
Combretaceae	*Terminalia arjuna*
Euphorbiaceae	*Mallotus philippinensis**
Fabaceae	*Albizia lebbeck, Bauhinia vahlii**, *Cassia fistula**, *Dalbergia sissoo*
Flacourtiaceae	*Flacourtia ramontchi*
Lamiaceae	*Colebrookia oppositifolia**
Lythraceae	*Woodfordia floribunda*
Myrtaceae	*Eucalyptus umbellata**
Poaceae	*Saccharum arundinaceum**, *Zea mays**
Rhamnaceae	*Zizyphus jujuba*
Rosaceae	*Prunus amygdalus**, *Prunus domestica, Pyrus pashia*
Salicaceae	*Populus nigra**
Sapindaceae	*Dodonaea viscosa*
Tiliaceae	*Grewia oppositifolia*
Verbenaceae	*Lantana camara**

*New Records of Fuel Wood Plants.

livestock ailments, namely, loss of appetite (*Adhatoda vasica*), stomach disorders (*Aloe vera*), enhancing lactation (*Bauhinia variegata*, *Eruca sativa, Mucuna pruriens, Tinospora cordifolia*), diarrhoea (*Cannabis sativa*), skin eruptions (*Capsicum annuum*), constipation (*Cassia fistula, Oroxylum indicum, Verbascum thapsus*), dysentery (*Cissampelos pareira*), stimulating ovulation and tail infection (*Euphorbia royleana*), cud (*Grewia oppositifolia*), prolapsed uterus (*Holarrhena antidysenterica*), febrifuge (*Jasminum multiflorum*), cooling effect against sun stroke (*Mirabilis jalapa*), toe infection (*Randia dumetorum*), flatulence, mouth ulceration, poisonous teats (*Stephania glabra*), expelling leeches (*Wendlandia heynei*) (Table XVI) whose well-being is considered to be very important by the people of this region. The most important reason for their inclination towards the use of herbal veterinary medicines is their proved therapeutic efficacy and availability in vicinity. Interestingly, the use of plants *Adhatoda vasica* for loss of appetite, *Cissampelos pareira* for checking dysentery, *Eruca sativa* for enhancing lactation are hitherto unreported in ethnoveterinary practices (Jain, 1991). Clinical validation of these, of course, be needed to confirm the claims so that these may be used as safe medicines for the use of communities of animal husbandry practitioners.

As remedies against various types of bites, the rural communities of Pragpur have been using 7 spp available in the vicinity, e.g. *Boerhaavia diffusa* (Nyctaginaceae), *Cannabis sativa* (Cannabinaceae), *Catharanthus roseus* (Apocynaceae), *Martynia annua* (Martyniaceae), *Papaver rhoeus* (Papaveraceae), *Tagetes erecta* (Asteraceae) and *Verbascum thapsus* (Scrophulariaceae). Their topical application is believed to help in neutralizing some of the inflammatory agents present in toxins and checks pain and swelling. The vast potential of such data needs to be screened on modern scientific lines so that maximum benefit can be derived from them in future.

Economic valuation of all the presently studied 204 spp was carried out to calculate the total importance values (TIV) as per Belal & Springuel (1996) (Table XVII). Evidently, the spp

TABLE XVI

ETHNOBOTANICAL IMPORTANT PLANTS USED FOR VETERINARY AILMENTS

Plant Species	*Ailments*
*Adhatoda vasica**	Loss of appetite
Aloe vera	Stomach disorders
Bauhinia variegata	Enhancing lactation
Cannabis sativa	Skin eruptions
Cassia fistula	Constipation
*Cissampelos pareira**	Dysentery
*Eruca sativa**	Enhancing lactation
Euphorbia royleana	Stimulating ovulation, Tail infection
Grewia oppositifolia	Cud
Holarrhena antidysenterica	Prolapsed uterus
Jasminum multiflorum	Febrifuge
Mirabilis jalapa	Cooling effect against sun stroke
Mucuna pruriens	Enhancing lactation
Oroxylum indicum	Constipation
Randia dumetorum	Toe infection
Stephania glabra	Flatulence, Mouth ulceration, Poisonous teats
Tinospora cordifolia,	Enhancing lactation
Verbascum thapsus	Constipation
Wendlandia heynei	Expelling leeches

*New Ethnoveterinary Uses.

like *Emblica officinalis* (Euphorbiaceae), *Brassica campestris* (Brassicaceae), *Mangifera indica* (Anacardiaceae), *Flacourtia ramontchi* (Flacourtiaceae), *Grewia oppositifolia* (Tiliaceae), *Aegle marmelos* (Rutaceae), *Prunus amygdalus* (Rosaceae), *Mallotus philippinensis* (Euphorbiaceae), *Morus alba* (Moraceae), *Bambusa nutans* (Bambusaceae), *Eucalyptus umbellata* (Myrtaceae), *Eugenia jambolana* (Myrtaceae), *Melia azadirachta* (Meliaceae), *Prunus persica* (Rosaceae), *Punica granatum* (Punicaceae), *Sesamum indicum* (Pedaliaceae), *Albizia lebbeck* (Fabaceae), *Bauhinia vahlii* (Fabaceae), *Dalbergia sissoo* (Fabaceae), *Musa sapientum* (Musaceae), *Solanum tuberosum* (Solanaceae), *Cassia fistula* (Fabaceae), *Terminalia arjuna* (Combretaceae), *Melia azedarach* (Meliaceae), *Pyrus pashia* (Rosaceae), *Bombax ceiba* (Bombacaeae), *Carissa spinarum* (Apocynaceae), *Cicer arietinum* (Fabaceae), *Zea mays* (Poaceae), *Cedrela toona* (Meliaceae), *Citrus medica* (Rutaceae), *Opuntia dillenii* (Cactaceae), *Saccharum officinarum* (Poaceae), *Tagetes erecta* (Asteraceae), *Juglans regia* (Juglandacae), *Murraya koenigii* (Rutaceae), *Amaranthus spinosus* (Amaranthaceae), *Amaranthus viridis* (Amaranthaceae), *Bauhinia variegata* (Fabaceae), *Chenopodium album* (Chenopodiaceae), *Cucurbita hispida* (Cucurbitaceae), *Lantana camara* (Verbenaceae), *Luffa aegyptica* (Cucurbitaceae), *Raphanus sativus* (Rutaceae), *Zizyphus jujuba* (Rhamnaceae), etc. (45 spp) are found to have high economic potential. Based on each importance value, plant spp such as *Gossypium hirsutum, Adhatoda vasica, Brassica campestris, Ficus religiosa, Triticum aestivum, Bambusa nutans, Eugenia jambolana, Zea mays, Amaranthus spinosus, A. viridis, Bauhinia vahlii, B. variegata, Chenopodium album, Cicer arietinum, Dalbergia sissoo, Eriobotrya japonica, Eugenia cerasoides, Ficus roxburghii, Fumaria indica, Ipomoea cairica, Leucaena leucocephala, Pennisetum purpureum, Pistacia integerrima, Populus nigra, Premna latifolia, Pyrus pashia, Trifolium alexandrianum,* etc. (27 spp) rank high in **fodder** value; *Zea mays, Albizia lebbeck, Bambusa* *nutans, Bauhinia vahlii, Eucalyptus umbellata, Flacourtia ramontchi, Lantana camara, Mallotus philippinensis, Parthenium hysterophorus, Prunus amygdalus, Pyrus pashia, Saccharum arundinaceum,* etc. (12 spp) in **fuel**; *Abrus precatorius, Aegle marmelos, Allium cepa, Aloe vera, Bambusa nutans, Cannabis sativa, Capsicum annuum, Curcuma domestica, Cynodon dactylon, Elettaria cardamomum, Emblica officinalis, Ficus benghalensis, F. religiosa, Foeniculum vulgare, Gossypium hirsutum, Mangifera indica, Mentha piperita, M. spicata, Ocimum sanctum, Phoenix sylvestris, Pinus roxburghii, Punica granatum, Saccharum arundinaceum, S. officinarum, S. spontaneum, Sapindus mukorossi, Sesamum indicum, Solanum tuberosum, Trigonella foenum-graecum, Triticum aestivum, Vitex negundo, Asparagus adscendens, Cucurbita pepo, Duranta repens, Euphorbia royleana, Murraya paniculata, M. koenigii, Musa sapientum, Ocimum basilicum, Opuntia dillenii, Putranjiva roxburghii, Tagetes erecta, Tamarindus indica, Bauhinia vahlii, Bougainvillea spectabilis, Butea monosperma, Cassia fistula, Datura innoxia, Eruca sativa, Eugenia jambolana, Jatropha curcas, Luffa aegyptica, Melia azadirachta, Morus alba, Raphanus sativus, Rosa damascena, Thuja orientalis, Zanthoxylum armatum, Zingiber officinale, Acacia nilotica, Albizia lebbeck, Bombax ceiba, Canna indica, Cedrela toona, Cestrum nocturnum, Catharanthus roseus, Cinnamomum camphora, Citrus grandis, C. jambhiri, C. limon, C. medica, Eucalyptus umbellata, Eugenia cerasoides, Flacourtia ramontchi, Grevillea robusta, Hibiscus rosa-sinensis, Holarrhena antidysenterica, Ipomoea cairica, I. carnea, Juglans regia, Lagerstroemia indica, Lantana camara, Malvaviscus conzatii, Mimosa pudica, Mirabilis jalapa, Nerium indicum, Pedialanthes tithymaloides, Plumbago zeylanica, Polyalthia longifolia, Populus nigra, Salvia coccinea, Saraca asoca, Tabernaemontana divaricata, Terminalia arjuna, Thevetia neriifolia,* etc. (93 spp) in **medicines**; *Amorphophallus campanulatus, Cicer arietinum, Punica granatum, Solanum tuberosum, Terminalia arjuna, Brassica campestris, Citrus jambhiri, C. limon, C. medica, Cucurbita pepo, Dioscorea bulbifera, Emblica officinalis, Lycopersicum esculentum, Mangifera indica, Musa sapientum, Zizyphus jujuba, Carica papaya, Cassia occidentalis, Colocasia esculenta, Cucurbita hispida, Juglans regia, Prunus amygdalus, Zea mays, Aegle marmelos, Amaranthus spinosus, A. viridis, Bauhinia variegata, Carissa inermis, C. spinarum, Chenopodium album, Citrus aurantium, C. grandis, Eugenia jambolana, Ficus glomerata, F. carica, F. roxburghii, Flacourtia ramontchi, Holoptelea integrifolia, Litchi chinensis, Luffa aegyptica, Momordica charantia, M. dioica, Opuntia dillenii, Prunus domestica, Psidium guajava, Pyrus communis, Raphanus sativus, Rumex hastatus, Saccharum officinarum, Tagetes erecta, Vitis jacquemontii,* etc. (50 spp) in **food**; *Bombax ceiba, Cedrela toona, Dalbergia sissoo, Eucalyptus umbellata, Mangifera indica, Melia azedirachta, Pinus roxburghii, Albizia lebbeck, Grewia oppositifolia, Melia azedarach* (10 spp) in **timber**; *Emblica officinalis, Lawsonia alba, Juglans regia, Mallotus philippinensis, Terminalia chebula, Cucurbita hispida* (6 spp) in **dye**; *Aegle marmelos, Cordia dichotoma, Prunus persica* (3 spp) in **gum**; *Brassica campestris, Helianthus annuus, Prunus amygdalus, Ricinus communis, Sesamum indicum* (5 spp) in **oil**; *Acacia catechu, Adhatoda vasica, Aegle marmelos, Allium cepa, A. sativum, Aloe vera, Asparagus adscendens, Coriandrum sativum, Curcuma domestica, Cuscuta reflexa, Elettaria cardamomum, Emblica officinalis, Melia azadirachta, Momordica charantia, Musa sapientum, Ocimum sanctum, Punica granatum, Ricinus communis, Rosa damascena, Saraca asoca, Sesamum indicum, Terminalia arjuna, T. bellirica, T. chebula, Tinospora cordifolia, Trigonella foenum-graecum, Vitex negundo, Woodfordia floribunda, Zanthoxylum armatum, Zingiber officinale, Argemone mexicana, Brassica campestris, Celastrus paniculatus, Celosia argentea, Centella asiatica, Cynodon dactylon, Dioscorea deltoidea, Momordica dioica, Ocimum basilicum, Pistacia integerrima, Prunus* *amygdalus, Solanum surattense, Spilanthes acmella, Stephania glabra, Abrus precatorius, Aloe maculata, Amorphophallus campanulatus, Artemisia japonica, Boerhaavia diffusa, Bombax ceiba, Bryophyllum calycinum, Capsicum annuum, Carica papaya, Catharanthus roseus, Cinnamomum camphora, Cissampelos pareira, Datura innoxia, Eugenia jambolana, Flacourtia ramontchi, Lannea coromandelica, Lycopersicum esculentum, Mucuna pruriens, Pogostemon benghalense, Randia dumetorum, Solanum tuberosum, Acacia nilotica, Achyranthes bidentata, Acorus calamus, Amaranthus spinosus, A. viridis, Artocarpus integrifolia, Bauhinia vahlii, B. variegata, Butea monosperma, Cannabis sativa, Carissa inermis, Cassia fistula, C. occidentalis, Chenopodium album, Cicer arietinum, Citrus aurantium, C. medica, Cucurbita hispida, Celastrus paniculatus, Eucalyptus umbellata, Euphorbia royleana, Ficus benghalensis, F. carica, Foeniculum vulgare, Fumaria indica, Helianthus annuus, Hibiscus rosasinensis, Holarrhena antidysenterica, Jasminum multiforium, Jatropha curcas, Lantana camara, Litchi chinensis, Mallotus philippinensis, Malvaviscus conzatii, Mangifera indica, Martynia annua, Melia azedarach, Mentha piperita, M. spicata, Mimosa pudica, Mirabilis jalapa, Murraya koenigii, Nerium indicum, Opuntia dillenii, Oroxylum indicum, Parthenium hysterophorus, Phyllanthus niruri, Plumbago zeylanica, Premna latifolia, Psidium guajava, Pyrus communis, Rumex hastatus, Solanum nigrum, S. viarum, Tabernaemontana divaricata, Tagetes erecta, Thevetia neriifolia, Verbascum thapsus, Vitis jacquemontii, Xanthium strumarium* etc., (125 spp) in **others.**

Nevertheless, the following data on the total importance values (TIV) calculated for the first time for the ethnic plants of Himachal Pradesh, for sure, can be profitably made use of for the welfare of mankind.

It is evident from the foregoing discussion that these people have a self-sufficient and self-reliant subsistence system which is agro-forestry based and offer the most significant evidence that how wonderfully they maintain a balanced ecological rhythm in the Himalaya. Considering the vast economic potential of ethnobotanical information generated presently, the traditional knowledge can be judiciously made use of by using modern technologies for improving and uplifting the life and economy of the inhabitants of this region of N.W. Himalaya.

Table XVII
Total Importance Values (TIV) of Ethnobotanically Important Plants of Pragpur, Distt. Kangra (H.P.)

Sr. No.	*Botanical Name*	*Dye*	*Fibre*	*Fodder*	*Fuel*	*Medicine*	*Oil*	*Tannin/ Gum*	*Timber*	*Others**	*Edible*	*TIV*
1	*2*	*3*	*4*	*5*	*6*	*7*	*8*	*9*	*10*	*11*	*12*	*13*
1.	*Abrus precatorius*					8				10		18
2.	*Acacia catechu*					10						10
3.	*Acacia nilotica*					7				7		17
4.	*Achras zapota*										6	6
5.	*Achyranthes bidentata*					7				6		13
6.	*Acorus calamus*					7						7
7.	*Adhatoda vasica*			9		10						19
8.	*Aegle marmelos*					10		6		10	7	33
9.	*Agave americana*		6	6							6	18
10.	*Ageratum conyzoides*									6		6
11.	*Albizia lebbeck*			6	7				9	7		29

(*Contd.*)

1	2	3	4	5	6	7	8	9	10	11	12	13
12.	*Allium cepa*					10				10		20
13.	*Allium sativum*					10						10
14.	*Aloe maculata*					8						8
15.	*Aloe vera*					10				10		20
16.	*Amaranthus spinosus*			7		7					7	21
17.	*Amaranthus viridis*			7		7					7	21
18.	*Amorphophallus campanulatus*					8					10	18
19.	*Annona squamosa*					6					6	12
20.	*Argemone mexicana*					9						9
21.	*Artemisia japonica*					8						8
22.	*Artocarpus integrifolia*					7					6	13
23.	*Artocarpus lakoocha*					6					6	12
24.	*Asparagus adscendens*					10				9		19
25.	*Bambusa nutans*			8	7					10	6	31
26.	*Bauhinia vahlii*			7	7	7				8		29
27.	*Bauhinia variegata*			7		7					7	21
28.	*Boerhaavia diffusa*					8						8
29.	*Bombax ceiba*					8			10	7		25
30.	*Bougainvillea spectabilis*									8		8
31.	*Brassica campestris*			9		9	10				9	37
32.	*Bryophyllum calycinum*					8						8
33.	*Butea monosperma*					7				8		15
34.	*Callistemon lanceolatus*									6		6
35.	*Canna indica*					6				7		13
36.	*Cannabis sativa*					7				10		17
37.	*Capsicum annuum*					8				10		18
38.	*Carica papaya*					8					8	16
39.	*Carissa inermis*			6		7					7	20
40.	*Carissa spinarum*			6	6	6					7	25
41.	*Cassia fistula*			6	6	7				8		27
42.	*Cassia occidentalis*				6	7					8	15
43.	*Catharanthus roseus*					8				7		15
44.	*Cedrela toona*			6					10	7		23
45.	*Celastrus paniculatus*					9						9
46.	*Celosia argentea*					9						9
47.	*Centella asiatica*					9						9
48.	*Cestrum nocturnum*					6				7		13
49.	*Chenopodium album*			7		7					7	21
50.	*Cicer arietinum*			7						7	10	24
51.	*Cinnamomum camphora*					8				7		15
52.	*Cissampelos pareira*					8						8
53.	*Citrus aurantium*					7					7	14
54.	*Citrus grandis*									7	7	14
55.	*Citrus jambhiri*									7	9	16
56.	*Citrus limon*									7	9	16
57.	*Citrus medica*					7				7	9	23
58.	*Coccinia grandis*					6					6	12
59.	*Colebrookia oppositifolia*				6	6						12
60.	*Colocasia esculenta*										8	8
61.	*Cordia dichotoma*					6		7			6	19
62.	*Coriandrum sativum*					10				6		16
63.	*Cucumis sativus*					6					6	12
64.	*Cucurbita hispida*	6				7					8	21
65.	*Cucurbita pepo*									9	9	18
66.	*Curcuma domestica*					10				10		20
67.	*Cuscuta reflexa*					10						10
68.	*Cynodon dactylon*					9				10		19

1	*2*	*3*	*4*	*5*	*6*	*7*	*8*	*9*	*10*	*11*	*12*	*13*
69.	*Dalbergia sissoo*			7	6				10	6		29
70.	*Datura innoxia*					8				8		16
71.	*Dendrocalamus strictus*			6	6	6						18
72.	*Dioscorea bulbifera*										9	9
73.	*Dioscorea deltoidea*					9						9
74.	*Dodonaea viscose*			6	6	7						19
75.	*Duranta repens*					6				9		15
76.	*Elettaria cardamomum*					10				10		20
77.	*Emblica officinalis*	10				10				10	9	39
78.	*Eriobotrya japonica*			7		6					6	19
79.	*Eruca sativa*			6		6				8		20
80.	*Eucalyptus umbellata*				7	7			10	7		31
81.	*Eugenia cerasoides*			7						7		14
82.	*Eugenia jambolana*			8		8				8	7	31
83.	*Euphorbia royleana*					7				9		16
84.	*Ficus benghalensis*					7				10		17
85.	*Ficus carica*			6		7					7	20
86.	*Ficus glomerata*			6							7	13
87.	*Ficus religiosa*			9						10		19
88.	*Ficus roxburghii*			7							7	14
89.	*Flacourtia ramontchi*			6	7	8				7	7	35
90.	*Foeniculum vulgare*					7				10		17
91.	*Fragaria chiloensis*										6	6
92.	*Fumaria indica*			7		7						14
93.	*Gossypium hirsutum*			10						10		20
94.	*Grevillea robusta*									7		7
95.	*Grewia oppositifolia*		7	6	6				9	6		34
96.	*Helianthus annuus*					7	10					17
97.	*Hibiscus rosasinensis*					7				7		14
98.	*Holarrhena antidysenterica*					7				7		14
99.	*Holoptelea integrifolia*			6						6	7	19
100.	*Impatiens balsamina*									6		6
101.	*Ipomoea cairica*			7						7		14
102.	*Ipomoea carnea*									7		7
103.	*Jasminum multiforium*					7				6		13
104.	*Jasminum sambac*					6				6		12
105.	*Jatropha curcas*					7				8		15
106.	*Juglans regia*	7								7	8	22
107.	*Lagenaria siceraria*					6					6	12
108.	*Lagerstroemia indica*					6				7		13
109.	*Lannea coromandelica*					8						8
110.	*Lantana camara*				7	7				7		21
111.	*Lawsonia alba*	10				6						16
112.	*Leucaena leucocephala*			7								7
113.	*Litchi chinensis*					7					7	14
114.	*Luffa aegyptica*					6				8	7	21
115.	*Lycopersicum esculentum*					8					9	17
116.	*Mallotus philippinensis*	7		6	7	7				6		33
117.	*Malvaviscus conzatii*					7				7		14
118.	*Mangifera indica*					7			10	10	9	36
119.	*Martynia annua*					7						7
120.	*Melia azadirachta*					10			10	8		31
121.	*Melia azedarach*			6		7			7	6		26
122.	*Mentha piperita*					7				9		16
123.	*Mentha spicata*					7				9		16
124.	*Mimosa pudica*					7				7		14

(*Contd.*)

1	2	3	4	5	6	7	8	9	10	11	12	13
125.	*Mirabilis jalapa*					7				7		14
126.	*Momordica charantia*					10					7	17
127.	*Momordica dioica*					9					7	16
128.	*Morus alba*		6	6		6				8	6	32
129.	*Mucuna pruriens*					8						8
130.	*Murraya koenigii*					7				9	6	22
131.	*Murraya paniculata*									9	6	15
132.	*Musa sapientum*					10				9	9	28
133.	*Nerium indicum*					7				7		14
134.	*Ocimum basilicum*					9				9		18
135.	*Ocimum sanctum*					10				10		20
136.	*Opuntia dillenii*					7				9	7	23
137.	*Oroxylum indicum*			6		7						13
138.	*Papaver rhoeus*					6				6		12
139.	*Parthenium hysterophorus*				7	7						14
140.	*Pedialanthes tithymaloides*									7		7
141.	*Pennisetum purpureum*			7								7
142.	*Phoenix sylvestris*									10	6	16
143.	*Phyllanthus niruri*					7						7
144.	*Pinus roxburghii*								10	10		20
145.	*Pistacia integerrima*			7		9						16
146.	*Plumbago zeylanica*					7				7		14
147.	*Pogostemon benghalense*					8						8
148.	*Polyalthia longifolia*									7		7
149.	*Populus nigra*			7	6					7		20
150.	*Premna latifolia*			7		7						14
151.	*Prunus amygdalus*				7	9	10				8	34
152.	*Prunus domestica*				6	6					7	19
153.	*Prunus persica*				6	6		6		6	6	30
154.	*Psidium guajava*					7					7	14
155.	*Punica granatum*					10				10	10	30
156.	*Putranjiva roxburghii*									9		9
157.	*Pyrus communis*					7					7	14
158.	*Pyrus pashia*			7	7	6					6	26
159.	*Randia dumetorum*					8						8
160.	*Raphanus sativus*					6				8	7	21
161.	*Ricinus communis*					10	10					20
162.	*Rosa damascena*					10				8		18
163.	*Rumex hastatus*					7					7	14
164.	*Saccharum arundinaceum*				7					10		17
165.	*Saccharum officinarum*					6			10	7		23
166.	*Saccharum spontaneum*									10		10
167.	*Salvia coccinea*									7		7
168.	*Sapindus mukorossi*									10		10
169.	*Saraca asoca*					10				7		17
170.	*Sechium edule*					6					6	12
171.	*Sesamum indicum*					10	10			10		30
172.	*Solanum viarum*					7						7
173.	*Solanum erianthum*					6						6
174.	*Solanum melongena*					6					6	12
175.	*Solanum nigrum*					7						7
176.	*Solanum surattense*					9						9
177.	*Solanum tuberosum*					8				10	10	28
178.	*Sonchus asper*					6						6
179.	*Spilanthes acmella*					9						9
180.	*Stephania glabra*											
181.	*Tabernaemontana divaricata*											

1	*2*	*3*	*4*	*5*	*6*	*7*	*8*	*9*	*10*	*11*	*12*	*13*
182.	*Tagetes erecta*					7				9	7	23
183.	*Tamarindus indica*									9		9
184.	*Taraxacum officinale*					6						6
185.	*Tecoma radicans*									6		6
186.	*Terminalia arjuna*					10				7	10	27
187.	*Terminalia bellirica*					10						10
188.	*Terminalia chebula*	7				10						17
189.	*Thevetia neriifolia*					7				7		14
190.	*Thuja orientalis*									8		8
191.	*Tinospora cordifolia*					10				6		16
192.	*Trifolium alexandrinum*			7								7
193.	*Trigonella foenum-graecum*						10			10		20
194.	*Triticum aestivum*			9						10		19
195.	*Verbascum thapsus*					7						7
196.	*Vitex negundo*					10				10		20
197.	*Vitis jacquemontii*					7					7	14
198.	*Wendlandia heynei*			6						6		12
199.	*Woodfordia floribunda*				6	10						16
200.	*Xanthium strumarium*					7				7		14
201.	*Zanthoxylum armatum*					10				8		18
202.	*Zea mays*			8	8						8	24
203.	*Zingiber officinale*					10				8		18
204.	*Zizyphus jujuba*			6	6						9	21

Others*: *House building, musical instruments, plates, bowls, roofing, torch light, plastering walls, thatching, water container, ladder, mats, measure of wt., shuttering purposes, writing pen, ornamental, fencing, bites (scorpion, wasp, insects), death rituals, fibre (cordage, chicks, basketry, stuffing), facial cosmetics, massage, bedding for animals ('pattri'), insect repellent, magico-religious beliefs, chewing gum, expelling leeches, games, grafting purposes, incense, rearing silkworm, scrubber, shampoo and detergent, souring, suppressing hunger, etc.*

APPENDIX I
NOTIFICATION REGARDING HERITAGE STATUS OF VILLAGE PRAGPUR

GOVERNMENT OF HIMACHAL PRADESH
DEPARTMENT OF TOURISM & C.A.

....

No.TSM-F(2)2/93 Dated 9th Dec., 1997.

NOTIFICATION

Pragpur in Dehra Tehsil of Kangra District in the State is a village having distinctive Architectural features of forts, villas, houses, streets etc. with medieaval ambience, which are of immense heritage and tourism interest. In view of its asthetic heritage value, the Governor, Himachal Pradesh is pleased to notify this village, which is located in the red line of revenue estate of Pragpur, as Heritage Village with immediate effect.

By Order,

DC-cum-Secretary (Tourism) to the
Government of Himachal Pradesh.

Endst.No.TSM-F(2)2/93 Dated 9th Dec., 1997.

Copy forwarded for information and necessary action to :-

1. The Divisional Commissioner, Shimla Division / Kangra Division / Mandi Division, Himachal Pradesh.
2. All the Deputy Commissioners / Superintendent of Police.
3. The Director Tourism & Civil Aviation, H.P., Shimla.
4. To all the Heads of Departments in Himachal Pradesh.

Additional Secretary (Tourism) to the
Government of Himachal Pradesh.

APPENDIX II
INDEX TO FAMILIES

Name of the Family	*Total Number of Spp in the Genera*	*Total Number of Genera and Spp*
1	2	3
Acanthaceae	1/1	*Adhatoda* (1)
Agavaceae	1/1	*Agave* (1)
Amaranthaceae	3/4	*Amaranthus* (2), *Achyranthes* (1), *Celosia* (1)
Anacardiaceae	3/3	*Lannea* (1), *Mangifera* (1), *Pistacia* (1)
Annonaceae	2/2	*Annona* (1), *Polyalthia* (1)
Apiaceae	3/3	*Centella* (1), *Coriandrum* (1), *Foeniculum* (1)
Apocynaceae	6/7	*Carissa* (2), *Catharanthus* (1), *Holarrhena* (1), *Nerium* (1), *Tabernaemontana* (1), *Thevetia* (1)
Araceae	3/3	*Acorus* (1), *Amorphophallus* (1), *Colocasia* (1)
Asteraceae	9/9	*Ageratum*(1), *Artemisia* (1), *Helianthus* (1), *Parthenium* (1), *Sonchus* (1), *Spilanthes* (1), *Tagetes* (1), *Taraxacum* (1), *Xanthium* (1)
Balsaminaceae	1/1	*Impatiens* (1)
Bambusaceae	2/2	*Bambusa* (1), *Dendrocalamus* (1)
Bignoniaceae	2/2	*Oroxylum* (1), *Tecoma* (1)
Bombacaceae	1/1	*Bombax* (1)
Brassicaceae	3/3	*Raphanus* (1), *Brassica* (1), *Eruca* (1)
Cactaceae	1/1	*Opuntia* (1)
Canabinaceae	1/1	*Cannabis* (1)
Cannaceae	1/1	*Canna* (1)
Caricaceae	1/1	*Carica* (1)
Celastraceae	1/1	*Celastrus* (1)
Chenopodiaceae	1/1	*Chenopodium* (1)
Combretaceae	1/3	*Terminalia* (3)
Convolvulaceae	1/3	*Ipomoea* (2)
Cordiaceae	1/1	*Cordia* (1)
Crassulaceae	1/1	*Bryophyllum*
Cucurbitaceae	7/9	*Coccinia* (1), *Cucumis* (1), *Cucerbita* (2), *Lagenaria* (1), *Luffa* (1), *Momordica* (2), *Sechium* (1)
Cupressaceae	1/1	*Thuja* (1)
Cuscutaceae	1/1	*Cuscuta* (1)
Dioscoreaceae	1/2	*Dioscorea* (2)
Euphorbiaceae	8/8	*Emblica* (1) *Euphorbia* (1), *Jatropha* (1), *Mallotus* (1), *Pedilanthes* (1), *Phyllanthus* (1), *Putranjiva* (1), *Ricinus* (1)
Fabaceae	15/18	*Abrus* (1), *Acacia* (2), *Albizia* (1), *Bauhinia* (2), *Butea* (1), *Cassia* (2), *Cicer* (1), *Dalbergia* (1), *Leucaena* (1), *Mimosa* (1), *Mucuna* (1), *Saraca* (1), *Tamarindus* (1), *Trigonella* (1), *Trifolium* (1)
Flacourtiaceae	1/1	*Flacourtia* (1)
Fumariaceae	1/1	*Fumaria* (1)
Juglandaceae	1/1	*Juglans* (1)
Lamiaceae	5/6	*Colebrookia* (1), *Mentha* (1), *Ocimum* (2), *Pogostemon* (1), *Salvia* (1)
Lauraceae	1/1	*Cinnamomum* (1)
Liliaceae	3/5	*Allium* (2), *Aloe* (2), *Asparagus*(1)
Lythraceae	3/3	*Lagerstroemia* (1), *Lawsonia* (1), *Woodfordia* (1)
Malvaceae	3/3	*Malvaviscus* (1), *Gossypium* (1), *Hibiscus* (1)
Martyniaceae	1/1	*Martynia* (1)
Meliaceae	3/4	*Melia* (2), *Cedrela* (1), *Mirabilis* (1)

1	2	3
Menispermaceae	3/3	*Cissampelos* (1), *Stephania* (1), *Tinospora* (1)
Moraceae	3/8	*Artocarpus* (2), *Ficus* (5), *Morus* (1)
Musaceae	1/1	*Musa* (1)
Myrtaceae	4/5	*Eucalyptus* (1), *Psidium* (1), *Eugenia* (2), *Callistemon* (1)
Nyctaginaceae	2/2	*Bougainvillea* (1), *Boerhaavia* (1)
Oleaceae	1/2	*Jasminum* (2)
Palmaceae	1/1	*Phoenix* (1)
Papaveraceae	2/2	*Argemone* (1), *Papaver* (1)
Pedaliaceae	1/1	*Sesamum* (1)
Pinaceae	1/1	*Pinus* (1)
Plumbaginaceae	1/1	*Plumbago* (1)
Poaceae	5/7	*Cynodon* (1), *Pennisetum* (1), *Saccharum* (3), *Triticum* (1), *Zea* (1)
Polygonaceae	1/1	*Rumex* (1)
Proteaceae	1/1	*Grevillea* (1)
Punicaceae	1/1	*Punica* (1)
Rhamnaceae	1/1	*Zizypus* (1)
Rosaceae	5/7	*Fragaria* (1), *Eriobotrya* (1), *Prunus* (3), *Pyrus*(2), *Rosa* (1)
Rubiaceae	2/2	*Randia* (1), *Wendlandia* (1)
Rutaceae	4/9	*Aegle* (1), *Citrus* (5), *Murraya* (2), *Zanthoxylum* (1)
Salicaceae	1/1	*Populus* (1)
Sapindaceae	3/3	*Dodonaea* (1), *Litchi* (1), *Sapindus* (1)
Sapotaceae	1/1	*Achras* (1)
Scrophulariaceae	1/1	*Verbascum* (1)
Solanaceae	5/10	*Capsicum* (1), *Cestrum* (1), *Datura* (1), *Lycopersicum* (1), *Solanum* (6)
Tiliaceae	1/1	*Grewia* (1)
Ulmaceae	1/1	*Holoptelea* (1)
Verbenaceae	4/4	*Duranta* (1), *Lantana* (1), *Premna* (1), *Vitex* (1)
Vitaceae	1/1	*Vitis* (1)
Zingiberaceae	2/3	*Curcuma* (1), *Elettaria* (1), *Zingiber* (1)

APPENDIX III
INDEX TO LOCAL NAMES

Botanical Name	*Local Name/s*	*Page*
1	2	3
Abrus precatorius L.	Rati	33
Acacia catechu Willd.	Khair	34
Acacia nilotica Linn.	Babul, Kikkar	36
Achras zapota L.	Chiku	37
Achyranthes bidentata L.	Puthkanda	37
Acorus calamus Linn.	Barae, Barya, Barre	38
Adhatoda vasica Nees	Safed basuti	40
Aegle marmelos (L.) Correa	Bel	41
Agave americana L.	Kuer Chatyan, Koaed	43
Ageratum conyzoides Linn.	Phulanu	44
Albizia lebbeck Benth.	Sirin	45
Allium cepa Linn.	Piaz	46
Allium sativum Linn.	Lasun	47

(Contd...)

1	2	3
Aloe maculata All.	Kuarian	48
Aloe vera (L.) Webb. & Benth	Kuarian, Ghi-kawer	49
Amaranthus spinosus Linn.	Chelri, Cholyai	51
Amaranthus viridis Linn.	Chelri, Cholyai	51
Amorphophallus campanulatus (Roxb.) Blume *ex* Decne.	Zamikand	52
Annona squamosa Linn.	Sitaphal	53
Argemone mexicana Linn.	Bhindi	54
Artemisia japonica Thunb.	Bhangru	55
Artocarpus integrifolia Linn.	Katal	56
Artocarpus lakoocha Roxb.	Dheu	57
Asparagus adscendens Roxb.	Sansban	57
Bambusa nutans Wall.	Bans	58
Bauhinia vahlii Wight & Arn.	Tor	59
Bauhinia variegata L.	Karale, Karayalae, Kachnar	60
Boerhaavia diffusa Linn.	Itsat	62
Bombax ceiba Linn.	Simal, Simbal	63
Bougainvillea spectabilis Willd	Garden-glory	64
Brassica campestris Linn.	Sarson	65
Bryophyllum calycinum Salisb.	Gillar Pattar, Patharchat	65
Butea monosperma (Lam.) Kuntze.	Palah, Palash, Plah	66
Callistemon lanceolatus DC.	Bottle Brush	68
Canna indica L.	Aeran, Hardarshan, Sudarshan	68
Cannabis sativa Linn.	Bhang	69
Capsicum annuum Linn.	Mirach	70
Carica papaya Linn.	Papita	71
Carissa inermis Vahl.	Badda Garnae	73
Carissa spinarum Hook. f.	Garnae	73
Cassia fistula L.	Aahali, Allia, Kanyar	74
Cassia occidentalis Linn.	Bada-rellu, Alliuna	76
Catharanthus roseus G. Don	Sadabahar	77
Cedrela toona Roxb.	Tooni	78
Celastrus paniculatus Willd.	Sankheere, Sankhiryan	79
Celosia argentea L.	Kukkar phul	80
Centella asiatica (Linn.) Urban	Brahmi	81
Cestrum nocturnum L.	Rat-Ki-Rani	82
Chenopodium album Linn.	Kunnah, Bathu	82
Cicer arietinum Linn.	Chollae	83
Cinnamomum camphora (L.) Nees & Ebesm	Kapoor, Kapur	84
Cissampelos pareira Linn.	Bhatindu, Pathindu	85
Citrus aurantium L.	Sangtara, Santra	87
Citrus grandis (L.) Osbeck.	Chakotra	88
Citrus jambhiri Lush.	Jambirdi, Jhamirdi, Kimb	89
Citrus limon (L.) Burm. f.	Khatta	89
Citrus medica L.	Nimbu	90
Coccinia grandis (L.) Voigt.	Kanduri	91
Colebrookia oppositifolia Smith.	Dashane	92
Colocasia esculenta (Linn.) Schott.	Kachalu	93
Cordia dichotoma Forst.	Lasora, Lasura	94
Coriandrum sativum Linn.	Dhania	95
Cucumis sativus L.	Kakri, Kheera	96
Cucurbita hispida Thunb.	Dhuda Petha, Petha	96
Cucurbita pepo L.	Kaddu	97

1	2	3
Curcuma domestica (Medik.) Valeton.	Haldi	98
Cuscuta reflexa Roxb.	Akashbel, Amarvela, Halwan	100
Cynodon dactylon (Linn.) Pers.	Doob, Doorva, Dhrub	101
Dalbergia sissoo Roxb.	Shisham, Tahali	102
Datura innoxia Mill.	Datura, Dhatura	103
Dendrocalamus hamiltonii Nees & Arn.	Baingh, Maggar	104
Dioscorea bulbifera Linn.	Tardi	105
Dioscorea deltoidea Wall.	Singli-Mingli	106
Dodonaea viscosa (Jacq.)	Mehandru	106
Duranta repens L.	Neel Kanthi (Neel Kanta)	107
Elettaria cardamomum Maton	Elaichi	108
Emblica officinalis Gaertn.	Ambla, Amla	109
Eriobotrya japonica Lindl.	Laukhat, Lukat	111
Eruca sativa Lam.	Taramira	111
Eucalyptus umbellata Domin.	Safeda	112
Eugenia cerasoides Roxb.	Jaamun, Jaman	113
Eugenia jambolana Lam.	Chparchoo, Chu, Chui	113
Euphorbia royleana Boiss.	Kathman	115
Ficus benghalensis Linn.	Bad, Barh, Bare	116
Ficus carica Linn.	Dura, Anji	117
Ficus glomerata Roxb.	Umara	118
Ficus religiosa Linn.	Pipal	119
Ficus roxburghii Wall. *ex* Miq.	Rumal, Trayambalu, Triamble	120
Flacourtia ramontchi L. Herit	Kangu	121
Foeniculum vulgare Mill.	Meethi-Saunf	122
Fragaria chiloensis (L.) Duchesne.	Strawberry	123
Fumaria indica (Haussk.) Pugsley.	Papra, Pitpapda, Pitpapra	124
Gossypium arboreum Linn.	Kapah	125
Grevillea robusta A. Cunn. *ex* R. Br.	Silver oak	126
Grewia oppositifolia Roxb.	Beul	126
Helianthus annuus L.	Sooraj mukhi	127
Hibiscus rosa-sinensis L.	China Rose, Shoe Flower	128
Holarrhena antidysenterica (Buch.-Ham.) Wall.	Pardesi	129
Holoptelea integrifolia (Roxb.) Planch.	Inderjaun, Keur	130
Impatiens balsamina Linn.	Teur	131
Ipomoea cairica (L.) Sweet	Railway creeper	132
Ipomoea carnea Jacq.	Basuti	132
Jasminum multiflorum Roth.	Chamaeli, Malti	133
Jasminum sambac (L.) Ait.	Motia	134
Jatropha curcas Linn.	Rattamjot, Jamnotta	134
Juglans regia L.	Akhrot	135
Lagenaria siceraria (Molina) Standl.	Ghiya	137
Lagerstroemia indica L.	Har-Shingar	137
Lannea coromandelica (Houtt.) Merrill.	Kehmble	138
Lantana camara Linn.	Phulnu	139
Lawsonia alba Lam.	Mehendi	140
Leucaena leucocephala (Lam.) de Wit.	Alseenia	141
Litchi chinensis Sonner.	Litchi	142
Luffa aegyptica Mill.	Ghiya Tori	142
Lycopersicum esculentum Mill.	Tamatar	143
Mallotus philippinensis (Lam.) Muell-Arg.	Kamal	144

(Contd...)

1	2	3
Malvaviscus conzattii Greenm.	Gudhal	146
Mangifera indica Linn.	Amb, Aam	146
Martynia annua L.	Kaun	148
Melia azedarach Linn.	Draek	149
Melia azadirachta Linn.	Neem	150
Mentha piperita Linn.	Pipermint	151
Mentha spicata Linn.	Pudina	152
Mimosa pudica Linn.	Lajalu	153
Mirabilis jalapa Linn.	Gulal	153
Momordica charantia L.	Karela	154
Momordica dioica Roxb. *ex* Wall.	Kakron	155
Morus alba Linn.	Toot	156
Mucuna pruriens L.	Gazal-Bel	158
Murraya koenigii Spreng.	Gandhla	159
Murraya paniculata (Linn.) Jack.	Gandhla	160
Musa sapientum Linn.	Kela	161
Nerium indicum Mill.	Kaner	162
Ocimum basilicum Linn.	Bhabhri	163
Ocimum sanctum Linn.	Baramasi Tulsi, Ram Tulsi	164
Opuntia dillenii Haw.	Chu, Thor	165
Oroxylum indicum Vent.	Tat-palanga	166
Papaver rhoeus L.	Gulla	168
Parthenium hysterophorus Linn.	Congress grass	168
Pedilanthes tithymaloides (L.) Poit.	Nag Bel	169
Pennisetum purpureum Schum.	Chari	169
Phoenix sylvestris Roxb.	Khajoor	170
Phyllanthus urinaria Linn.	Bhumi-Amla	171
Pinus roxburghii Sarg.	Chil	172
Pistacia integerrima Stewart *ex* Brand.	Kakad-singi, Kakarsingi	173
Plumbago zeylanica Linn.	Chitra, Chitrak	174
Pogostemon benghalensis Burm. f	Dusyan, Kali Basuti	175
Polyalthia longifolia (Sonn.) Thwaites	Ashok	176
Populus nigra L.	Poplar	176
Premna latifolia Roxb.	Bankar	177
Prunus amygdalus Baill.	Badam	178
Prunus domestica L.	Aloo Bukhara, Plum	178
Prunus persica Benth. & Hook. f.	Adu, Aru	179
Psidium guajava Linn.	Amrood	180
Punica granatum Linn.	Daadu, Daārdn	181
Putranjiva roxburghii Wall.	Putranjan	183
Pyrus communis L.	Nashpati	184
Pyrus pashia Ham.	Kainth, Kaenth	184
Randia dumetorum (Retz.) Lam.	Rara	185
Raphanus sativus Linn.	Mooli	186
Ricinus communis Linn.	Airand, Arind, Erand	187
Rosa damascena Mill.	Gulab	189
Rumex hastatus Don.	Maroli	189
Saccharum arundinaceum Reitz.	Sarphar	190
Saccharum officinarum L.	Ganna, Kamandi	191
Saccharum spontaneum Linn.	Kans	192
Salvia coccinea Linn.	Phultulsi	192
Sapindus mukorossi Gaertn.	Doddae, Doda, Reetha, Ritha	193

APPENDIX IV
INDEX TO NUTRITIONAL VALUE OF ETHNOBOTANICALLY IMPORTANT PLANTS OF PRAGPUR

Part used/Category	*Moisture*	*Protein*	*Fat*	*Minerals*	*Fibres*	*Carbohydrates*	*Calories*
Leaves							
Amaranthus spinosus	85.0	3.0	0.7	3.6	1.1	7.0	43
Amaranthus viridis	81.8	5.2	0.3	2.8	6.1	3.8	38
Bauhinia variegata	78.1	3.6	1.0	2.1	5.5	9.7	62
Chenopodium album	89.6	3.7	0.4	2.6	0.8	2.9	30
Colocasia esculenta	82.7	3.9	1.5	2.2	2.9	6.8	56

(Contd...)

Part used/Category	*Moisture*	*Protein*	*Fat*	*Minerals*	*Fibres*	*Carbohydrates*	*Calories*
Fruits							
Aegle marmelos	61.5	1.8	0.3	1.7	2.9	31.8	137
Annona squamosa	69.3	1.7	0.6	1.1	3.7	23.6	107
Artocarpus integrifolia	76.2	1.9	0.1	0.9	1.1	19.8	88
Artocarpus lakoocha	82.1	0.7	0.1	0.8	2.0	13.3	66
Cordia dichotoma	82.5	1.8	1.0	2.2	0.3	12.2	65
Eugenia jambolana	83.7	0.7	0.3	0.4	0.9	14.0	62
Flacourtia indica	67.8	1.7	1.8	1.3	4.7	22.7	114
Holoptelea integrifolia	—	13.71	—	—	—	69.91	—
Mangifera indica	86.1	0.6	0.1	0.3	1.1	11.8	51
Momordica dioica	84.1	3.1	1.0	1.1	3.0	7.7	52
Morus alba	87.5	1.5	0.4	0.9	8.3	1.4	
Phoenix sylvestris	59.2	1.2	0.4	1.7	3.7	33.8	144
Prunus persica	80.4	0.3	0.3	0.4	5.2	13.4	58
Tamarindus indica	20.9	3.1	0.1	2.9	5.6	67.4	283
Zizyphus jujuba	81.6	0.8	0.3	0.3	—	17.0	74
Underground parts							
Allium cepa	86.8	1.2	—	0.4	0.6	11.0	—
Allium sativum	62.0	0.5	0.1	1.0	0.8	29	—
Amorphophallus campanulatus	73.7	1.2	0.1	0.8	0.8	18.4	79
Colocasia esculenta	73.1	3.0	0.1	1.7	1.0	21.1	97
Dioscorea bulbifera	67.6	4.5	0.8	1.3	1.1	24.7	124
Raphanus sativus	94.4	0.07	0.01	0.6	—	3.4	—
Spices							
Capsicum annuum	82.60	2.90	0.60	—	6.80	6.10	1.0
Coriandrum sativum (leaves)	87.0	3.3	0.6	—	—	6.5	—
Coriandrum sativum (seed)	6.3	1.3	—	—	19.6	31.5	—
Curcuma domestica	5.8	8.6	8.9	—	6.9	63.0	—
Elettaria cardamomum	7-10	7.0-14.0	—	—	6.7-12.8	—	3.8-6.9
Foeniculm vulgares	6.30	9.5	10	—	18.5	42.3	—
Murraya paniculata	66.3	6.1	1.0	—	6.4	16.0	—
Ocimum basilicum	6.1	1.9	3.6	—	20.5	41.2	16.7
Punica granatum (*Anardana*)	78.0	1.6	0.1	—	5.1	14.5	—
Tamarindus indica	18.2	28.2		—	—	—	—
Trigonella foenum-graecum	6.3	9.5	10.0	—	18.5	42.3	—
Zingiber officinale	10.85	—	—	—	7.17	—	6.64

	Crude protein (%)	*Crude fibre*	*Nitrogen free extract*	*Ether extract*	*Total ash*	*Calcium*	*Phosphorus*
Fodder							
Acacia catechu	13.03	21.88-22.55	46.9-50.56	3.05-4.55	9.66-9.80	2.46-2.74	0.14-0.17
Acacia nilotica	18.72-19.75	6.49-9.81	—	2.16-4.20	4.79-11.23	1.11-2.80	0.25-0.52
Bauhinia vahlii	2.94	—	—	14.55	4.30	—	—
Bauhinia variegata	10.73-15.91	25.28-32.97	40.87-51.83	1.33-3.93	6.27-12.37	1.76-4.13	0.20-0.38
Cedrela toona	—	—	—	—	9.45-10.86	2.63-4.23	Mg-0.60 N-2.24-2.49
Cynodon dactylon	10.47	28.17	47.81	1.80	11.75	—	—
Dalbergia sissoo	15.00	34.10 neutral detergent fibre	—	—	—	2.00	0.17
Ficus benghalensis	7.74-11.17	13.91-29.95	37.61-51.59	2.46-5.08	9.30-16.92	2.53-4.12	0.08-0.42

	Crude protein (%)	*Crude fibre*	*Nitrogen free extract*	*Ether extract*	*Total ash*	*Calcium*	*Phosphorus*
Ficus glomerata	11.06-15.20	12.27-16.50	55.52-59.00	2.23-2.87	11.88-18.39	1.73-2.96	0.16-0.45
Ficus roxburghii	12.28-13.35	7.71-17.79	50.02-64.87	4.49-4.65	9.42-15.42	1.31-2.19	0.17-0.22
Flacourtia indica	—	—	—	—	6.60	2.21	—
Grewia optiva	15.6-20.0	18.08-22.5	41.7-49.0	2.1-6.7	2.5-4.0	0.15-0.30	—
Holoptelea integrifolia	13.71	—	—	1.86	14.52	4.25	0.15
Lannea coromandelica	11.44-16.17	16.12-19.76	53.10-59.05	2.47-4.20	8.50-9.19	1.75-2.41	1.75-2.41
Leucaena leucocephala	22.22	46.30 neutral detergent fibre	—	3.56	—	—	—
Mallotus philippinensis	13.37	29.05	44.64	3.65	8.69	1.34	0.24
Melia azedarach	—	13.30	—	—	—	2.55	0.23
Morus alba	15.0-27.64	9.07-15.27	47.98-49.70	2.30-8.04	14.32-22.87	2.42-4.71	0.23-0.97
Pennisetum purpureum	1.0	7.4	10.2	—	—	0.12	0.07
Zizyphus jujuba	12.86-16.93	13.47-17.07	55.33-57.90	1.52-3.23	10.24-11.72	1.42-3.59	0.21-0.33

**Source:* Bennet *et al.*, 1991; Dwivedi, 1993; Pruthi, 1976.

APPENDIX V
GLOSSARY OF WORDS COMMONLY USED BY THE LOCAL POPULACE OF PRAGPUR

English	*Kangri*
Ailment and State of the Body	
Asthma	Dama
Belching	Dkar
Bladder stone	Patthri
Blind	Anna
Bodly pains	Pira; Darda
Boil	Phora
Cold	Najla or Jkhwam
Colic	Sul
Constipation	Kabji
Cough	Khang
Deaf	Tona
Diarrhoea	Atisar
Disk slip	Cas
Dumb	Gunga
Epilepsy	Mirgi
Fever	Tap or Bkhwar
Fit	Dora
Jaundice	Pusar rog or peolia
Leprosy	Kor
Loose motion	Jlab or dast
Measles	Trasal
Mumps	Knedde
Nose bleeding	Naksir
Odema, swelling	Phara or Sojas
One eyed	Kana
Paralysis	Trang
Phlegm	Kapph or Balgum
Phobia	Alkh
Pneumonia	Nomunia
Pus	Pak
Reddishness	Lalgi
Sciatica	Ringan
Sick	Bmar or Rogi
Small pox	Mata
Sneez	Chik
Stool	Cara
Sweating	Psinna
Throat irritation	Khriri
Tonsilitis	Kandi
Toothache	Klwanda
Tuberculosis	Chairog
Twisting Ankle	Mos
Urine	Psab or Muttar
Vomit	Ulti
Parts of Body	
Arm	Ban
Ankle	Gitta
Armpit	Kacch
Back	Pith
Belly	Pet
Breast/Nipple	Ciccu
Cheek	Khakh
Chest	Chatti
Chin	Chunni or Thodi
Calf	Pilli
Ear	Kan
Eye	Akh
Eye ball	Della
Face	Munh
Finger	Ongli or Gutthi
Foot	Per
Forehead	Kaudi
Gum	Masude
Hair	Kes
Head	Sir
Hand	Atth
Heart	Kalja
Heel	Addi
Knee	Godda
Liver	Jiggar
Lip	Labbar
Lung	Phephra
Head	Mattha
Mole	Tilla
Mouth	Muhn
Nail	Na
Neck	Kyari
Nose	Nakk
Nostril	Nasan
Shoulder	Munda
Skull	Khopri

(Contd...)

English	*Kangri*
Soft head	Talu
Sole	Toe
Testicle	And
Thigh	Pat
Throat	Kandu
Thumb	Guttha
Tongue	Jib
Tooth	Dand
Waist	Lakk
Wrist	Maunna
Kinship (Relation)	
Aunt	Cacci
Brother	Pau or Pai
Child	Bacca
Daughter	Di
Daughter' husband	Jwai
Daughter-in-law	Nu
Father	Bab, bappu or babbul
Father's elder brother	Taya
Grandfather	Dadda or daddu
Great-grandfather	Pardadda
Grandson	Potru
Husband	Lara
Mother	Ma or Amma
Sister	Ben or bebbe
Son	Puttar
Son's wife	Nu
Uncle	Cacca
Wife	Jnana
Wife's brother	Sala
Clothes (Kapre)	
Cap	Topu
Clothes	Kapre
Duppata	Cadru
Gloves	Dastane
Lengha	Ghagru
Pocket	Khisa
Shirt	Jaggu
Socks	Jurab
Thread	Taga
Trouser	Suthunu
Turban	Safa
Wool	Un
Fruits, Vegetable, Cereals, Eatables	
Apple	Seb, Seaun
Butter	Nauni
Curd	Dehu
Food	Khana

English	*Kangri*
Grass	Ka
Liquor	Gurangi
Meat	Skaar
Milk	Dud
Maize	Chali
Potato	Alu
Rice	Chaul
Sugar	Khand
Tea	Cha
Wheat	Kanak
Tree and its Parts	
Branch	Tehani
Flower	Phul
Leaf	Patta
Root	Jarh
Thorn	Kanda
Tree	Rukh or butta
Wood	Lakri
Housing and Housing Compartments	
Courtyard	Baranda
Door	Dwar
House	Kar or Makan
Stable	Tabella
Ground Floor Common Room	Owan
Room	Kamra
Room on First Floor	Bor
Roof Top	Ladda
Sitting Room	Bethak
Wall	Kand
Window	Dwari
Animals and Related Terms	
Ass	Khota
Bull	Bald
Calf	Bacchu
Cat	Billie
Crow	Ko
Dog	Kutta
Egg	Anda
Fish	Machi
Goat	Bakri
Hare	Serru
Hen	Kukri
Jackal	Gidder
Lion	Ser
Lice	Gun
Monkey	Bander
Mule	Kachar
Nest	Alana

English	*Kangri*
Rat	Chua
Tail	Poonch
Domestic Article	
Almirah	Almari
Balance	Trakri
Blow pipe	Painpla
Bowl	Katori
Box	Dabba
Broom	Caru
Comb	Kangi
Cot	Manja
Key	Chabbi
Lid	Dakan
Lock	Janda
Mirror	Sisa
Needle	Sui
Plate	Thalu
Spoon	Chamcha
Table	Mej
Thread	Dhaga
Tools	
Axe	Kulhari
Digger	Chabal
Hammer	Thora
Plough	Al
Sickle	Drat or drati
Stick	Sothi
Days of the Week	
Sunday	Twar
Monday	Swar
Tuesday	Mangal
Wednesday	Bud
Thursday	Bir bar
Friday	Sukkar
Saturday	Saniccar
Meals and Dishes	
Breakfast	Datyallu
Butter	Noni
Chapati	Rotti
Clarified butter	Kiu
Cooked rice	Pat or col
Curd	Dau or dai
Fried Fermented Sweet Wheat Chapati	Babru
Halwa	Kra
Milk Pudding	Khir
Rice Pan Cake	Enklu
Snack	Tukru
Sour dish	Mani or khatta
Sweet dish	Mittha

English	*Kangri*
Tea	Ca
Water Milk Dish	Reru or kari or karu
Wheat Fermented Chapati	Baturu
Heavenly Bodies	
Air	Oa
Clouds	Baddal
Earth	Tart
Earthquake	Hillan
Hailing	Aan
Lighting	Bijli
Moon	Chandarma
Moonlight	Chanani
Rain	Barkha
Sky	Ambar
Stars	Tare
Sun	Suraj
Sunshine	Tuppa
Customs, Rites and Ceremonies	
Dowry	Daj
Engagement	Karmai
Fast	Barat

English	*Kangri*
Marriage	Bya
Marriage Party	Brat
Married Women	Pyoki
Marriage Procession	Jnet
Marriage Party Member	Jneti
Day of Death	Tith
10th Day After Death	Kapre
13th Day After Death	Sog or tera or kirya
30th Day After Death	Mkwan
1st Death Anniversary	Barkhi
4th Death Anniversary	Cabarkh
Numerals	
One	Ik
Two	Do
Three	Tin
Four	Car
Five	Panj
Six	Che

English	*Kangri*
Seven	Sat
Eight	Atth
Nine	Naun
Ten	Das
Eleven	Gyara
Twelve	Bara
Thirteen	Tera
Fourteen	Coda
Fifteen	Pandara
Sixteen	Sola
Seventeen	Stara
Eighteen	Thara
Ninteen	Unni
Twenty	Bi
Thirty	Ti
Forty	Chali
Fifty	Panja
Sixty	Sath
Seventy	Sattar
Eighty	Assi
Ninety	Nabbe
Hundred	So
Thousand	Zahar

REFERENCES

Agarwal, V.S. 1986. *Economic Plants of India.* Bishen Singh Mahendra Pal Singh, Dehradun & Kailash Prakashan, Calcutta.

Agarwal, V.S. & Ghosh, B. 1985. *Drug Plants of India (Root Drugs)*. Kalyani Publishers, New Delhi.

Agrawal, S.R. 1981. Trees, Flowers and Fruits in Indian Folk Songs, Folk Proverbs and Folk Tales, 3-12. *In:* Jain, S.K. (Ed.) *Glimpses of Indian Ethnobotany.* Oxford & IBH, New Delhi.

Ahluwalia, K.S. 1952. Medicinal Plants of Kangra Valley. *Indian Forester* **78**(4): 181-194.

Ahuja, B.S. 1965. *Medicinal Plants of Saharanpur*. Survey of Medicinal Plants, Central Council of Ayurvedic Research, Gurukul Kangri Vishwavidyalaya, Haridwar.

Alcorn, J.B. 1996. Is Biodiversity Conserved by Indigenous People?: 233-238. *In:* Jain, S.K. (Ed.) *Ethnobiology in Human Welfare*. Deep Publ., New Delhi.

Ambasta, S.P. (Ed.) 1986. *The Useful Plants of India*. CSIR, New Delhi.

Aminuddin & Girach, R.D. 1991. Ethnobotanical Studies on Bodo Tribe of District Koraput (Orissa), India. *Ethnobotany* **3**: 15-19.

_______ 1993. Observations on Ethnobotany of the Bhujia—A Tribe of Sonabera Plateau. *Ethnobotany* **5**: 83-86.

_______ 1996. Native Phytotherapy Among the Paudi Bhuinya of Bonai Hills. *Ethnobotany* **8**: 66-70.

Anonymous, 1948-1976. *The Wealth of India—Raw Materials,* Vols. 1-11. CSIR, Delhi.

_______, 1990. *Phytochemical Investigation of Certain Medicinal Plants Used in Ayurveda*. Central Council for Research in Ayurveda and Siddha, New Delhi.

_______, 1994. *Ethnobiology in India—A Status Report*. Ministry of Environment and Forest, Government of India.

_______, 1998a. *An Indian Coordinated Project on Ethnobotany.* Final Technical Report 1992-1998. Ministry of Environment & Forests, Government of India.

_______, 1998b. *The Wealth of India. A Dictionary of Indian Raw Material and Industrial Products.* Vols. I-XIV. CSIR, New Delhi.

Ansari, A.A. 1991. Ethnobotanical Notes on Some Plants of Khirsu, Pauri-Garhwal, U.P. *Ethnobotany* **3**: 105-107.

Arizpe, L. 1996. Culture and Environment. *Nature and Resources* 32: 1-30.

Arora, R.K. 1995. Ethnobotanical Studies on Plant Genetic Resources—National Efforts and Concern. *Ethnobotany* **1**: 125-136.

Arora, R.K. & Mehra, K.L. 1977. Special Exploration for the Collection of Cultivated Plants and their Wild Relations in India. Bot. Surv. India **19** (1-4): 139-144.

Arora, R.K. & Pandey, A. 1996. *Wild Edible Plants of India—Diversity, Conservation and Use*. NBPGR, New Delhi.

Arora, R.K., Maheswari, M.L., Chandel, K.P.S. & Gupta, R. 1980. Mano (*Inula racemosa*): Little Known Aromatic Plants of Lahaul Valley, India. *Eco. Bot.* **34** (2): 175-180.

Asolkar, L.V., Kakkar, K.K. & Chakre, O.J. 1992. *Second Supplement to Glossary of Indian Medicinal Plants with Active Principles*. Part 1 (A-K). CSIR, New Delhi.

Aswal, B.S. 1996. Conservation of Ethno-Medicinal Plants Diversity of Garhwal Himalaya in India. 133-135. *In:* Jain, S.K. (Ed.) *Ethnobiology in Human Welfare.* Deep Publ., New Delhi.

Aswal, B.S. & Mehrotra, B.N. 1994. *Flora of Lahaul Spiti (A Cold Desert in North-West Himalaya).* B.S.M.P.S., Dehradun.

Atkinson, E.T. 1882. *Economic Botany of the Himalayan Region.* Cosmo Publ., New Delhi.

Augustine, Jomy & Sivadasan, M. 2004. Ethnomedicinal Plants of Periyar Tiger Reserve, Kerala India. *Ethnobotany* **16**: 44-49.

Badola, H.K. 2001. Medicinal Plant Diversity in Himachal Pradesh, 407-430. *In:* Samant, S.S., Dhar, U. and Palani, L.M.S. (Eds.) *Himalayan Medicinal Plants: Potential and Prospects.* Himvikas Occasional Publication, No. 14, Gyanodya Prakashan.

Badola, H.K. & Pal, M. 2003. Threatened Medicinal Plants and Their Conservation in Himachal Himalayas. *Indian Forester* **129** (1): 55-68.

Bajpayee, K.K. & Dixit, G. 1996. Ethnobotanical Studies on Food Stuffs of Tribals of Tarai-Region, Uttar Pradesh. *J. Econ. Taxon. Bot.* Additional Series **12**: 128-132.

Balasubramaniam, P. & Prasad, Narendra 1996. Medicinal Plants Among the Irulars of Attapady and Bolumampatti Forests in the Nilgiri Biosphere Reserve. *J. Econ. Taxon. Bot.* Additional Series **12**: 253-259.

Banerjee, Archana, 1996. Medicinal Uses of Some Flowers by Santhals of West Bengal. *J. Econ. Taxon. Bot.* Additional Series **12**: 314-317.

Banerjee, D.K. & Pal, D.C. 1996. Plant Used by the Tribals of North Indian Plains for Hair and Scalp Protection, 187-188. *In*: Jain, S.K. (Ed.) *Ethnobiology in Human Welfare*. Deep Publ., New Delhi.

Baruah, P. & Sarma, G.C. 1984. Studies on the Medicinal Uses of Plants by the Boro Tribals of Assam II. *J. Econ. Tax. Bot.* **5**: 599-604.

_______ 1987. Studies on the Medicinal Uses of Plants by the North-East Tribes. Pt. 3. *J. Econ. Tax. Bot.* **11**: 71-76.

Belal, Ahmed Esmat & Springuel, Irina, 1996. Economic Value of Plant Diversity in Arid Environments. *Nature & Resources* **32** (1): 33-39.

Bennet, S.S.R. 1985. Ethnobotanical Studies in West Sikkim. *J. Econ. Tax. Bot.* **7:** 317-321.

_______ 1986. *Name Changes in Flowering Plants of India and Adjacent Regions*. Triseas Publishers, Dehradun, India.

Bennet, S.S.R., Biswas, S., Chandra, V., Singh, P. J., Chandra, S. & Dayal, R. 1991. *Food from Forest.* F.R.I., I.C.F.R.E., Dehradun.

Bhalla, N.P., Sahu, T.R., Mishra, G.P. & Dakwale, R.N. 1982. Traditional Plant Medicines of Sagar Distt., Madhya Pradesh, India. *J. Econ. Tax. Bot.* **3**: 23-32.

Bhandary, M.J., Chandrashekar, K.R. & Kaveriappa, K.M. 1996. Ethnobotany of Gowlis of Uttar Kannada District, Karnataka. *J. Econ. Taxon. Bot.* Additional Series **12**: 244-249.

Bhardwaj, S. & Gakhar, S.K. 2003. Ethnomedicinal Plants Used by the Tribals of Mizoram to Cure Dysentery. *Ethnobotany* **15**: 51-54.

Bhargava, N. 1981. Plants in Folk Life and Folklore in Andaman & Nicobar Islands, 329-344. *In:* Jain, S.K. (Ed.) *Glimpses of Indian Ethnobotany*. Oxford & IBH, New Delhi.

Bhatnagar, P. 1997. Marketing and Trade of Medicinal Plants. *Sustain. Forester* **11**: 4-11.

Bhatt, D.C., Mitaliya, K.D., Mehta, S.K. & Joshi, P.N. 2002. Notes on Some Ethnomedicinal Plants of Pachchham Hills of Kachchh Distt. Gujarat. *Ethnobotany* **14**: 34-35.

Bhatt, D.C., Mitaliya, K.D. & Mehta, S.K. 2001. Some Sacred Plants of Gujarat and Their Medicinal Uses. *Ethnobotany* **13**: 146-149.

Bhatt, R.P. & Sabnis, S.D. 1987. Contribution to the Ethnobotany of Khedbrahma Region of North Gujarat. *J. Econ. Tax. Bot.* **9**: 139-145.

Bhattacharjee, S., Tiwari, K.C., Majumdar, R. & Misra, A.K. 1980. Folklore Medicine from District Kamrup (Assam). *Bull. Medico-Ethno-bot. Res.* **1**: 441-460.

Bilochi, N.C. & Datta, K.K. 1993. Global Overview of Diarrhoeal Diseases and Cholera. *Swastha Hind.* **36** (9 & 10): 250-253.

Binoj, K.M.S. & Balakrishnan, N.P. 1996. Ethnobotanical Studies of One Genus Euphorbial (Euphorbiaceae). *J. Econ. Taxon. Bol.* Additional Series **12**: 46-49.

Binu, S., Nayar, T.S. & Pushpangadan, P. 1992. An Outline of Ethnobotanical Research in India. *J. Econ. Tax. Bot.* Additional Series **10**: 405-428.

Bondya, S.L., Khanna, K.K. & Singh, K.P. 2006. Ethnomedicinal Used of Leafy Vegetable from the Tribals Folklore of Achanakmar-Amarkantak Biosphere Reserve (M.P. & Chattishgarh). *Ethnobotany* **18**: 145-148.

Bondya, S.L. & Sharma, H.P. 2003. Ethnobotanical Studies on Plants Used in Diabetes (Madumeha) under the Bharagora Block of Jharkhand. *Ethnobotany* **16**: 139-140.

Borthakur, S.K. 1993. Native Phytotherapy for Child and Women Diseases from Assam. *Ethnobotany* **5**(87): 87-91.

_______ 1996a. Post-Natal Care of Women in Traditional System in Assam. *Ethnobotany* **8**: 51-55.

_______ 1996b. Wild Edible Plants in Markets of Assam, India—An Ethnobotanical Investigation, 31-34. *In*: Jain, S.K. (Ed.) *Ethnobiology in Human Welfare*. Deep. Publ., New Delhi.

Borthakur, S.K., Chaudhary, B.T. & Gogoi, R. 2004. Folklore Hepato-Protective Herbal Recipes from Assam in Northeast India. *Ethnobotany* **16**: 76-82.

Borthakur, S.K., & Gogoi, P. 1994. Indigenous Technology of Making Writing Materials Among the Tai Khamtis. *Ethnobotany* **6**: 5-8.

Borthakur, S.K. & Sarma, U.K. 1996. Ethnoveterinary Medicine with Special Reference to Cattle Prevalent Among the Nepalis of Assam, India, 209-210. *In*: Jain, S.K. (Ed.) *Ethnobiology in Human Welfare.* Deep Publ., New Delhi.

Brahmam, M. & Saxena, H.O. 1990. Ethnobotany of Gardamardan Hills—Some Noteworthy Folk-Medicinal Uses. *Ethnobotany* **2**: 71-79.

Brahmam, M., Dhal, N.K. & Saxena, H.O. 1996. Ethnobotanical Studies Among the Tanla of Malayagiri Hills in Dhenkanal District, Orissa, India, 393-396. *In*: Jain, S.K. (Ed.) *Ethnobiology in Human Welfare.* Deep Publ., New Delhi.

Brierly, J.S. 1985. West Indian Kitchen Gardens: A Historical Perspective with Current Insights from Grenada. *Food & Nutrition Bull.* **7**: 52-60.

Chak, I.M. & Sharma, J.N. 1965. Effect of Asarone on Experimentally Induced Conflict Neurosin in Rats. *Indian J. Exp. Biol.* **3**: 252.

Chandra, S. 1990. *Foundations of Ethnobotany* (*Pre-1900 Ethnobotany*). A Review and Bibliography. Deep Publ., New Delhi.

Chandra Sekar, K. & Srivastava, S.K. 2003. Ethnomedicinal Studies in Pin Valley National Park, Lahaul-Spiti, Himachal Pradesh. *Ethnobotany* **15**: 44-47.

_______ 2005. New Reports on Aphrodisiac Plants from Pin Valley National Park, Himachal Pradesh. *Ethnobotany* **17**: 189-190.

Chatterjee, A. & Pakrashi, S.C. (Eds.) 1997. *The Treatise on Indian Medicinal Plants.* Vols. 1-5. CSIR, New Delhi.

Chaudhari, R.H.N., Molla, H.A., Pal, D.C. & Roy, B. 1982. Plants Used in Traditional Medicine by Some Tribals of Jalpaiguri District, West Bengal. *Bull Bot. Surv. India* 24: 87-90.

Chauhan, N.S. 1974. Medicinal Plants of Una Forest Division, Una District, H.P. lying in Shivalik Range. *Nagarjun* 27: 31-39.

_______ 1983. Some Rare and Interesting Medicinal Plants of H.P. All India Summer Inst. On Agroforestry in the Western Himalaya, July 22–August 7, 1983.

_______ 1987. Collection, Identification, Cultivation and Utilization of Medicinal and Aromatic Plants in H.P. Status Paper Presented to the Sub-Committee on Medicinal Plants Development in H.P. Under the State Science, Technology and Enviornment Council, April 4, Shimla.

_______ 1988a, Exploitation and Utilization of Herbal Resouurces of H.P. Invited Lecture Delivered "Workshop on Opportunities Identification in Agro-based and Food-Products, pp. 23-32, STEP, Holiday Home, Shimla, November.

_______ 1988b. Ethnobotanical Study of Medicinal Plants of Himachal Pradesh. 187-198. *In*: Kaushik, P. (Ed.) *Indigenous Medicinal Plants*. Today and Tomorrow Printers & Publ., New Delhi.

_______ 1989a. Medicinal and Aromatic Plants of Himachal Pradesh and Their Role in Social Forestry

Programme, IIIrd Forestery Conference (Proceeding) Vol. II: 89 (1) 89 (9): FRI, Dehradun.

_______ 1989b. Potential of Aromatic Plants Flora in Himachal Pradesh. *Indian Perfumer* **33**(2): 118-122.

_______ 1989c. Endangered Ayurvedic Pharmacopoeial Plant Resources of H.P. in P. Kaushik (Ed.) Indiginous Medicinal Plants. Including Microbes and Fungi: 199-205. Today & Tommorow Printers and Publ., New Delhi.

_______ 1989d. Herbal Wealth of Cold Desert Area of Himachal Pradesh. *Proceedings, National Seminar-cum-Workshop on Cold Desert Area in India*. 65-68, Government Press, Shimla, H.P.

_______ 1990. Medicinal Orchids of Himachal Pradesh. *J. Orchid Soc. India* **4**(2): 99-105.

_______ 1996a. Ethno-Medicobotany of Pabbar Valley in District Shimla, Himachal Pradesh (India). *In*: *UHF-IUFRO International Workshop in Prospects of Medicinal Plants.* 4-9 Nov., p. 30.

_______ 1996b. Plant Resources of Economic Use in Himachal Pradesh. DEE-UH & F, Solan, 77 pp.

_______ 1999. *Medicinal and Aromatic Plants of Himachal Pradesh*. Indus Publ. Co., New Delhi.

_______ 2003. Important Medicinal and Aromatic Plants of Himachal Pradesh. *Indian Forester* 129(8): 979-998.

Chauhan, N.S. & Khosla, P. K. 1988. Commercially Important Medicinal and Aromatic Plants of H.P., 81-89. *In* Khosla, P.K. (Ed.) *Trends in Tree Science.* ISTS Publication.

Chauhan, N.S. & Negi, Y.S. 1988. Production Potential of Medicinal and Aromatic Plants in Himachal Pradesh, 71-80. *In* Khosla, P.K. (Ed.) *Trends in Tree Science.* ISTS Publication.

Chaurasia, O.P., Singh, B. & Ballabh, B. 2001. Potential Aromatic Floras of Himalayan Cold District: Ladakh & Lahaul-Spiti. *J. Econ. Taxon. Bot.* **25** (1): 91-97.

Chhetri, D.R. 2005. Ethnomedicinal Plants of Khangchendzonga National Park, Sikkim, India. *Ethnobotany* **17**: 176-178.

Chhetri, R.B. 1994. Further Observations on Ethnomedicobotany of Khasi Hills in Meghalaya, India. *Ethnobotany* **6**: 33-36.

Chopra, I.C., Abrolo, B.K. & Handa, K.L. 1983. *Medicinal Plants of the Arid Zones.* Today & Tomorrow's Printers & Publ., New Delhi.

Chopra, R.N., Nayar, S.L. & Chopra, I.C. 1956. *Glosary of Indian Medicinal Plants*. CSIR, New Delhi.

Chowdhery, H.J. 1996. Arunachal Pradesh: Its People, Ethnobotanical Diversity and Conservation, 323-328. *In*: Jain, S.K. (Ed.) *Ethnobiology in Human Welfare.* Deep. Publ., New Delhi.

Chowdhery, H.J & Wadhwa, B.M. 1984. *Flora of Himachal Pradesh.* Vols. 1-3. Bot. Sur. India, Calcutta.

Chowdhury, J.U., Alam, M.K., & Hassan, M.A. 1996. Some Traditional Folk Formularies Against Dysentery and Diarrhoea in Bangladesh. *J. Econ. Taxon. Bot.* Additional Series **12**: 20-23.

Collett, H. 1902. *Flora Simlensis*. Thacker Spink and Co. Calcutta and Shimla, Reprinted 1971. B.S.M.P.S., Dehradun.

Conklin, H.C. 1980. *Ethnographic Atlas of I fugo: A Study of Environment, Culture and Society in Noron Luzon.* Pugguwon Lupaih and Miklos, New Haren.

Dagar, H.S. 2003. Ethnobotanical Evaluation of Some Endemic Plants in Andaman and Nicobar Islands. 134-147. *In*: *Vistas in Ethnobotany*, Vol. I, *Indian Journal of Pure and Applied Biology*.

Dagar, H.S. & Dagar, J.C. 1996. Ethnobotanical Studies of the Nicobarese of Chowra Island of Nicobar Group of Islands. *J. Econ. Tax. Bot.* Additional Series **12**: 381-388.

_______ 1999. *Ethnobotany of Aborigines of Andaman-Nicobar Islands*. Surv. of India, Port Blair, Andaman & Nicobar Islands, India.

Dandiya, P.C. & Sharma, J.D. 1961. Pharmacological Effects of a-asarone and b-asarone. *Nature* **192**: 1299.

_______ 1962. Studies on *Acorus Calamus* V. Pharmacological action of a-asarone and b-asarone in C.N.S. *Indian J. Med. Res.* **50**: 46.

Dar, G.H., Vir Jee, Kachroo, P. & Buth, G.M. 1984. Ethnobotany of Kashmir-I. Sind Valley. *J. Econ. Tax. Bot.* 5: 668-675.

Dastur, J.F. 1970. *Medicinal Plants of India and Pakistan*. D.B. Taraporevala Sons & Co., Private Ltd. 2nd ed. Bombay, p. 212.

Datt, Bhaskar 1996. Ethnobotanical Resources of Chhatarpur District (Madhya Pradesh), *In*: Jain, S.K. (Ed.)

Ethnobiology in Human Welfare. Deep Publ., New Delhi.

Dey, K.L. 1973. *The Indigenous Drugs of India*. The Chronica Bot., New Delhi.

Dhaliwal, D.S. & Sharma, M. 1999. *Flora of Kullu District (Himachal Pradesh)*. Bishen Singh Mahender Pal Singh, Dehradun.

Dhiman, D.R. 1976. *Himachal Pradesh ki Vanoshdiya Sampada*. Imperial Printing Press, Dharmshala. H.P.

Dixit, R.D., & Pal, D.C. 1996. Role of Household Remedies in Primary Health-Care, 108-109. *In*: Jain, S.K. (Ed.) *Ethnobiology in Human Welfare*. Deep Publ., New Delhi.

Dobriyal, R.M., Singh, G.S., Rao, K.S. & Saxena, K.G. 1997. Medicinal Plant Resources in Chhakinal Watershed in North-Western Himalaya. *J. Herbs Spices & Medicinal Plants* 5: 15-27.

Dobson, A. 1995. Biodiversity and Human Health. *Trends Ecol. & Enol.*: **10**: 390-391.

Drury, H. 1978. *Useful Plants of India*. Periodical Experts Book Agency, Delhi.

Duhoon, S.S., Kopper, M.N., and Chandra, U. 1996. Seabuckthorn (*Hippophae spp*)—A Less Known Wonder Plant of Ethno-Medico-Botanical Importance in Cold Desert of India. *J. Econ. Tax. Bot.* Additional Series **12**:43-45.

Duthie, J.E. 1888. *Fodder Grasses of Northern India*. Avishkar Publishers Distributors, Jaipur (Raj.) India.

Dwarakan, P. & Ansari, A.A. 1996. Less Known Uses of Plants of Kollimalai (Salem District, Tamil Nadu) in South India. *J. Econ. Taxon. Bot.* Additional Series **12**: 284-286.

Dwivedi, A.P. 1993. *Forest—The Non-Wood Resources.* International Book Distributors, Dehradun (India).

Dwivedi, S.W., Dwivedi, Sangetta, Patel, P.C. 2005. Herbal Remedies Used by Rural People in Treatment of Joint Diseases. *Ethnobotany* **17**: 193-196.

Etkin, N. 1993. Anthropological Methods in Ethnophramacology: *J. Ethnopharmacol.* **38**: 93-104.

Farnsworth, N.R. 1985. A Computerized Data Base for Medicinal Plants. *East. Pharmst.,* New Delhi **28**: 53-55.

Farnsworth, N.R. & Soejarto, D.D. 1991. Global Importance of Medicinal Plants: 25-51. *In:* Akerele, O., Heywood, V. & Synge, H. (Ed.) *The Conservation of Medicinal Plants*. Cambridge University Press, Cambridge, New York.

Faulks, P.J. 1958. *An Introduction to Ethnobotany*. Moredale, London.

Ford, R.I. (Ed.) 1978. *The Nature and Scope of Ethnobotany.* Anthropological Paper No. 67. Museum of Anthro. Univ. of Michigan, Ann. Arbor.

Gangwar, A.K. & Ramkrishnan, P.S. 1990. Ethnobiological Notes on Some Tribes of Arunachal Pradesh, Northeastern India. *Econ. Bot.* **44**: 94-105.

Garcia da Orta. 1563. *Coloquious Des Simples* Drogas—The Consassedicinai da India, Goa.

Gbile, Z. 1986. Ethnobotany Taxonomy and Conservation of Medicinal Plants: 231-259. *In*: Sofowora, A. (Ed.) The *State of Medicinal Plants in Nigeria*. Univ. of Ibadan Press, Nigeria.

Geetha, S., Lakshmi, G. & Ranjithakani, P. 1996a. Ethnobotanical Review: Wild Fibre Yielding Plants of Kolli Hills, Tamil Nadu. *J. Eco. Taxon. Bot.* Additional Series. **12**: 250-252.

_______ 1996b. Ethnoveterinary Medicinal Plants of Kolli Hills, Tamil Nadu. *J. Econ. Taxon. Bot.* Additional Series. **12**: 289-191.

Ghosh, A, Maity, S. & Maity, M. 1996. Ethnomedicine in Bankura District, West Bengal. *J. Econ. Taxon. Bot.* Additional Series **12**: 318-320.

Gill, L.S., Idu, M. & Ogbor, D.N. 1997. Folk Medicinal Plants; Practices and Beliefs of the Bini People in Nigeria. *Ethnobotany* **9**: 1-5.

Gill, L.S., Nyawuame, H.G.K., Esezobor, E.I. & Osagie, I.S. 1993. Nigerian Folk Medicine: Practices and Beliefs of Esan People. *Ethnobotany* 5: 129-142.

Gill, L.S. & Nyawuame, H.G.K. 1994. Leguminosae in Ethnomedical Practices in Nigeria. *Ethnobotany* **6**: 51-64.

Girach, R.D. 1992. Medicinal Plants Used by Kondh Tribe of District Phulbani (Orissa) in Eastern India. *Ethnobotany* **4**: 53-66.

Girach, R.D. & Aminuddin, 1995. Ethnomedicinal Uses of Plants Among the Tribals of Singhbhum District, Bihar, India. *Ethobotany* **7**: 103-107.

Girach, R.D., Aminuddin, Ahmed, M., Brahmam, M. & Misra, M.K. 1996. Native Phytotherapy Among Rural

Population of District Bhadrak, Orissa, 162-164. *In*: Jain, S.K. (Ed.) *Ethnobiology in Human Welfare.* Deep Publ., New Delhi.

Girach, R.D., Aminudhin, Brahmam, M. & Mishra, M.K. 1997. Observations on Ethnomedicinal Plants of Bhadrak District, Orissa, India. *Ethnobotany* **9**: 44-46.

Girach, R.D., Brahmam, M. & Mishra, M.K. 1998. Folk Veterinary Herbal Medicine of Bhadrak District, Orissa, India. *Ethnobotany* **10**: 85-88.

Goel, A.K., Sahoo, A.K. & Mudgal, V. 1984. *A Contribution to the Ethnobotany of Santal Pargana, Bihar.* Bot. Surv. India, Howrah.

Gogoi, J.C., Sharma, D.K. & Borthakur, S.K. 2003. Folklore Lepatoprotective Herbal Recipes for Assam in Northeastern India. 1-21. *In*: *Vistas in Ethnobotany* Vol. 1. *Indian Journal of Applied and Pure Biology*.

Gogoi, P. & Borthakur, S.K. 1991. Plants in Religio-Cultural Beliefs of Tai Khamtis of Assam (India). *Ethnobotany* **3**: 89-95.

Gogoi, P. & Borthakur, S.K. 1996. Ethnic Plant Classification and Nomenclature of the Tai Khamtis of North-eastern India, 482-484. *In*: Jain, S.K. (Ed.) *Ethnobotany in Human Welfare.* Deep Publi., New Delhi.

Gogoi, R. & Borthakur, S.K. 2001. Notes on Herbal Recipes of Bodo Tribe in Kamrup District, Assam. *Ethnobotany* **13**: 15-23.

Goud, P. Saiprasad & Pullaiah, T. 1996a. Ethnobotany of Kurnool District—Some Wild Plants Used as a Food. *J. Econ. Taxon. Bot.* Additional Series. **12**: 224-227.

_______ 1996b. Folk Veterinary of Kurnool District, Andhra Pradesh. *Ethnobotany* **8**: 71-74.

Gupta, M.P. 1988. Present Status and Future Prospects for the Cultivation and Collection of Medicinal Plants of Himachal Pradesh. *Indian Forester* 114(1): 19-25.

Gupta, R. 1961. Flora of Lam Dal. *Ind. Forester* **87**(5): 316-324.

_______ 1964. Survey Record of Medicinal and Aromatic Plants of Chamba Forest Division, Himachal Pradesh. *Ind. Forester* **90**: 454-468.

_______ 1971. Medicinal and Aromatic Plants of Bhandal Ranges and Churah Forest Division, Chamba District, Himachal Pradesh. *J. Bomb. Nat. Hist. Soc.* **68** (3): 791-803.

_______ 1981. Plants in Folk Medicine of the Himalaya, 83-90. *In*: Jain, S.K. (Ed.) *Glimpses of Indian Ethnobotany*. Oxford & IBH, New Delhi.

Hajra, P.K. 1987. Plants in Magico-Religious Beliefs and in Sanskrit Literature, 98-103. *In:* Jain S.K. (Ed.) *A Manual of Ethobotany.* Scientific Publ., Jodhpur.

Hajra, P.K. & Baishya, A.K. 1981. Ethnobotanical Notes on the Miris (Mishings) of Assam Plains, 161-169. *In:* Jain S.K. (Ed.) *Glimpses of Indian Ethnobotany.* Oxford & IBH, New Delhi.

Hamilton, A. 1955. The People and Plant Initiative: X-XI. *In*: Martin, G.J. (Ed.) *Ethnobotany.* Chapman & Hall, London.

Harshberger, J.W. 1895. Some New Ideas. *Philadelphia Evening Telegraph* (December 5).

Hedberg, I. 1993. Botanical Methods in Ethnopharmacology and the Need for Conservation of Medicinal Plants. *J. Ethnopharmacol.* **38**: 121-128.

Heiser, C.B. 1995. Edgar Anderson Botanist and Curator of Useful Plants. *Ann. Miss. Bot. Gard.* **82**(1): 54-60.

Hembrom, P.P. 1991. Tribal Medicine in Chotanagpur and Santhap Parganas of Bihar, India. *Ethnobotany* **3**: 97-99.

_______ 1996. Contact Therapy Practised by Mundas of Chotanagpur (Bihar). *Ethnobotany* **8**: 36-39.

Hembrom, P.P. & Goel, A.K. 2005. Horopathy: Ethnomedicine of Mundas. *Ethnobotany* **17**: 89-95.

Henry, A.N., Hosagoudar, V.B. & Ravi Kumar, K. 1996. Ethno-Medico-Botany of the Southern Western Ghats of India, 173-180. *In*: Jain, S.K. (Ed.) *Ethnobiology in Human Welfare*, Deep Publ., New Delhi.

Hooker, J.D. 1872-1897. *The Flora of British India,* Vols. I-VIII. Lalit Mohan Basu, Allahabad.

Hosagoudar, V.B. & Henry, A.N. 1996a. Ethnobotany of Soligas in Biligiri Rangana Betta, Karnataka, Southern India. *J. Econ. Taxon. Bot.* Additional Series **12**: 228-243.

_______ 1996b. Ethnobotany of Kadars, Malasars and Kuthuvans of the Anamalais in Coimbatore District, Tamil Nadu, India. *J. Econ. Taxon. Bot.* Additional Series 12: 260-267.

_______ 1996c. Ethnobotany of Tribes Irular, Kurumban and Paniyan of Nilgiris in Tamil Nadu, Southern

India. *J.Econ. Taxon. Bot.* Additional Series **12**: 364-366.

Huidrom, B.K.S. 1996. Plants Used in Medico-Sexual Purposes by Meitein Community in Manipur State, India. *J. Econ. Taxon. Bot.* Additional Series **12**: 364-366.

Iverson, S.T. 1988. Conservation of the Usambara Mountains, Botanical Values at the Species and Phytogeographical Level, 218-243. *In:* Hedberg, I. (Ed.) *Systematic Botany—A Key Science for Tropical Research and Documentation.* Almqvist & Wiksell Int., Stockholm.

Jadhav, D. 2006. Ethnomedicinal Survey of Maalgamdi in Ujjain District, Madhya Pradesh, India. *Ethnobotany* **18**: 157-159.

Jain, A.K. 1992. Clans of Sor Tribals of Madhya Pradesh and Their Role in Conservation. *Ethnobotany* 4: 67-69.

Jain, A.K. & Patole, S.N. 2001. Less Known Medicinal Uses of Plants Among Some Tribal and Rural Communities of Panchmarhi Forest (M.P.). *Ethnobotany* **13**: 96-100.

Jain, A.K. & Sharma, H.O. 1996. Ethnobiological Studies of Shariya Tribe of Central India, 397-399. *In*: Jain, S.K. (Ed.) *Ethnobiology in Human Welfare*. Deep Publ., New Delhi.

Jain, S.K. 1968. *Medicinal Plants*. National Book Trust, New Delhi.

_______ (Ed.) 1981. *Glimpses of Indian Ethnobotany*. Oxford & IBH, New Delhi.

_______ 1986. Ethnobotany. *Interdiscipl. Sci. Rev.* II: 285-292.

_______ 1991. *Dictionary of Indian Folk Medicine and Ethnobotany*. Deep Publ., New Delhi.

_______ 1995. Ethnobotanical Diversity in Zingibers of India. *Ethnobotany* 1: 83-88.

_______ (Ed.) 1996. *Ethnobiology in Human Welfare*. Deep Publ., New Delhi.

_______ 2002. *Bibliography of Indian Ethnobotany*. Scientific Publ., Jodhpur.

_______ 2004. Objective Ethnobotany—Knowledge Traditional Approaches Modern. *Ethnobotany* **16**: 1-9.

Jain, S.K. & Borthakur, S.K. 1980. Ethnobotany of the Mikiris of India. *Econ. Bot.* **34**: 264-272.

Jain, S.K. & De Filipps, R.A. 1991. *Medicinal Plants of India* (2 Vols.) Ref. Publ., Algonac.

Jain S.K. & Goel, A.K. 1995. Workshop Exercise – 1. Proforma for Field Work: 142-147. *In*: Jain, S.K. (Ed.) *A Manual of Ethnobotany*. Scientific Publ., Jodhpur.

Jain, S.K., Mudgal, V., Banerjee, D.K., Guha, A., Pal, D.C. & Das, D. 1984. *Bibliography of Ethnobotany*. Bot. Survey India, Calcutta.

Jain, S.K., Ranjan, Vinay, Sikarwar, R.K.S., & Saklani, A. 1994. Botanical Distribution of Psychoactive Plants of India. *Ethnobotany* **6:** 65-75.

Jain, S.K. & Rao, R.R. (Eds.) 1977. *A Handbook of Field and Herbarium Methods.* Today & Tomorrow Printers & Publ., New Delhi.

Jain, S.K. & Saklani, Arvind, 1992. *Cross-Cultural Ethnobotanical Studies in North-East India. Ethnobotany* **4:** 25-38.

Jain, S.K., Sinha, B.K. & Gupta, R.C. (Eds.) 1991. *Notable Plants in Ethnomedicine of India*. Deep Publ., New Delhi.

Jain, S.K. & Srivastava, S. 1999. *Directory of Ethnoveterinary Plants of India.* Deep Publ., New Delhi.

_______ 2001. Indian Ethnobotanical Literature in Last Two Decades—A Graphic Review and Future Directions. *Ethnobotany* **13:** 1-8.

Jain, S.K. & Tarafder, C.R. 1970. Medicinal Plant Lore of the Santhals. A Revival of P.O. Boddings's Work. *Econ. Bot.* 24: 241-278.

Jain, S.P. & Puri, H.S. 1994. An Ethno-Medico-Botanical Survey of Parvati Valley in Himachal Pradesh (India). *J. Econ. Tax. Bot.* **10:** 231-222.

Janaki Ammal, E.K. 1955. *An Introduction to the Subsistence Economy of India. Background Paper No. 10.* Winner Gref Foundation International Symposium on 'Man's Role in Changing the Face of the Earth'. Princeton Inn. Princeton, N.J. June 16-22.

Jha, R.R. & Varma, S.K. 1996. Ethnobotany of Sauria Paharias of Santhal Pargana, Bihar: I Medicinal Plants. *Ethnobotany* **8:** 31-35.

Jha, V. & Basak, P.K. 1994. Ethnobotanical Aspects of Mithila Paintings. *Ethnobotany* **6:** 9-18.

Jha, V., Choudhary, U.N. & Saraswati, K.C. 1991. Botanical Aspects of an Ethnoveterinary Prescription in

Mithila, North Bihar (India). *Ethnobotany* **3**: 101-104.

Jha, V., Mishra, S., Kargupta, A.N. & Jha, A. 1996. Leaves and Flowers Utilised as Supplementary Vegetables in Darbhanga (North Bihar) and their Ethnobotanical Significance. *J. Econ. Taxon. Bot.* Additional Series **12**: 395-402.

Jones, V. 1941. The Nature and Scope of Ethnobotany. *Chronica Bot.* **4**: 219-221.

Joseph, J. & Kharkonger, P. 1981. A Preliminary Ethnobotanical Survey in the Khasi and Jaintia Hills, Meghalaya, 115-123. *In:* Jain, S.K. (Ed.) *Glimpses of Indian Ethnobotany*. Oxford & IBH, New Delhi.

Joshi, P. 1995. *Ethnobotany of the Primitive Tribes in Rajasthan*. Rupa Books Pvt. Ltd., Jaipur.

Kadel, C. & Jain, A.K. 2006. Plant Used in Ethnoveterinary Practices in Jhabua District, Madhya Pradesh. *Ethnobotany* **18**: 149-152.

Kala, C.P. & Manjrekar, N. 1999. Ethnomedicobotany of India Trans-Himalaya: A Case Study from Spiti. *J. Econ. Tax. Bot.* **23** (1): 177-183.

Kalpesh, I., Rao, V.R., Parabia, F.M., Kothari I.L. & Parabia M.H. 2004. Ethnobotany of Little Rann of Kachh, Gujarat. *Ethnobotany* **16**: 40-43.

Kapur, S.K. 1993. Ethnomedico Plants of Kangra Valley (Himachal Pradesh). *J. Econ. Tax. Bot.* **17(2)**: 395-408.

_______ 1996. Traditionally Important Medicinal Plants of Bhaderwah Hills (Jammu Province) Part–III. *J. Econ. Taxon. Bot.,* Additional Series **12**: 62-69.

Kapur, S.K. & Nanda, S. 1996. Traditionally Important Medicinal Plants of Bhaderwah Hills (Jammu Province) Part-II. *J. Econ. Taxon. Bot.* Additional Series **12**: 56-61.

Kapur, S.K. & Singh, P. 1996. Traditionally Important Medicinal Plants of Udhampur District (Jammu Province) Part-I. *J. Econ. Taxon. Bot.,* Additional Series **12**: 82-88.

Kapur, S.K., Nanda, S. & Srivastava, T.N. 1996. Ethnobotanical Uses of PRL-Herbarium-III. *J. Econ. Taxon. Bot*., Additional Series **12**: 50-55.

Karnick, C.R., Tiwari, R., Majumdar, R. & Bhattacharjee, S. 1981. Newer Ethnobotanical and Folklore Studies of Some Medicinal Plants of Gauhati and Surrounding Areas. *Nagarjun* **24**: 240-245.

Katewa, S.S. & Arora, Asha 1997. Some Plants in Folk Medicine of Udaypur District (Rajasthan). *Ethnobotany* **9**: 48-51.

Katewa, S.S., Choudhary, B.L., Jain, Anita, Takhar, H.K. 2001. Some Plants in Folk Medicine of Rajsamand District (Rajasthan). *Ethnobotany* **13**: 129-134.

Katewa, S.S. & Sharma, Rajesh 1998. Ethnomedicinal Observation from Certain Watershed Areas of Rajasthan. *Ethnobotany* **10**: 46-49.

Kaushik, P. 1988. *Indigenous Medicinal Plants* (Including Microbes and Fungi). Today and Tomorrow Printers & Publ., New Delhi.

Khan, A.A. 2005. Ethnobotany in Twenty-first Century in India. *Ethnobotany* **17**: 71-78.

Khanna, K.K. 2002. Unreported Ethnomedicinal Uses of Plants from the Tribal and Rural Folklores of Gonda District, Uttar Pradesh. *Ethnobotany* **14**: 52-56.

Khanna, K.K. & Kumar, R. 2000. Ethnomedicinal Plants Used by the Gujjar Tribe of Saharanpur District, Uttar Predesh. *Ethnobotany* **12**: 17-22.

Khanna, K.K., Shukla, G. & Mudgal, V. 1996a. New Traditional Medicinal Uses of Plants from Jalaun District, Uttar Pradesh. *J. Econ. Taxon. Bot*. Additional Series **12**: 108-111.

Khanna, K.K., Srivastava, P.K. & Mudgal, V. 1996b. Noteworthy Medicinal Plants Uses from Rural Folklore of Raebareli District, Uttar Pradesh. *J. Econ. Taxon. Bot.* Additional Series **12**: 118-122.

Kharkonger, P. & Joseph, J. 1981. Folklore Medicobotany of Rural Khasi and Jaintia Tribes in Meghalaya, 124-136. *In*: Jain, S.K. (Ed.) *Glimpses in Indian Ethnobotany.* Oxford & IBH, New Delhi.

Khoshoo, T.N. 1995. Census of India's Biodiversity: Tasks Ahead. *Curr. Sci.* **69**: 14-17.

Kirtikar, K.R. & Basu, B.D. 1935. *Indian Medicinal Plants.* Vols. I-IV. Lalit Mohan Basu, Allahabad, India.

Koelz, W.N. 1979. Notes on the Ethnobotany of Lahul, A Province of the Punjab. *Quar. J. Crude Drug Res.* **17**: 1-56.

Kothari, M.J. & Moorthy, S. 1996. Ethnobotany in Human Welfare of Raigad District in Maharashtra State, India, 403-407. *In*: Jain, S.K. (Ed.) *Ethnobiology in Human Welfare*. Deep Publ., New Delhi.

Kshirsagar, R.D. & Singh, N.P. 2000. Less Known Ethnomedicinal Use of Plants in Coorg District of Karnataka State, Southern India. *Ethnobotany* **12**: 12-16.

——— 2007. *Ethnobotany of Mysore and Coorg Districts, Karnataka State*. Bishen Singh Mahender Pal Singh, Dehradun.

Kshirsagar, S.R., Parobia, M.H. & Reddy M.N. 2003. Ethnobotany of Coastal Areas in Gujarat. *Ethnobotany* **15**: 60-63.

Kulkarni, D.K. & Kumbhojkar, M.S. 1993. Kitchen Garden Plants of Mahadeokali Tribe in Maharashtra. *Ethnobotany* **5**: 119-127.

——— 1996. Pest Control in Tribal Areas of Western Maharashtra—An Ethnobotanical Approach. *Ethnobotany* **8**: 56-59.

Kumar, A. 1996. Some Ethnomedicinal Plants of the Murias of the Indravati Tiger Reserve, Bastar (Madhya Pradesh). *J. Econ. Taxon. Bot.* Additional Series **12**: 201-205.

Kumar, K. & Goel, A.K. 1998. Little Known Ethnomedicinal Plants of Santhal and Paharia Tribes in Santhal Pargana, Bihar, India. *Ethnobotany* **10**: 66-67.

Kumar, R. Vijaya & Pullaiah, T. 1998. Medicinal Plants Used by Tribals of Prakasam District (Andhra Pradesh), *Ethnobotany* **10**: 97-102.

——— 2003. A Survey of Wild Edible Plants of Chenchu Tribes of Prakasam District (Andhra Pradesh), 22-31. *In*: *Vistas in Ethnobotany* Vol. 1. *Indian Journal of Applied and Pure Biology*.

Kumar, S. & Chauhan, A.K.S. 2005. Medicinal Plant Used by Local Inhabitants in Bharatpur District (Rajasthan). *Ethnobotany* **17**: 179-183.

——— 2006. Less Known Medicinal Use of Plant Species in Koeladeo National Park, Bharatpur (Rajasthan). *Ethnobotany* **18**: 153-156.

Kumar, S., Gozal, S. & P. Farzana, 2004. Ethnoveterinary Plants in Indian Arid Zone. *Ethnobotany* **16**: 91-95.

Kumar, V. & Jain, S.K. 1998. A Contribution to Ethnobotany of Surguja District in Madhya Pradesh, India. *Ethnobotany* **10**: 86-89.

Kurian, J.C. (Ed.) 1995. *Plants that Heals*. Oriental Watchman Publ. House, Pune, India.

Lal, B. & Dube, V.P. 1984. Ethnobotanical Studies on *Grewia optiva* Drum. ex Berret Among the Jaunsaris and the People of the Adjoining Areas of Tehri Garhwal, U.P. *Himal. Res. Dev.* **3**: 6-8.

Lal, B., Vats, S.K., Singh, R.D. & Gupta, A.K. 1996. Plants Used as Ethnomedicine and Supplementary Food by Gaddis of Himachal Pradesh, 384-387. *In:* Jain, S.K. (Ed.) *Ethnobiology in Human Welfare*. Deep Publ., New Delhi.

Lalramnghinglova, H. & Jha, L.K. 1999. Ethnobotany: A Review. *J. Econ. Taxon. Bot.* **23**: 1-27.

Lalramnghinglova, J.H. 1996. Ethnobotany of Mizoram—A Preliminary Survey. *J. Econ. Taxon. Bot.* Additional Series **12**: 439-459.

Lata, Sneh 1996. *Grewia Optiva* Drumm. 'Bhimal': A Boon for Rural People in Kumaun Hills, India, 471-471. *In*: Jain, S.K. (Ed.) *Ethnobiology in Human Welfare,* Deep Publ., New Delhi.

Lipp, F.J. 1989. Methods for Ethnophramacology Field Work. *J. Ethnopharmacol.* **25**: 139-150.

Mahato, A. Kumar & Mahato, P. 1996. Ethnobotanical Wealth of Chhotanagpur Plateau–IV. Some Medicinal Plants Used Against Intestinal Worms. *J. Econ. Taxon. Bot.* Additional Series **12**: 389-391.

Maheshwari, J.K. (Ed.) 1992. *Ethnobotany in India*. Scientific Publ., Jodhpur.

——— 1996. *Ethnobotany in South Asia*. Scientific Publ., Jodhpur.

Maheshwari, J.K. & Singh, J.P. 1987. Contribution to the Ethnobotany of Bhoxa Tribe of Bijnor and Pauri Garhwal District, U.P., *J. Econ. Taxon. Bot.* **5**(2): 251-259.

Maheshwari, J.K., Singh, K.K. & Saha, S. 1981. *The Ethnobotany of the Tharus of Kheri District, Uttar Pradesh*. Nation. Bot. Res. Inst. Lucknow.

——— 1986. *Ethnobotany of Tribals of Mirzapur District, U.P.* Nation. Bot. Res. Inst., Lucknow.

Malik, S. 1996. Belief and Health Care: A Study Among the Saoras of Koraput District, Orissa, 151-152. *In*: Jain S.K. (Ed.) *Ethnobiology in Human Welfare*, Deep. Publ., New Delhi.

Maji, S. & Sikdar, J.K. 1982. A Taxonomic Survey and Systematic Census on the Edible Wild Plants of Midnapore District, West Bengal. *J. Econ. Taxon. Bot.* **3**: 717-737.

Maliya, S.D. 2004. Some New or Less Known Folk Medicines of District Bahraich, U.P., India. *Ethnobotany* **16**: 113-115.

Maliya, S.D. & Singh, K.K. 2003. Some New or Less Known Folk Medicines of Bahraich District, Uttar Pradesh. *Ethnobotany* **15**: 132-135.

Manandhar, N.P. 1980. *Medicinal Plants of Nepal Himalaya.* Ratna Pustak Bhandar, Kathmandu, Nepal.

_______ 1985. Ethnobotanical Notes on Certain Medicinal Plants Used by Tharus of Dang-Deokhuri District, Nepal. *Int. J. Crude Drug Res.* **25(4)**: 153-159.

_______ 1987. Traditional Medicinal Plants Used by Tribals of Lamjung District, Nepal. *Int. J. Crude Drug Res.* **25(4)**: 236-240.

_______ 1989a. Medicinal Plants Used by Chepang Tribes of Makawanpur District, Nepal. *Fitotreapia* **60(1)**: 61-68

_______ 1989b. *Useful Wild Plants of Nepal.* Franz Steiner Verlag Wiesbaden GMBH, Stuttgart, W. Germany.

_______ 1990. Folklore Medicine of Chitwan District, Nepal. *Ethnobotany* **2**: 31-38.

_______ 1996. Traditional Practice for Oral Health Care in Nepal. *J. Econ. Taxon. Bot.* Additional Series **12**: 408-413.

Mandal, S.K. & Basu, S.K. 1996. Ethnobotanical Studies Among Some Tribals of Niligiri District, Tamil Nadu. *J. Econ. Taxon. Bot.* Additional Series **12**: 268-271.

Manilal, K.S. 1989. Linkages of Ethnobotany with other Sciences and Discipline. *Ethnobotany* **1**: 15-24.

Masih, S.K. 2003, Diversity of Ethnomedicinal Wealth in Amar Kantak Plateau Region (India), 204-214. *In*: *Vistas in Ethnobotany,* Vol. I, *Indian Journal of Applied and Pure Biology*.

Maya, S., Kumari, S.P. & Menon, V.S. 2003. Ethnobotanical Notes on the Flora of Sacred Tanks of Kerala. *Ethnobotany* **15**: 55-59.

Megoneitso, K. & Rao, R.R. 1983. Ethnobotanical Studies in Nagaland-4. Sixty-two Medicinal Plants Used by the Angami Nagas. *J. Econ. Taxon. Bot.* **4**: 167-172.

Mehrotra, S., Rawat, A.K.S., Singh, H.K. & Shome, U. 1995. Standardization of Popular Ayurvedic Adaptogenic Preparation 'Chyavanaprasha' and Ethnobotany of its Ingredients. *Ethnobotany* **7**: 1-15.

Mishra, D.K., Samanta, G. & Mishra, T.K. 1996a. Ethnomedicine of the Tribe Kharia of Midnapore District, West Bengal. *J. Econ. Taxon. Bot.* Additional Series **12**: 329-331.

Mishra, S., Jha, V. & Jha, S. 1996b. Plants in Ethnoveterinary Practices in Darbhanga (North Bihar). *In*: Jain, S.K. (Ed.) *Ethnobiology in Human Welfare*, Deep Publ., New Delhi.

Mishra, T.K., Bhunia, A., Sahoo, T.K. & Roy, S.B. 1997. Ethnobotany *vis-à-vis* Joint Forest Management Movement of Midnapore (West Bengal), India. *Ethnobotany* **9**: 52-55.

Mittre, V. 1981. Wild Plants in Indian Folk Life—A Historical Perspective, 37-58. *In:* Jain, S.K. (Ed.) *Glimpses of Indian Ethnobotany.* Oxford & IBH, New Delhi.

Mohanty, R.B. & Padhy, S.N. 1996. Traditional Phytotherapy for Diarrhoeal Diseases in Ganjam and Phulbani Districts of South Orissa, India. *Ethnobotany* **8**: 60-65.

Mohanty, S.K., Dash, S.K. & Padhy, S.N. 1998. Traditional Phytotherapy for Diarrhoea Disease in India—A Review. *Ethnobotany* **10**: 103-111.

Mokat, D.K. & Deokule, S.S. 2004. Plants Used as Veterinary Medicine in Ratnagiri District of Maharashtra. *Ethnobotany* **16**: 131-135.

Molla, H.A. & Roy, B. 1996. Some Ethnobotanical Claims from the Jalpaiguri District of West Bengal. *J. Econ. Taxon. Bot.* Additional Series. **12**: 322-324.

Mudgal, V. 1987. Literature on Ethnobotany—Indian and Foreign, 45-57. *In*: Jain, S.K. (Ed.) *A Manual of Ethnobotany.* Scientific Publ., Jodhpur.

Mudgal, V., Pal, D.C., Kayal, R.N. & Saha, S. 1999. *Ethnobotany of Totopora.* Bishen Singh Mahender Pal Singh, Dehradun.

Mukherjee, A., Mukherjee, S. & Mukherjee, M. 1986. Less Known Medicinal Plants of Rubiaceae in Darjeeling Hills, West Bengal (India). *Indian J. Appl. & Pure Bot.* **1**: 86-91.

Mukherjee, A. & Namhata, D. 1990. Medicinal Plants of Tribals of Sundargarh District, Orissa. *Ethnobotany* **2**: 57-60.

Mume, J.O. 1976. *Tradomedicalism*. Jom Nature Cure Centre Ajbaro, Nigeria.

Murthy, K.S., Sharma, P.C. & Kishore, P. 1986. Tribal Remedies for Snake-Bite from Orissa. *Anc. Sci. Life* **6**: 122-123.

Nair, N.C. 1964. On a Botanical Tour to Lahaul and Spiti (Punjab Himalaya). *Bull. Bot. Surv. India.* **6**: 219-235.

Nair, R.C. 1977. *Flora of Bashahr Himalaya*. Int. Bioscience Publ., Hisar.

Nath, S.C. & Borodoloi, D.N. 1989. Ethnobotanical Observation on Some Medicinal Folklore of Tirap District, Arunachal Pradesh. *J. Econ. Taxon. Bot.* **13**: 321-325.

Nautiyal, S. 1981. Some Medicinal Plants of Garhwal Hills—A Traditional Use. *J. Sci. Res. Pl. Med.* **2**: 12-17.

Nayar, M.P., Ramamurthy, K. & Agarwal, V.S. 1989. *Economic Plants of India*. Vol. I & II. Bot. Survey India, Calcutta.

Nazarudeen, A., Seeni, S., Koshy, K.C. & Pushangadan, P. 1996. Folk Plants of Food, Medicine, Adornment and Repellent Used by the Paniyar Community in North Kerala. *J. Econ. Taxon. Bot.* Additional Series **12**: 299-305.

Negi, K.S. & Gaur, R.D. 1991. A Contribution to the Edible Wild Fruits of Uttar Pradesh Hills. *Bull. Bot. Surv. India* **43**: 233-266.

Negi, K.S. & Pant, K.C. 1990. Notes on Ethnobotany of the Gangwal—A Tribe of Garhwal Himalaya. *Ethnobotany* **2**: 81-85.

Negi, K.S., Tiwari, J.K. & Gaur, R.D. 1985. Economic Importance of Some Common Trees in Garhwal Himalaya: An Ethnobotanical Study. *Indian J. Forestry* **8**: 276-289.

Negi, K.S., Tiwari, J.K., Gaur, R.D. & Pant, K.C. 1993. Notes on Ethnobotany of Five Districts of Garhwal Himalaya, Uttar Pradesh, India. *Ethnobotany* **5**: 73-81.

Negi, P.S. & Subramani, S.P. 2002. Ethnobotanical Study in the Village Chhitkul of Sangla Valley, District Kinnaur, Himachal Pradesh. *J. Non-Timber Forest Prod.* **9** (3-4): 113-120.

Negi, Y.S. & Bhalla, P. 2002. Collection and Marketing of Important Medicinal and Aromatic Plants in Tribal Areas of Himachal Pradesh. *Indian Forester* **128**(6): 641-649.

Padoch, C. & de Jong, W. 1991. The House Gardens of Santa Rosa: Diversity and Variability in an Amazaonian Agricultural System. *Econ. Bot.* **45**(2): 166-175.

Painuli, R.M. & Maheshwari, J.K. 1996. Some Interesting Ethnomedicinal Plants Used by Sahariya Tribe of Madhya Pradesh. *J. Econ. Taxon. Bot.* Additional Series. **12**: 179-185.

Pal, D.C. 1981. Plants Used in Treatment of Cattle and Birds Among Tribals of Eastern India, 245-257. *In:* Jain, S.K. (Ed.) *Glimpses of Indian Ethnobotany*. Oxford & IBH, New Delhi.

Pal, D.C. & Jain, S.K. 1998. *Tribal Medicine.* Nayaprakash, Calcutta.

Panda, S. 1996. Plant Use—A Recent Perspective from Eastern Himalayas. *J. Econ. Taxon. Bot.* Additional Series **12**: 332-337.

Pandey, A.K. & Rout, S.D. 2006. Ethnobotanical Uses of Plants of Similipal Biosphere Reserve (Orissa). *Ethnobotany*: **18**: 102-106.

Pandey, D.N. 1998. *Ethno Forestry* (*Local Knowledge for Sustainable Forestry and Livelihood Security*). Himanshu Publ., Udaipur, New Delhi.

Pandey, H.P. & Verma, B.K. 2002. Plants in Oral Healthcare Among the Aboriginals of Gonda and Balrampur Regions, U.P., India. *Ethnobotany* **14**: 81-86.

——— 2003. Magico-Therapeutic Plants of Uttar Pradesh, India, 178-187. *In*: *Vistas in Ethnobotany,*Vol. I, *Indian Journal of Pure and Applied Biology*.

Parmar, C. & Kaushal, M.K. 1982. *Wild Fruits of the Sub-Himalayan Region*. Kalyani Publishers, New Delhi, Ludhiana.

Parrotta, J.A. 2000. *Healing Plants of Peninsular India*. CAB International, U.K.

Patel, N.B., Sidaha, B.S. & Jain, B.K. 2003. Tribal Artifacts of Dholwani Forest of Sabarkantha (Gujarat). *Ethnobotany* **15**: 40-43.

Patil, D.A. 2004. Origin of Medicines vis-à-vis Doctrine of Signatures. *Ethnobotany* **16**: 52-58.

Patil, S.H. & Merat, M.M. 2003. Ethnoveterinary Practices of Satpudas of Nandurbar District of Maharashtra. *Ethnobotany* **15**: 103-106.

Patole, S.N. & Jain, A.K. 2002. Some Wild Edible Plants of Pachmarhi Biosphere Reserve (M.P.). *Ethnobotany* **14**: 48-51.

Pawar, S. & Patil, D.A. 2004. Herbal Folk Remedies for Urinogenital Complaints in Jalgaon District of Maharashtra. *Ethnobotany* **16**: 86-90.

Pei, S.J. 2001. Ethnobotanical Approaches of Traditional Medicinal Studies: Some Experience from Asia. *Pharmacent Biol.* **39** (Supplement): 74-79.

Pei, S.J. & Long, C.L. 1998. *The Applied Ethnobotany.* Yunnan Ethnology Press, Kunmung.

Polunin, O. & Stainton, A. 1984. *Flowers of the Himalaya*. Oxford University Press, Delhi.

Prasad, P., Nagrajan, R., Singh, A.J.A., Narayanan, L.M. & Natarajan, C.R. 1996. Ethnobotany of the Kanikkars of South Tamil Nadu-1. *J. Econ. Taxon. Bot*. Additional Series **12**: 292-298.

Pruthi, J.S. 1976. *Spices and Condiments.* National Book Trust, New Delhi.

Pullaiah, T. & Kumar, T. Dharma Chandra, 1996. Herbal Plants in Mannanur Forest Mahaboobnagar District, Andhra Pradesh. *J. Econ. Taxon. Bot*. Additional Series **12**: 218-220.

Punjani, B.K. 1998. Role of Plants in Field Fencing in Tribal Area of District Sabarkantha (North Gujarat). *Ethnobotany* **10**: 56-60.

Pushpangadan, P., 1990. Health Status of Tribals of India. Keynote Address, National Seminar on Health of Tribal People. Indian Society for Health Administration (ISHA), May 21-26, pp. 57-61.

Pushpangadan, P. & Kumar, B. 2005. Ethnobotany, CBD, WTO and the Biodiversity Act of India. *Ethnobotany* **17**: 2-12.

Radhakrishnan, K., Pandurangan, A.G. & Pushpangadan P. 1996. Ethnobotany of Wild Edible Plants of Kerala, India, 48-51. *In*: Jain, S.K. (Ed.) *Ethnobotany in Human Welfare*. Deep Publ., New Delhi.

_______ 2000. Tribal Artifacts of Kerala. *Ethnobotany* **12**: 67-71.

Ragupathy, S. & Mahadevan, A. 1991. Ethnobotany of Kodiakkarai Reserve Forest, Tamil Nadu, South India. *Ethnobotany* **3**: 79-82.

Rajwar, G.S. 1983. Low Altitude Medicinal Plants of South Garwhal. *Bull. Medico-Ethnobot. Res.* **4**: 14-28.

Ram, A., Jeewan, R.K., Reddy, K.E. & Venkata, R.R. 2002. Medicinal Plant-Lore of Sugails of Gooty Forests Andhra Pradesh. *Ethnobotany* **14**: 37-42.

Rama Rao, N. & Henry, A.N. (Ed.) 1996. *Ethnobotany of Eastern Ghats in Andhra Pradesh, India*. Bot. Surv. of India, Calcutta.

Ramachandran, V.S. & Nair, N.C. 1981. Ethnobotanical Conservations of Irulars of Tamil Nadu (India). *J. Econ. Taxon. Bot*. **2**: 183-190.

Rana, J.C. 2004. *Lesser Known Food Plants of Himachal Pradesh*. NBPGR, Regional Station, Phagli, Shimla.

Rana, T.S., Dutt, B. & Rao, R.R. 1996. Strategies of Sustainable Utilization of Plant Resources by the Tribals of the Tons Valley, Western Himalaya. *Ethnobotany* **8**: 96-104.

Ranjan, V. 1996. Some Ethnomedicinal Plants of Lalitpur District, Uttar Pradesh, India, 149-150. *In*: Jain, S.K. (Ed.) *Ethnobotany in Human Welfare*. Deep Publ., New Delhi.

Rao, D.M. & Pullaiah, T. 2001. Ethno-Medico-Botanical Studies in Guntur District of Andhra Pradesh, India. *Ethnobotany* **13**: 40-44.

Rao, N.S.B., Rajasekhar, D., Raju, D.C. & Nagaraju, J. 1996. Ethnomedicinal Notes on Some Plants of Tirumala Hills for Dental Disorders. *Ethnobotany* **8**: 88-91.

Rao, R.R. 1996. Traditional Knowledge and Sustainable Development: Key Role of Ethnobiologists. *Ethnobotany* **8**: 14-24.

Rastogi, M.A. 1960. Medicines from the Wild. A Case Study of the Great Himalayan Park. *The Indian Magazine of Her People and Culture*. pp. 74-75.

Rastogi, R.P. & Mehrotra, B.N. 1990. *Compendium of Indian Medicinal Plants*. Vol. 1 (1960-1969). CDRI & CSIR.

_______ 1991. *Compendium of Indian Medicinal Plants*. Vol. 2 (1970-1979). CDRI & CSIR.

_______ 1993. *Compendium of Indian Medicinal Plants*. Vol. 3 (1980-1984). CDRI & CSIR.

_______ 1995a. *Compendium of Indian Medicinal Plants*. Vol. 4 (1985-1989). CDRI & CSIR.

_______ 1995b. *Compendium of Indian Medicinal Plants*. Vol. 5 (1990-1994). CDRI & CSIR.

Rau, M.A. 1961. A Recent Survey of Medicinal Plant Resources of Lahaul Valley (Panjab India). *Pakistan J. Soc. & Ind. Res. Spl. Nr. International Symp.* 4: 201-205.

Ravishankar, T. & Henry, A.N. 1992. Ethnobotany of Aditlabad District, Andhra Pradesh, India. *Ethnobtany* **4**: 45-52.

Reddy, H.H., Vijayalakshmi, K. & Venkata Raju, R.R. 1996. Native Phytotherapy for Snake Bite in Nallamalais Eastern Ghats, India. *J. Econ. Taxon. Bot.* Additional Series. **12**: 214-217.

Reddy, K.N., Bhanja, R., & Vatsavaya. S. Raju, 1998. Plant Used in Ethnoveterinary Practices in Warangal District of Andhra Pradesh, India. *Ethnobotany* **10**: 75-84.

Redford, K. & Stearman, A. 1993. Forest Dwelling Native Amazonians and the Conservation of Biodiversity: Interest in Common or Collision? *Conservation Biology* **I**: 248-255.

Retnam, K.R. & Martin, P. 2006. *Ethnomedicinal Plants*, Agrobios Jodhpur.

Robbins *et al.*, 1916. *Ethnobotany of the Tenn Indians.* Bureau of American Ethnobotany. Bulletin No. 5, USA.

Ruddle, K. 1974. *The Yuka Cultivation System: A Study of Shifting Cultivation in Colombia and Venezuela.* Indero-American 52 University California Press, Berkeley.

Sahoo, A.K. & Mudgal, V. 1995. Less Known Ethnobotanical Uses of Plants of Phulbani District, Orissa, India. *Ethnobotany* **7**: 63-67.

Sahu, T.R. 1996. Life Support Promising Food Plants Among Aboriginals of Bastar (M.P.), India, 26-30. *In*: Jain, S.K. (Ed.) *Ethnobiology in Human Welfare*, Deep Publ. New Delhi.

Saini, D.C. 1996a. Ethnobotany of Tharus of Basti District, Uttar Pradesh. *J. Econ. Taxon. Bot.* Additional Series **12**: 138-153.

Saini, V.K. 1996b. Plants in the Welfare of Tribal Women and Children in Certain Areas of Central India, 140-144. *In*: Jain, S.K. (Ed.) *Ethnobiology in Human Welfare*. Deep Publ., New Delhi.

Saklani, A. & Jain, S.K. 1994. *Cross-Cultural Ethnobotany of North-East India.* Deep Publ. New Delhi.

Samati, H. 2004. Kitchen Garden Plants of Pnar Tribe in Jaintia Hills District of Meghalaya. *Ethnobotany* **16**: 125-130.

Samwatsar, S. & Diwanji, V.B. 1996. Plants Used for Skin Diseases, Cuts, Wounds and Bruises by the Tribals of Western M.P. *J. Econ. Taxon. Bot.* Additional Series. **12**: 192-195.

Sarangi, N. & Sahu R.K. 2004. Ethnomedicinal Plants Used in Veneral and Gynaecological Disorders in Kalahandi Orissa. *Ethnobotany* **16**: 16-20.

Sarin, Y.K. 1967. A Survey of Vegetable Raw Resources of Lahaul. *Indian Forester* **93**(7): 489-499.

Sarin, Y.K. & Chopra, C.L. 1985. Medicinal and Aromatic Plant Resources of Himachal Pradesh, Present Status and Future Prospects. *Workshop on Sci. and Tech.*, Shimla, Oct. 29-30, 1985.

Satapathy, K.B. & Brahmam, M. 1996. Some Medicinal Plants Used by Tribals of Sundargarh District, Orissa, India, 153-155. *In*: Jain, S.K. (Ed.) *Ethnobiology in Human Welfare*, Deep Publ., New Delhi.

Saxena, A.P. & Vyas, K.M. 1981. Ethnobotanical Records on Infectious Diseases from Tribals of Banda District. *J. Econ. Taxon. Bot.* **2**: 191-194.

Saxena, H.O. 1986. Observations on the Ethnobotany of Madhya Pradesh. *Bull. Bot. Surv. India* **28**: 149-156.

Saxena, H.O., Brahmam, M. & Dutta, P.K. 1981. Ethnobotanical Studies in Orissa, 232-244. *In*: Jain, S.K. (Ed.) *Glimpses of Indian Ethnobotany*. Oxford & IBH, New Delhi.

Schultes, R.E. & Von Reis, S. 1995. *Ethnobotany: Evolution of a Discipline.* Chapman & Hall, London, UK.

Schultes, R.E. 1960. Tapping our Heritage of Ethnobotanical Lore. *Econ. Bot.* **8**: 224-227.

Sen, S. & Batra, A. 1997. Ethno-Medico-Botany of Household Remedies of Phagli Tehsil of Jaipur District (Rajasthan). *Ethnobotany* **9**: 122-128.

Sen, S.K. & Behera, L.M. 2003. Ethnomedicinal Plants Used Against Skin Diseases of Bargarh District in Orissa. *Ethnobotany* **15**: 90-96.

Sen, S., Batra, A. & Ali, Z.D. 2002. Some Herbal Home Remedies from Jaipur District, Rajasthan. *Ethnobotany* **14**: 116-122.

Sensarma, P. 1989. *Plants in the Indian Puranas*. An Ethnobotanical Investigation. Naya Prakash, Calcutta.

_______ 1991. Herbal Veterinary Medicines in an Ancient-Sanskrit Work—The Garuda Purana. *Ethnobotany* **3**: 83-87.

Shabnam, S.R. 1964. Medicinal Plants of Chamba. *Indian Forester* 90: 50-63.

Shah, G.L. 1984. Some Economically Important Plants of Salsette Island Near Bombay. *J. Econ. Taxon. Bot.* 5: 753-765.

Shah, N.C. & Joshi, M.C. 1971. An Ethnobotanical Study of Kumaon Region of India. *Econ. Bot.* **25**: 414-422.

Shah, N.C. & Singh, S.C. 1990. Hitherto Unreported Phytotherapeutical Uses from Tribal Pockets of Madhya Pradesh (India). *Ethnobotany* **2**: 91-95.

Sharma, 2003. *Medicinal Plants of India—An Encyclopaedia.* Daya Publishing House, New Delhi.

Sharma, B. & Maheshwari, S. 2005. Traditional Medical Practices of Gaddi Tribes in Kangra District, Himachal Pradesh, *Indian J. Trad. Knowledge* **4** (2): 173-188.

Sharma, B.D. & Rana, J.C. 1999. Traditional Medicinal Uses of Plants of Himachal Hills. *J. Econ. Taxon. Bot.* **23**(1): 173-176.

Sharma, D. & Misra, D.D. 1972. Scope of Phytochemical Industry in Himachal Pradesh. Nagarjun **15**(II): 26-27.

Sharma, K.R. & Sood, M. 1997. *Important Medicinal Plants of Himachal Pradesh*. Vol. I & II. Dr. Y.S. Parmar University of Horticulture and Forestry, Nauni, Solan, H.P.

Sharma, P.K., Chauhan, N.S. & Lal, B. 2004. Observation on the Traditional Phytotherapy Among the Inhabitants of Parvati Valley in Western Himalaya, India. *J. Ethnopharmacology*. **92**: 167-176.

Sharma, P.K., Dhyani, S.K. & Shanker, V. 1979. Some Useful and Medicinal Plants of the District Dehradun and Siwalik. *J. Sci. Res. Pl. Med.* **1**: 17-43.

Sharma, P.P. & Singh, N.P. 2001. Ethnomedicinal Uses of Some Edible Plants in Dadra, Nagar Haveli and Daman (U.T.). *Ethnobotany* **13**: 121-125.

Sharma, S.C. 1996. A Medicobotanical Study in Relation to Veterinary Medicines of Shahjahanpur District (Uttar Pradesh). *J. Econ. Taxon. Bot.* Additional Series **12**: 123-127.

Sharma, S.K., Bhattacharya, D.K. & Devi B. 2001. Medicinal Plants Used by Bodo Tribe of Nalbari District in Assam. *Ethnobotany* **13**: 135-139.

——— 2002. Traditional Use of Herbal Medicines by Madai Tribe of Nalbari District of Assam. *Ethnobotany* **14**: 103-111.

Sharma, T.R. & Borthakur, S.K. 2003. Inquire into Old Lead-II: Ethnoveterinary Medicine for Treatment of Horse in Assam, 69-85. *In*: *Vistas in Ethnobotany*, Vol. I, *Indian Journal of Pure and Applied Biology*.

Shekhawat, G.S. & Anand, S. 1984. An Ethnobotanical Profile of Indian Desert. *J. Econ. Taxon. Bot.* **5**: 591-598.

Shome, U., Rawat, A.K.S. & Mehrotra, S., 1996. Time Tested Household Remedies, 96-100. *In*: Jain, S.K. (Ed.) *Ethnobiology in Human Welfare*, Deep Publ., New Delhi.

Shrestha, P. 1985. Research Note: Contribution to the Ethnobotany of the Palpa Area. *Contribution to the Nepalese Studies* **12**(2): 63-74.

Shukla, G. & Verma, B.K. 1996. Roots—A Vital Plant Part to Cure Body Ailments Among Tribals/Rural Folklore of Western Bihar. *J. Econ. Taxon. Bot.* Additional Series **12**: 392-394.

Sikarwar, R.L.S. 1996. Ethnoveterinary Herbal Medicines in Morena District of Madhya Pradesh, India, 194-196. *In*: Jain, S.K. (Ed.) *Ethnobiology in Human Welfare,* Deep Publ., New Delhi.

——— 2002. Ethnogynaecological Uses of Plants New to India. *Ethnobotany* **14**: 112-115.

Sikarwar, R.L.S. & Kaushik, J.P. 1992. Some Less Known Medicinal Uses of Trees Among the Sahariyas of Morena District, M.P., India. *Ethnobotany* **4**: 71-74.

Silori, C.S. & Rana, A.R. 2000. Indigenous Knowledge of Medicinal Plants and Their Uses in Narayan Sarover Sanctuary Kakhch. *Ethnobotany* **12**: 1-7.

Singh, A.K., Singh, R.N. & Singh, S.K. 1987. Some Ethnobotanical Plants of Terai Region of Gorakhpur District, I. *J. Econ. Taxon. Bot.* 9: 407-410.

Singh, B., Chaurasia, O.P. & Ballabh, B. 2001. Edible Wild Plants of Trans-Himalayan Cold Desert: 483-512. *In:* Pande, P.C. & Samat, S.S. (Eds.) *Plant Diversity of the Himalaya*. Gyanodya Prakashan, Nanital.

Singh, B., Chaurasia, O.P. & Jadhav, K.L. 1996. An Ethnobotanical Study of Indus Valley (Ladakh). *J. Econ.*

Taxon. Bot. Additional Series 12: 92-101.

Singh, G.S. 1990. Ethnobotanical Studies of Useful Plants of the Kullu District in North-Western Himalaya, India. *J. Econ. Taxon. Bot.* **23**: 185-188.

_______ 1999a. Ethnobotanical Study of Useful Plants of Kullu District in North-Western Himalaya, India. *J. Econ. Taxon. Bot.* **23** (1): 185-198.

Singh, G., Singh, H.B. & Mukherjee, T.K. (Eds.). 2003b *Ethnomedicine of North-East India.* NISCIS, New Delhi.

Singh, G.S., Rao, K.S. & Saxena, K.G. 1998a. Eco-Cultural Analysis of Sacred Species and Ecosystem in Chhakinal Watershed, Himachal Pradesh: 301-313. *In*: Ramakrishnan, P.S., Saxena, K.G. & Chandrashekara, V.M. (Eds.): *Conserving the Sacred for Biodiversities Management.* UNESCO. Oxford & IBH, New Delhi.

_______ 1998b. Environment, Ecological and Socio-Economic Impact of Introduced Crops in Western Himalaya: A Case Study of Kullu Valley. *J. Human Ecol.* **9**: 63-72.

Singh, H. 2003. Herbal Recipes for Spermatoorrhoea by Bhoxa Tribe of Uttaranchal. *Ethnobotany* **15**: 115-117.

Singh, H., Kumar, B., Hyniewta, T.M. & Bora, P.J. 1997. Ethno-Medico-Botanical Studies in Tripura, India. *Ethnobotany* **9**: 56-58.

Singh, H. & Sharma, M. 2006. *Flora of Chamba District (Himachal Pradesh).* Bishen Singh Mahender Pal Singh, Dehradun.

Singh, H.B., Singh, R.S. & Sandhu, J.S. (Eds.) 2003a. *Herbal Medicine of Manipur: A Colour Encyclopaedia.* Daya Publ. House, New Delhi.

Singh, J., Bhuyan, T.C. & Ahmed, A. 1996b. Ethnobotanical Studies on the Mishing Tribes of Assam with Special Reference to Food and Medicinal Plants–I. *J. Econ. Taxon. Bot.* Additional Series 12: 350-356.

Singh, J.P. & Shankar, V. 1996. Traditional Forage Resources of India: Exploitation Potential: 307-309. *In*: Jain, S.K. (Ed.) *Ethnobiology in Human Welfare*. Deep Publ., New Delhi.

Singh, K.K. & Kumar, K. 2000a. *Ethnobotanical Wisdom of Gaddi Tribe of Western Himalayas*. Bishen Singh Mahender Pal Singh, Dehradun.

Singh, K.K. & Kumar, K. 2000b. Observation on Ethnoveterinary Medicine Among the Gaddi Tribe of Kangra Valley, H.P. *Ethnobotany* **12**: 145-152.

Singh, K.K. & Maheshwari, J.K. 1992. Some Less Known Medicinal Uses of Trees Among the Sahariyas of Morena District, M.P., India. *Ethnobotany* 4: 71-74.

Singh, K.K., Palvi, S.K. & Singh, H.B. 1980. Survey of Some Medicinal Plants of Dharchula Block-I Pithoragarh District of U.P. *Bull. Medico-Ethnobot. Res.* 1: 1-7.

Singh, K.K. & Prakash, A. 1994. Indigenous Phytotherapy Among the Gond Tribe of Uttar Pradesh, India. *Ethnobotany* 6: 37-41.

Singh, K.K. & Rao, R.R. 2003. Studies on Some Potential Ethnomedicine Among the Tribals of Uttar Pradesh, India, 42-57. *In*: *Vistas in Ethnobotany*, Vol. I, *Indian Journal of Applied & Pure Biology*.

Singh, P.B. 1993. Medicinal Plants of Ayurvedic Importance from Mandi District of Himachal Pradesh. *Bull. of Medico-Ethnobot. Res.* **14** (3-4): 126-136.

_______ 1996. Wild Edible Plants of Mandi District in Northwest Himalaya, 22-25. *In:* Jain, S.K. (Ed.) *Ethnobiology in Human Welfare.* Deep Publ., New Delhi.

_______ 1999b. *Illustrated Field Guide to Commercially Important Medicinal and Aromatic Plants of Himachal Pradesh* (With Special Reference to Mandi District). Published by Society for Herbal Medicine and Himalayan Biodiversity, Mandi, Himachal Pradesh.

Singh, P.B. & Aswal, B.S. 1992. Medicinal Plants of Himachal Pradesh Used in Indian Pharmaceutical Industry. *B.M.E.B.R.* **13**: 172-208.

Singh, S.K. & Rawat G.S. 2000. *Flora of Great Himalayan National Park—Himachal Pradesh.* Bishen Singh Mahendra Pal Singh, Dehradun.

Singh, S.R. & Singh, N.I. 1985. A Preliminary Ethnobotanical Study on Wild Edible Plants in the Markets of Manipur-I. *J. Econ. Taxon. Bot.* **6**: 699-703.

Singh, V. & Chauhan, N.S. 2005. Traditional Practices of Herbal Medicines in the Lahaul Valleys. Himachal Himalayas. *Indian J. Trad. Knowledge* **4**(2): 208-220.

Singh, V. & Pandey, R.P. 1996. Ethno-Medicinal Plants Used for Venereal and Gynaecological Diseases in Rajasthan (India). *J. Econ. Taxon. Bot.* Additional Series 12: 154-165.

——— 1998. *Ethnobotany of Rajasthan, India.* Scientific Publ., Jodhpur.

Singh, V.K. & Zaheer Anwar Ali, 1996. Folk Medicinal Plants Used for Family Planning in India, 184-186. *In*: Jain, S.K. (Ed.). *Ethnobiology in Human Welfare*. Deep Publ., New Delhi.

Sinha, R.K. 1996. *Ethnobotany—The Renaissance of Traditional Herbal Medicines*. INA Shree, Jaipur.

Sinha, S.C. 1986. *Ethnobotanical Study of Manipur*, Ph.D. Thesis, Manipur University, Imphal.

——— 1996. Wild Edible Plants of Manipur, India, 42-47. *In*: Jain, S.K. (Ed.) *Ethnobiology in Human Welfare*. Deep Publ., New Delhi.

Siwakoti, M. & Verma, S.K. 1996. Medicinal Plants of the Terai of Eastern Nepal. *J. Econ. Taxon. Bot.* Additional Series **12**: 423-438.

Sood, R.P., Kalia, N.K., Brada, B.L. 1978. Scope for Commercial Exploitation of Various Aromatic Plants in Palampur (H.P.). *Ind. Perf.* **22**(2): 136-137.

Sood, S.K. & Prakash, V. 2007. *Edible Roots and Underground Stems of Ethnic India*. Satish Serial Publ., Delhi.

Sood, S.K. & Thakur, S. 2004. *Ethnobotany of Rewalsar Himalaya (Dist. Mandi, Himachal Pradesh, India)*. Deep Publi., New Delhi.

Sood, S.K., Bhardwaj, R. & Lakhanpal, T.N. 2005a. *Ethnic Indian Plants in Cure of Diabetes.* Scientific Publ., Jodhpur, India.

Sood, S.K., Nath, R., & Kalia, D.C. 2001. *Ethnobotany of Cold Desert Tribes of Lahaul-Spiti (N.W. Himalaya)*. Deep Publi., New Delhi.

Sood, S.K., Parmar, S. & Lakhanpal, T.N. 2005b. *Ethnic Plants of India Used in Cancer Cure—A Compendium.* Bishen Singh Mahendra Pal Singh, Dehradun.

Sood, S.K., Rana, S. & Lakhanpal, T.N. 2005c. *Ethnic Aphrodisiac Plants.* Scientific Publ., Jodhpur, India.

Sood, S.K., Thakur, V. & Lakhanpal, T.N. 2005d. *Sacred and Magico-Religious Plants of India.* Scientific Publ., Jodhpur, India.

Srivastava, S.K. & Chandra Sekar, K. 2004. Ethnomedicine of the Pin Valley National Park. Himachal Pradesh: Plant Used in Treatment of Dysentery. *Ethnobotany* **16**: 62-63.

Srivastava, S.K., Mehrotra, S., Srivastava, G.K. & Chauhan, D.K. 2003. Folk Uses of Herbs from Mahabaleshwar Region of Sahyadri Hills. *J. Eco. Taxon. Bot.* **27**: 857-863.

Stainton, A. 1988. *Flowers of Himalaya: A Supplement.* Oxford University Press, Delhi.

Swami, A. & Gupta, B.K. 1996. A Note on Some Commonly Occurring Medicinal Weeds of Udhampur District (J&K). *J. Econ. Taxon. Bot.* Additional Series **12**: 89-91.

Swaminathan, M.S. 1993. Genetic Diversity—Safetynet for Smack Farmers and Food Stocks for Biotechnologists. *Paper Presented to ISPGR Dialogue 1993*, New Delhi.

Tirkey, A., 2004. Some Ethnomedicinal Plant Species of Chattisgarh State. *Ethnobotany* **16**: 118-124.

Tirkey, A., Khan, F. & Khan, S.S. 2003a. Medicinal Plants Used in the Treatment of Hair in Raijarh District of Madhya Pradesh (India), 104-110. *In*: *Vistas in Ethnobotany,* Vol. I, *Indian Journal of Pure and Applied Biology.*

——— 2003b. Medicinal Plants Used by the Tribals of Raijarh District (M.P.) for Jaundice, 163-170. *In*: *Vistas in Ethnobotany*, Vol. I, *Indian Journal of Pure and Applied Biology*.

Tiwari, K.C. & Tiwari, V.P. 1996. Some Important Medicinal Plants of the Tropical, Subtropical and Temperate Region of Siang, Subansiri and Tirap District of Arunachal Pradesh. *J. Econ. Taxon. Bot*. Additional Series **12**: 359-363.

Tosh, J. 1996. Ethnobotanical Study of Western Maharashtra. *J. Econ. Taxon. Bot.* Additional Series **12**: 169-174.

Uniyal, M.R. 1968. Medicinal Plants of Bhagirathi Valley Lying in Uttarkashi Forest Division. *Indian Forester* **94**: 407-420.

Uniyal, B. 2003. Utilization of Medicinal Plants by the Rural Women of Kullu, Himachal Pradesh. *Indian J. Trad. Knowledge* **2**(4): 366-370.

Uniyal, M.R., Bhat, A.V. & Chaturvedi, 1973a. Preliminary Observations on Medicinal Plants of Lahaul Spiti Division in Himachal Pradesh. *Bull. Medico-Ethnobot. Res.* **3**(1): 1-26.

Uniyal, M.R. & Chauhan, N.S. 1971. Medicinal Plants of Uhal Valley in Kangra Forest Division. H.P. *Jour. Res. Ind. Med.* **6**(3): 287-299.

_______ 1972. Commercially Important Medicinal Plants of Kullu Forest Division. H.P. *Nagarjun* **4**(2): 28-32.

_______ 1973b. Traditionally Important Medicinal Plants of Kullu Forest Division, H.P. *Jour. Res. Ind. Med.* **7**(1): 76-85.

_______ 1973c. Traditionally Important Medicinal Plants of Kangra Valley in Dharamsala Forest Circle, Himachal Pradesh. *J. Res. Indian Med.* **8**(1): 76-85.

Uniyal, S.K., Singh, K.N., Jamwal, P. & Lal, B. 2006. Trditional Uses of Medicinal Plants Among the Tribal Communities of Chhota Bhangal, Western Himalaya. *J. Ethnobiology & Ethnomedicine* **2**: 2-11.

Upadhyay, R. & Singh, J. 2005. Ethno-Medicinal Uses of Plants in Tikri Forest of Gonda District (U.P.). *Ethnobotany* **17**: 167-170.

Upadhye, A.S. & Kulkarni, D.K. 2004. Traditional Phytotherapy for Insect Bites Among the Tribals in Western Maharashtra. *Ethnobotany* **16**: 59-61.

Upadhye, A.S., Vartak, V.D. & Kumbhojkar, M.S. 1994. Ethno-Medicobotanical Studies in Western Maharashtra, India. *Ethnobotany* 6: 25-31.

Uphof, J.C. Th. 1968. *Dictionary of Economic Plants*. H.R. Eugelmann (J. Cramer), Weinheim.

Varghese, E. 1996. *Applied Ethnobotany—A Case Study Among the Kharia of C. India.* Deep Publi., Delhi.

Varghese, E. & Hembrom, P.P. 2000. *Ethnomedicinal Lore of the Paharias.* Central Human Ecol., Mumbai.

Varma, S.K. 1997. Comparative Studies on Folk Drugs of Tribals of Chotanagpur and Santhal Pargana of Bihar, India. *Ethnobotany* **9**: 70-76.

Vedavathy, S., Mrudula, V. & Sudhakar, R. 1997. *Tribal Medicine of Chittoor Distt., A.P., Tirupati.*

Venkataratnam, K.R. & Venkata Raju, R.R. 2004. Folk Medicine from Gundlabrahmeswaram Wild-life Sanctuary. *Ethnobotany* **16**: 33-39.

Verma, P., Khan, A.A. & Singh, K.K. 1995. Traditional Phytotherapy Among the Baiga Tribe of Shahdol District of Madhya Pradesh, India. *Ethnobotany* **7**: 69-73.

Verma, S. & Chauhan, N.S. 2006. Studies on Ethno-Medico-Botany of Kunihar Forest Division, Distt. Solan, H.P. *Ethnobotany* **18**: 160-165.

Verma, S.K. & Chauhan, N.S. 2007. Indigenous Medicinal Plants Knowledge of Kunihar Forest Division. *Indian J. Trad. Knowledge* **6**(3): 498-507.

Vihari, V. 1995. Ethnobotany of Cosmetics of Indo-Nepal Border. *Ethnobotany* **7**: 89-94.

Virijee, G.H., Dhar, V., Kachroo, P. & Buth, G.M. 1984. Taxa-Ethnobotanical Studies of the Rural Areas in District Rajouri (Jammu). *J. Econ. Tax. Bot.* **5**: 831-834.

Vishwanathan, M.V. & Singh, H.B. 1996. Plants Used as Household Remedies in India. 105-107. *In*: Jain, S.K. (Ed.) *Ethnobiology in Human Welfare*. Deep Publ., New Delhi.

Vishwanathan, M.B. 1997. Ethnobotany of the Malyalis in North Arcot District, Tamil Nadu, India. *Ethnobotany* **9**: 77-79.

Viswanathan, M.B., Kumar, P.E.H. & Ramesh, N. 2001. Ethnomedicines of Kanis in Kalakkad-Mundanthurai Tiger Reserve, Tamil Nadu. *Ethnobotany* **13**: 60-66.

Viswanathan, M.V., Kashyapa, K. & Ramachandran, K. 1981. Less Known Fibre-Yielding Plants of Peninsular India. *Bull. Bot. Surv. India* **23**: 240-242.

Wambene, C.O.N. 1990. Natural Products in Developing Economy. *In:* Igbocchi, A.C. & Osisigu, I.U.W (Eds.) *National Workshop on Natural Products*. Univ. of Benin Press, Nigeria.

Watt, G. 1972. *A Dictionary of the Economic Products of India*, Vols. I-VI. Periodical Experts, Delhi (India).

Wickens, G.E. 1990. What is Economic Botany? *Econ. Bot.* **44**: 12-28.

Wielgorskaya, T. 1995. *Dictionary of Generic Names of Seed Plants*. Bishan Singh Mahendra Pal Singh, Dehradun (India).

PLATES

Plate 1: (A) A signpost indicating height above sea level of village Pragpur and the total number of people inhabiting it. (B) Same; notifying the heritage status of village Pragpur. (C) Distant view of Pragpur. (D) Dhauladhar hills as seen from the heritage village Pragpur.

Plate 2: (A) Famous Judges Court (side view). (B) Same; close-up view showing the beautiful architectural plans.

Plate 3: (A) Layout of a newly constructed two-storeyed house. (B) Layout of large-sized ventilators in one of the 2-storeyed old havelis. (C) An old house having iron sheets used for roofing and walls made up of unburnt bricks. (D) Same; roof made up of slates.

Plate 4: (A) and (B) Old havelis showing places for keeping earthen lamps, wooden and iron grilled windows and ventilators. Note the use of pressed clay for cementing the bricks.

Plate 5: (A-D) Patterns of wooden doors, windows and ventilators of old havelis of village Pragpur.

Plate 6: (A) Broom. (B) 'Battoi'—a large-sized metal vessel employed for cooking food on ceremonial occasions. (C) An earthern 'chullaha' in kitchen employed for cooking. (D) A stand for various articles for the removal of ash from earthen 'chullaha'.

Plate 3: Ornaments adorned by womenfolk of village Pragpur: (A) Silver bracelet. (B) A pair of large-sized silver ear-rings. (C) Necklace of gold. (D) 'Kantha'—A necklace of silver. (E) 'Pajeb'—An ornament for feet. (F) 'Chak'—An adornment for head of married woman. (G) Various types of gold ornaments.

Plate 8: (A) Radha-Krishna Temple. (B) Close view of monkey god—'Hanuman' resting under the canopy of a tree in Hanuman temple of village Pragpur.

Plate 9: (A-D) 'Naun'—water storage reservoirs for drinking and irrigation in Pragpur.

Plate 10: (A) Sickel. (B) Earthen Pitchers. (C) 'Chakla' and 'Belan'—a device for rolling of 'chapatis' and 'Madani' (wooden churner—a device for churning milk and curd. (D) Bamboo baskets of different sizes for drying and serving purposes.

Plate 11: (A) *Abrus precatorius* L. (B) *Acacia catechu* Willd. (C) *Acacia nilotica* Linn. (D) *Achras zapota* Linn.

Plate 12: (A) *Achyranthes bidentata* Blume. (B) *Acorus calamus* Linn. (C) *Adhatoda vasica* Nees. (D) *Aegle marmelos* (L.) Correa.

Plate 13: (A) *Agave americana* L. (B) *Ageratum conyzoides* Linn. (C) *Albizia lebbeck* Benth. (D) *Allium cepa* Linn.

Plate 14: (A) *Allium sativum* Linn. (B) *Aloe maculata* All. (C) *Aloe vera* (L.) Webb. & Benth. (D) *Amaranthus spinosus* Linn.

Plate 15: (A) *Amaranthus viridis* Linn. (B) *Amorphophallus campanulatus* (Roxb.) Blume *ex* Decne. (C) *Annona squamosa* Linn. (D) *Argemone mexicana* Linn.

Plate 16: (A) *Artemisia japonica* Thunb. (B) *Artocarpus integrifolia* Linn. (C) *Artocarpus lakoocha* Roxb. (D) *Asparagus adscendens* Roxb.

Plate 17: (A) *Bambusa nutans* Wall. (B) *Bauhinia vahlii* Wight and Arn. (C) *Bauhinia variegata* L. (D) *Boerhavia diffusa* Linn.

Plate 18: (A) *Bombax ceiba* Linn. (B) *Bougainvillea spectabilis* Willd. (C) *Brassica campestris* Linn. (D) *Bryophyllum calycinum* Salisb.

Plate 19: (A) *Butea monosperma* (Lam.) Kuntze. (B) *Callistemon lanceolatus* DC. (C) *Canna indica* L. (D) *Cannabis sativa* Linn.

Plate 20: (A) *Capsicum annuum* Linn. (B) *Carica papaya* Linn. (C) *Carissa inermis* Vahl. (D) *Carissa spinarum* Hook. f.

Plate 21: (A) *Cassia fistula* L. (B) *Cassia occidentalis* Linn. (C) *Catharanthus roseus* G. Don. (D) *Cedrela toona* Roxb. *ex* Rottl.

Plate 22: (A) *Celastrus paniculatus* Willd. (B) *Celosia argentea* L. (C) *Centella asiatica* (Linn.) Urban. (D) *Cestrum nocturnum* L.

Plate 23: (A) *Chenopodium album* Linn. (B) *Cicer arietinum* Linn. (C) *Cinnamomum camphora* (L.) Nees and Ebesm. (D) *Cissampelos pareira* Linn.

Plate 24: (A) *Citrus aurantium* L. (B) *Citrus grandis* (L.) Osbeck. (C) *Citrus jambhiri* Lush. (D) *Citrus limon* (L.) Burm. f.

Plate 25: (A) *Citrus medica* L. (B) *Coccinia grandis* (L.) Voigt. (C) *Colebrookia oppositifolia* Smith. (D) *Colocasia esculenta* (Linn.) Schott.

Plate 26: (A) Twig and fruits of *Cordia dichotoma* Forst. (B) *Coriandrum sativum* Linn. (C) *Cucumis sativus* L. (D) *Cucurbita hispida* Thunb.

Plate 27: (A) Plant, flower and fruit of *Cucurbita pepo* L. (B) *Curcuma domestica* (Medik.) Valeton. (C) *Cuscuta reflexa* Roxb. (D) *Cynodon dactylon* (Linn.) Pers.

Plate 28: (A) *Dalbergia sissoo* Roxb. (B) *Datura innoxia* Mill. (C) *Dendrocalamus hamiltonii* Nees and Arn. *ex* Munro. (D) *Dioscorea bulbifera* Linn.

Plate 29: (A) *Dioscorea deltoidea* Wall. (B) *Dodonaea viscosa* Jacq. (C) *Duranta repens* L. (D) *Elettaria cardamomum* (L.) Maton.

Plate 30: (A) *Emblica officinalis* Gaertn. (B) *Eriobotrya japonica* Lindl. (C) *Eruca sativa* Lam. (D) *Eucalyptus umbellata* Domin. (distant and close-up view).

Plate 31: (A) *Eugenia cerasoides* Roxb. (B) *Eugenia jambolana* Lam. (C) *Euphorbia royleana* Boiss. (D) *Ficus benghalensis* Linn.

Plate 32: (A) *Ficus carica* Linn. (B) *Ficus glomerata* Roxb. (C) *Ficus religiosa* Linn. (D) *Ficus roxburghii* Wall. *ex* Miq.

Plate 33: (A) *Flacourtia ramontchi* L. Herit. (B) *Foeniculum vulgare* Mill. (C) *Fragaria chiloensis* (L.) Duchesne. (D) *Fumaria indica* (Haussk.) Pugsley.

Plate 34: (A) *Gossypium arboreum* Linn. (B) *Grevillea robusta* A. Cunn. *ex* R. Br. (C) *Grewia oppositifolia* Roxb. (D) *Helianthus annuus* L.

Plate 35: (A) *Hibiscus rosa sinensis* L. (B) *Holarrhena antidysenterica* (Buch.-Ham.) Wall. (C) *Holoptelea integrifolia* (Roxb.) Planch. (D) *Impatiens balsamina* Linn.

Plate 36: (A) *Ipomoea cairica* (L.) Sweet. (B) *Ipomoea carnea* Jacq. (C) *Jasminum multiflorum* Roth. (D) *Jasminum sambac* (L.) Ait.

Plate 37: (A) *Jatropha curcas* Linn. (B) Woody branch and fruits of *Juglans regia* L. (C) *Lagenaria siceraria* (Molina) Standl. (D) *Lagerstroemia indica* L.

Plate 38: (A) *Lannea coromandelica* (Houtt.) Merrill. (B) *Lantana camara* Linn. (C) *Lawsonia alba* Lam. (D) *Leucaena leucocephala* (Lam.) de Wit.

Plate 39: (A) *Litchi chinensis* Sonner. (B) *Luffa aegyptica* Mill. (C) *Lycopersicum esculentum* Mill. (D) *Mallotus philippinensis* (Lam.) Muell.-Arg.

Plate 40: (A) *Malvaviscus conzatii* Greenm. (B) *Mangifera indica* Linn. (C) *Martynia annua* L. (D) *Melia azedarach* Linn.

Plate 41: (A) *Melia azadirachta* Linn. (B) *Mentha piperita* Linn. (C) *Mentha spicata* Linn. (D) *Mimosa pudica* Linn.

Plate 42: (A) *Mirabilis jalapa* Linn. (B) *Momordica charantia* L. (C) *Momordica dioica* Roxb. *ex* Wall. (D) *Morus alba* Linn.

Plate 43: (A) *Mucuna pruriens* L. (B) *Murraya koenigii* Spreng. (C) *Murraya paniculata* (Linn.) Jack. (D) *Musa sapientum* Linn.

Plate 44: (A) *Nerium indicum* Mill. (B) *Ocimum basilicum* Linn. (C) *Ocimum sanctum* Linn. (D) *Opuntia dillenii* Haw.

Plate 45: (A) *Oroxylum indicum* Vent. (B) *Papaver rhoeus* L. (C) *Parthenium hysterophorus* Linn. (D) *Pedilanthes tithymaloides* (L.) Poit.

Plate 46: (A) *Pennisetum purpureum* Schum. (B) *Phoenix sylvestris* Roxb. (C) *Phyllanthus urinaria* Linn. (D) *Pinus roxburghii* Sarg.

Plate 47: (A) *Pistacia integerrima* Stewart ex Brand. (B) *Plumbago zeylanica* Linn. (C) *Pogostemon benghalensis* Burn.f. (D) *Polyalthia longifolia* (Sonn.) Thwaites.

Plate 48: (A) *Populus nigra* L. (B) *Premna latifolia* Roxb. (C) *Prunus amygdalus* Batsch. (D) *Prunus domestica* L.

Plate 49: (A) *Prunus persica* Benth. and Hook. f. (B) *Psidium guajava* Linn. (C) A branch and fruits of *Punica granatum* Linn. (D) *Putranjiva roxburghii* Wall.

Plate 50: (A) *Pyrus communis* L. (B) *Pyrus pashia* Ham. (C) *Randia dumetorum* (Retz.) Lam. (D) *Raphanus sativus* Linn.

Plate 51: (A) *Ricinus communis* Linn. (B) *Rosa damascena* Mill. (C) *Rumex hastatus* Don. (D) *Saccharum arundinaceum* Reitz.

Plate 52: (A) *Saccharum officinarum* L. (B) *Saccharum spontaneum* Linn. (C) *Salvia coccinea* Linn. (D) *Sapindus mukorossi* Gaertn.

Plate 53: (A) *Saraca asoca* (Roxb.) DE Willd. (B) *Sechium edule* Sw. (C) *Sesamum indicum* DC. (D) *Solanum erianthum* Don.

Plate 54: (A) *Solanum melongena* L. (B) *Solanum nigrum* Linn. (C) *Solanum surattense* Burm. f. (D) *Solanum tuberosum* Linn.

Plate 55: (A) *Solanum viarum* Dun. (B) *Sonchus asper* Hill. (C) *Spilanthes acmella* Murr. (D) *Stephania glabra* (Roxb.) Miers.

Plate 56: (A) *Tabernaemontana divaricata* R. Br. *ex* Roem. and Sch. (B) *Tagetes erecta* Linn. (C) *Tamarindus indica* Linn. (D) *Taraxacum offinciale* Wigg.

Plate 57: (A) *Tecoma grandiflora* Loisel. (B) *Terminalia arjuna* Wt. and Arn. (C) *Terminalia bellirica* (Gaertn.) Roxb. (D) *Terminalia chebula* Retz.

Plate 58: (A) *Thevetia neriifolia* Juss. *ex* Steud. (B) *Thuja orientalis* L. (C) *Tinospora cordifolia* (Willd.) Miers. (D) *Trifolium alexandrinum* L.

Plate 59: (A) *Trigonella foenum-graecum* Linn. (B) *Triticum aestivum* Linn. (C) *Verbascum thapsus* Linn. (D) *Vitex negundo* L.

Plate 60: (A) *Vitis jacquemontii* Parker. (B) *Wendlandia heynei* (Roem. and Schult.) Sant and Merch. (C) *Woodfordia floribunda* Salisb. (D) *Xanthium strumarium* Linn.

Plate 61: (A) *Zanthoxylum armatum* DC. (B) *Zingiber officinale* Rose. (C) *Zea mays* L. (D) *Zizyphus jujuba* Lamk.